# Industrial Painting
## Principles and Practices
### Second Edition

Norman R. Roobol

PUBLICATIONS
Cincinnati, Ohio

Roobol, Norman R.
    Industrial painting : principles and practices / Norman R. Roobol
— 2nd ed.
      356 p. 17.8 × 25.2 cm.
    Includes index.
    ISBN 1-56990-215-1
    1. Painting, Industrial. I. Title
TT305.R56 1997
667'.9—dc20                                                       96-38488
                                                                                               CIP

Copyright © 1997 by Hanser Gardner Publications. All rights reserved. No part of this book may be reproduced in any form or by any means without the express written consent of the publisher.

Please direct all inquiries to:
    Hanser Gardner Publications
    6915 Valley Avenue
    Cincinnati, OH 45244


Printed in the United States of America.
1 2 3 4 5 6    01 00 99 98 97 96

# Preface to the Second Edition

I have designed this book on industrial painting to be useful and practical rather than deeply theoretical. It provides information that will be of value on the plant floor. No knowledge of chemical theory is necessary to understand any section, and everything is explained in a straightforward manner. Where it is instructive, I have included small amounts of basic science, so you can fully comprehend the principles involved and also to avoid a "do this, do that" style presentation. The book contains a wealth of information for everyone who is associated with painting; those new to paint and those veterans of many years' painting experience can learn from it.

The chapter order in this second edition follows even more closely to that of my three-day Industrial Painting Processes course. That instructional painting seminar has been presented over 200 times, both as a public offering and as private in-plant sessions. The actual course content has constantly been altered over the past twenty years. The methods and procedures of industrial painting operations undergo constant gradual shifts, but the topic order of this book is basically the same as the current course. This sequence has been found to be best for maximizing the learning experience.

For many years I had been looking for a strong and accurate industrial painting textbook. Even after I realized that I would probably need to write it myself, the hope remained that someone else would go through the long and arduous task of putting together a book such as this. Yet when the need for a text finally forced me to begin this book, it was often enjoyable despite the inevitable delays, frustrations, and difficulties. If this book helps people to improve paint quality and economy, my goal will be achieved.

I am always eager to hear from anyone who has comments on any part of this book. All suggestions for improvement will be incorporated into future editions of this work. Contact me at the address below.

Norman R. Roobol
507 Haddington Lane
Peachtree City, GA 30269
770-487-9133
fax 770-487-2792
e-mail drroobol@juno.com

# Dedication

This is for Joan, my wife and my life. It's for my kids, too, who show their love and support in so many ways. Thanks, you guys! Also, it's for Louise Ezinga, the most loving and loved mother-in-law imaginable.

# Acknowledgements

Many people helped me with this textbook by providing information, reviewing the coverage, suggesting topic arrangements, and all the other things that make a book better. I hereby express my deep gratitude to each of them, especially to the kind folks at Hanser Gardner, Jerry Poll, Tom J. Burke, Carl P. Izzo, and Glen Muir. But most of all, I am so happy and thankful for the incessant support and brilliant editing of Jennifer King. Unless you've worked with different editors, you cannot fully understand how much difference complete competence and total concern make in a book such as this paint text.

The motto below is that of my undergraduate alma mater, but it becomes the personal aim of most who attend, for "...the Calvin 'Spark' you once have caught, shall never die again."

"Cor Meum Tibi Offero Domine, Prompte Et Sincere"

# Credits

Our sincere appreciation to the following companies for the use of their photographs: Binks Manufacturing Co., pp. 67, 163, 170, 198, 220, 225, 227, 229, 230, 234, 260, 288, 290, 303; Dürr Industries, Inc., 112, 148, 307, 311; ITW-Ransburg Electrostatic Systems, 189, 204; Nordson Corporation, 68, 69, 71, 73, 172, 173, 196, 299, 300; Toyota Motor Manufacturing, USA, 217.

# Table of Contents

**Preface** ........................................................................................................ iii

**Dedication/Acknowledgements/Credits** ..................................................... iv

**Chapter 1: Paint Components and Their Functions** ................................. 1
Introduction to Paint ........................................................................................ 1
Paint Composition ........................................................................................... 2
    Resins ......................................................................................................... 3
    Pigments .................................................................................................... 6
    Solvents (Fluidizers) ................................................................................. 10
    Additives ................................................................................................... 13

**Chapter 2: Classifying Industrial Paints** ................................................... 17
Three Major Groups of Paint .......................................................................... 17
    Industrial Paint Types According to End Use Characteristics ................. 18
    Other End-use Classifications .................................................................. 25

**Chapter 3: Industrial Paints Categorized by Resin Category,**
    **Physical Makeup, and Cure Mechanism** ............................................. 27
Resin Categories ............................................................................................. 27
    Acrylics ..................................................................................................... 28
    Alkyds ....................................................................................................... 29
    Cellulosics ................................................................................................. 29
    Epoxies ...................................................................................................... 29
    Halogenated Resins .................................................................................. 30
    Oleoresinous (Oil-Based) Coatings .......................................................... 30
    Phenolics .................................................................................................. 30
    Polyesters ................................................................................................. 31
    Silicones ................................................................................................... 31
    Urethanes .................................................................................................. 31
    Vinyls ........................................................................................................ 33
    Resin Mixtures ......................................................................................... 34
Physical Makeup Categories ........................................................................... 34
    Liquid Fluidization Methods .................................................................... 34
Cure Mechanism Categories ........................................................................... 37
    Lacquers .................................................................................................... 37
    Enamels .................................................................................................... 37

## Chapter 4: Low- and High-solids Coatings ................................................. 39
The History of Low-solids Coatings ................................................. 39
High-solids Coatings ................................................................. 40
    High-solids Resins ............................................................. 41
    Surface Tension ................................................................ 42
    High-solids Advantages ....................................................... 43
    High-solids Disadvantages .................................................... 44

## Chapter 5: Waterborne Coatings ............................................................. 47
Definition of a Waterborne Coating ................................................ 47
    Solution Waterborne Coatings ................................................ 48
    Emulsion Waterborne Coatings ............................................... 49
    Dispersion Waterborne Coatings ............................................. 51
Solvent in Waterborne Coatings .................................................... 51
Waterborne Wood Finishes .......................................................... 53
WBC Considerations ................................................................. 54
    Operation Permits ............................................................. 54
    Pretreatment ................................................................... 55
    Cost ........................................................................... 55
    Resin Availability ............................................................ 55
    Modifications ................................................................. 55
    Process Commonality .......................................................... 55
    Electrostatics ................................................................ 56
    Atomization ................................................................... 57
    Application Problems ......................................................... 57
    Agitation ..................................................................... 58
    Odor and Cleanup ............................................................. 58
    Fire Hazard ................................................................... 59
    Storage ....................................................................... 59
    Dip Tanks ..................................................................... 59
    Foam .......................................................................... 59
WBC Advantages/Disadvantages ..................................................... 59

## Chapter 6: Powder Coating ................................................................ 61
Introduction to Powder Coating .................................................... 61
Powder Application Methods ........................................................ 64
    Electrostatic Spray .......................................................... 64
    Fluidized Bed ................................................................ 72
    Specialty Powder Coating Methods ........................................... 73
Powder Coating Advantages ......................................................... 74
    Cost .......................................................................... 74
    VOC Compliance ............................................................... 75
    Powder Coating Quality ...................................................... 75

  Powder Overspray Reuse ........................................................................... 76
  Ease of Application ................................................................................... 76
  Energy Savings ......................................................................................... 77
  Quick "Packageability" ............................................................................. 78
  Resin Availability ..................................................................................... 78
Powder Coating Disadvantages ........................................................................ 78
  Manufacturing Limitations ....................................................................... 78
  Application Problems ............................................................................... 79
  Coating Repairs ........................................................................................ 83

**Chapter 7: Cleaning the Surface** ................................................................. 85
Cleaning a Variety of Materials ........................................................................ 85
Metal Surface Cleaning ..................................................................................... 85
  Mechanical Cleaning ................................................................................ 86
  Solvent Cleaning ....................................................................................... 87
  Aqueous Cleaning ..................................................................................... 89
  Acid Cleaning ........................................................................................... 91
  Alkaline Detergent Cleaning .................................................................... 92
Emulsion Cleaners ............................................................................................. 93
  Cleaners for Aluminum ............................................................................ 93
Other Considerations ......................................................................................... 95
Rinsing After Cleaning ...................................................................................... 95
  Wastewater Treatment .............................................................................. 97

**Chapter 8: Conversion Coatings** .................................................................. 99
Conversion Coating Qualifications ................................................................... 99
Conversion Coatings for Metal ....................................................................... 100
  Conversion Coatings for Steel ................................................................ 101
  Conversion Coatings for Zinc ................................................................. 105
  Conversion Coatings for Aluminum ....................................................... 106
  Specialty and Mixed Metal Phosphates .................................................. 110
Spray and Dip Processes ................................................................................. 110
Phosphate Troubleshooting ............................................................................. 113
  Problem 1 Streaky, Blotchy, Thin, or Discontinuous Phosphate Coatings ............... 113
  Problem 2 Loose, Powdery, and Nonadherent Phosphate Coatings ............ 115
  Problem 3 Salt Deposits, Acid Spots, and Water Spots ......................... 115
  Problem 4 Flash Rust .............................................................................. 116
Sealing Phosphate Coatings ............................................................................ 117
Other Conversion Coatings ............................................................................. 119
Dryoff Ovens ................................................................................................... 119
Conversion Coating Waste Treatment ............................................................ 119

## Chapter 9: Paint Application I—Traditional Methods ... 121
Historical Applications ... 121
Modern Application Methods ... 121
    Rolling ... 121
    Dipping ... 121
    Flow Coating ... 122
    Continuous Coaters ... 123
    Dip-Spin Coaters ... 125
    Curtain Coating ... 125
    Roll Coating, Coil Coating ... 126
Safety Procedures in the Paint Shop ... 127
    Fire Safety Precautions ... 131
    Cleanliness ... 131
    Safe Spraying Techniques ... 131
    Waste Disposal ... 132

## Chapter 10: Paint Application II—Electrocoating and Autodeposition Coating ... 133
Coating by Immersion ... 133
Electrocoating ... 133
    Electrical Considerations ... 133
    E-Coat Paint Constituents ... 135
    Bath Chemical Reaction ... 137
    Rinsing ... 138
    Pigments ... 142
    Bath Parameters ... 146
    Tank Details ... 146
    E-coat Curing Cycle ... 148
    E-coat Advantages ... 149
    E-coat Disadvantages ... 150
Autodeposition ... 152

## Chapter 11: Paint Application III—Spray Guns ... 155
Components of Air-Atomizing Spray Guns ... 155
Compressed Air Supply ... 155
Paint Supply ... 157
Gun Operation ... 158
    Types of Guns ... 161
Spraying Techniques for Air-Atomized Guns ... 166
Air Spray Characteristics ... 167

## Chapter 12: Paint Application IV—Airless Spray and Air-Assisted Airless-Spray ... 169
Components of Airless Spray Guns ... 169
    Gun Operation ... 171
    Spraying Techniques for Airless Guns ... 172

Advantages of Airless Spray Guns .................................................................... 174
Disadvantages of Airless Spray Guns ................................................................ 174
Air-Assisted Airless Spray Guns ................................................................................ 176
Air-Assisted Spray Gun Advantages ................................................................. 177

## Chapter 13: Paint Application V—Electrostatic Painting ........................... 179
The Basics of Electrostatics .................................................................................... 179
Electrostatics in Painting ....................................................................................... 180
Grounding and Safety Precautions ................................................................... 182
The Effects of Humidity ................................................................................... 186
Advantages of Electrostatic Spray ................................................................... 186
Disadvantages of Electrostatic Spray ............................................................... 187

## Chapter 14: Paint Application VI—Rotary Atomizers ................................ 193
Introduction to Rotary Atomization ....................................................................... 193
Shape of the Device ......................................................................................... 195
Mounting Configuration ................................................................................... 195
Rotational Speed and the Degree of Atomization ................................................. 196
Paint Application ................................................................................................... 200
Configuration of the Disc System .......................................................................... 201
Disc and Bell Rotation ........................................................................................... 201
Rotary System Operation ....................................................................................... 202
Hand-held Bell ....................................................................................................... 203
High-speed Rotary Atomizer Advantages .............................................................. 204
Fine Atomization .............................................................................................. 204
High-solids, Waterborne Versatility .................................................................. 205
Viscosity Flexibility .......................................................................................... 205
Transfer Efficiency ............................................................................................ 205
High-Speed Rotary Atomizer Disadvantage .......................................................... 205

## Chapter 15: Coating Types and Curing Methods ........................................ 207
The Curing of Coatings .......................................................................................... 207
Coatings That Cure by Solvent Evaporation Only ................................................. 207
Coatings That Cure by Cross Linking .................................................................... 208
Oxidizing Coatings ........................................................................................... 208
Moisture-cure Coatings ..................................................................................... 208
Heat Cross-linking Coatings ............................................................................. 208
Reactive Catalytic Coatings .............................................................................. 209
Catalyst Vapor Coatings ................................................................................... 210
Radiation-cure Coatings ................................................................................... 211
Types of Ovens ...................................................................................................... 213
Convection Ovens ............................................................................................. 213
Infrared Ovens .................................................................................................. 215

Other Curing Methods ........................................................................................................216
    Induction Heating ........................................................................................................217
    Microwave and Radio Frequency (RF) Curing ..........................................................217
    Heat of Condensation Curing ......................................................................................218

## Chapter 16: Film Defects in Liquid Coatings ....................................................219
The Root Cause of Defects ................................................................................................219
Types of Defects ................................................................................................................219
    Blisters ........................................................................................................................219
    Bubbles and Craters ....................................................................................................221
    Color Mismatch ..........................................................................................................222
    Dirt ..............................................................................................................................223
    Fisheyes ......................................................................................................................225
    Gloss Variations ..........................................................................................................226
    Mottle ..........................................................................................................................228
    Orange Peel ................................................................................................................229
    Runs, Sags, and Curtains ............................................................................................230
    Paint Adhesion Loss ..................................................................................................232
    Soft Paint Films ..........................................................................................................233
    Solvent Pops, Boils, and Pinholes ..............................................................................234
    Solvent Wash ..............................................................................................................235

## Chapter 17: Paint-related Testing ......................................................................237
Categories of Paint Tests ..................................................................................................237
    Tests of Paint "in the Bulk" ........................................................................................237
    Tests on Parts and Paint Application Equipment ......................................................241
    Tests of Applied Paint Film ........................................................................................242
Which Tests to Run? ..........................................................................................................256

## Chapter 18: Stripping ..........................................................................................259
Reasons for Removing Unwanted Paint ............................................................................259
Types of Stripping ..............................................................................................................259
    Mechanical ..................................................................................................................259
    Blasting by Abrasive Grit ............................................................................................260
    Blasting by Water ........................................................................................................261
    Chemical ....................................................................................................................261
    Solvent ........................................................................................................................262
    Burn-off Ovens ..........................................................................................................265
    Molten Salt ..................................................................................................................265
    Hot Fluidized Sand ....................................................................................................266
Selecting a Stripping Process ............................................................................................266
    Cost and Volume of Items to be Stripped ..................................................................266

Substrate Characteristics .................................................................................. 266
Safety ............................................................................................................. 266
Environmental Regulations ............................................................................ 266

## Chapter 19: Special Considerations for Painting Plastics ............ 267
Understanding Plastics Chemistry ............................................................................ 267
Why Paint Plastic? .................................................................................................... 268
    Highlighting ................................................................................................... 269
    Texturing ........................................................................................................ 269
    Protection ....................................................................................................... 269
    Appearance Uniformity ................................................................................. 269
    Hiding Defects ............................................................................................... 269
    Functionality .................................................................................................. 270
    Conductivity .................................................................................................. 270
    Second Surface .............................................................................................. 270
Cleaning Plastic Before Painting .............................................................................. 270
    Fingerprints .................................................................................................... 270
    Dust and Lint ................................................................................................. 270
    Mold Release ................................................................................................. 271
Preparing Plastic Surfaces for Painting .................................................................... 272
Plastic Surface Peculiarities Affecting Painting ...................................................... 273
    Water Sensitivity ........................................................................................... 273
    Heat Sensitivity ............................................................................................. 274
    Sanding Sensitivity ........................................................................................ 274
    Nonconductivity ............................................................................................ 275
    Color Matching with Metal .......................................................................... 276
    Gloss Variations with Metal ......................................................................... 276
    Wicking ......................................................................................................... 276
Paint Film Laminates For Plastics ............................................................................ 276
Plastic Paintability Information Resources .............................................................. 278

## Chapter 20: Conveyors for Painting ............................................... 279
The Need for Conveyors ........................................................................................... 279
    Flat Belt, Roller, or Track Conveyors .......................................................... 280
    Chain-on-edge Conveyors ............................................................................. 281
    Chain-in-floor Conveyors ............................................................................. 281
    Overhead Conveyors ..................................................................................... 283
    Inverted Conveyors ....................................................................................... 284
The Importance of Good Conveyor Design ............................................................. 284

**Chapter 21: Finishing Robots** ............................................................................................. 285
Industrial Robot Programming ............................................................................................. 285
Rotary Bell and Spray Painting Robots ................................................................................ 286
Programming for Painting ..................................................................................................... 289
Robot Operation Requirements ............................................................................................. 292
Spray Painting Robot Advantages ......................................................................................... 294
Case History ........................................................................................................................... 294
Future Developments ............................................................................................................. 295

**Chapter 22: Spray Booths for Liquid Painting** ................................................................. 297
Spray Booth Basics ................................................................................................................ 297
    Dry-filter Booth ............................................................................................................... 300
    Water-wash Booths .......................................................................................................... 301
    Dry-filter or Water-wash Booths? ................................................................................... 305
High-solids Paint Overspray Recovery ................................................................................. 306
VOC/HAP Reduction Methods ............................................................................................. 306
    Incineration Devices ........................................................................................................ 306
    Carbon-bed Solvent Adsorption Systems ...................................................................... 310

**Glossary** ................................................................................................................................ 315

**Index** ...................................................................................................................................... 333

# Chapter 1

# Paint Components and Their Functions

## Introduction to Paint

In any field of study, the meaning of a group of related terms is required learning for anyone who wishes to understand the material. Industrial painting is no different in this regard; therefore, as the chapters progress, the needed terminology will be given. But three chemical terms are so basic and essential to the explanation of paint adhesion and material compatability that they are mentioned here. The terms atom, molecule, and polarity are crucial to explaining the topic, and they will be used repeatedly throughout the text.

At school, you were taught that all materials are composed of *atoms*, those unbelievably tiny bits of matter that make up absolutely everything in nature. Perhaps the names of some of the 98 natural atoms are familiar: carbon, sodium, iron, oxygen, aluminum, chlorine, lead, and neon. For our purposes, you only have to be aware that atoms of the elements exist; their individual names will be recognized as they occur in the text.

It is important that you recall how atoms can bond together to form limitless kinds of *molecules*. When you read "$H_2O$" or hear the sounds "aich two oh," these immediately cause the mental recognition of "water." The formula $H_2O$ is the chemist's shorthand for the molecules of the compound we call water, which is formed when two hydrogen atoms plus one oxygen atom physically merge and then chemically bond together. Of course, it takes billions of water molecules to make even a barely visible drop of water. Atoms of different elements can unite chemically into molecules of totally new substances. Small molecules of two such substances used in finishing are butyl alcohol and methyl ethyl ketone. These solvents have molecules with approximately 12 atoms each. Medium- and large-sized molecules are also possible. Paint resins frequently have molecules that are comprised of 100,000 or more atoms, for example.

Molecules of different substances may or may not mix together. Oil will dissolve in xylene, but when oil and water are mixed, even after vigorous shaking, the oil separates and floats to the top. Yet table salt, which dissolves quickly when stirred into water, absolutely will not dissolve in xylene. This behavior is explained by *polarity,* a magnet-like property of all molecules. Polarity is always constant for each kind of molecule, but among various molecules it ranges from very slight to high. Oil and xylene have low polarity; water and salt have rather high polarity.

The polarity of materials must be somewhat close for them to be compatible, that is, to dissolve or adhere together. When rain falls on a newly waxed car, the water beads up because the wax has only slight polarity. The highly polar water molecules act like clumps of tiny magnets, drawing themselves into little droplets that attempt to touch the wax as little as possible. By recognizing the significance of polarity matches, you can understand why a highly polar paint will have little or no chance of sticking to a very low-polarity surface such as polypropylene. Yet a polar paint will adhere firmly to a polar plastic such as polyacrylate.

## Paint Composition

The term *paint* includes all organic resin-containing materials used as decorative, protective, and functional coatings on any kind of surface. It does not matter how paint is cured, how it is applied, what other ingredients it contains, whether it is liquid or powder, or whether it is solventborne, waterborne, or solvent free. Some writers claim that powder coatings are not paint, but in this book we make no distinction between liquid paints and powder paints. Paint is paint in any of its myriad forms.

I'm frequently asked if the term coating means something other than paint. There are no legal or scientific reasons for such a distinction; however, to a degree at least, ordinary usage tends to apply the label *paint* to films that are about 25-30 mils (650 microns) thick or less and *coatings* to films that are more likely to be functional rather than decorative and that are thicker than roughly 10-15 mils (325 microns). The important point is that everyone should strive to make certain the terminology he or she uses is fully understood by others.

Four major classes of materials are commonly present in paints, as shown in Figure 1-1:

1. Resins, also called binders, vehicles, polymers, and plastics form the paint film.
2. Pigments provide opacity and color, but are omitted from varnishes and clearcoats.
3. Fluidizers, often termed solvents, are used in most, but not all, liquid paints; they are not used in powder paints.
4. Additives may be added to provide special properties to the uncured or the cured paint.

The resins, pigments, and additives make up the *solids* portion of a paint; the fluidizers or solvents that evaporate during curing are called the *volatiles* in the formulation. Paint is prepared by mixing together a particular resin or resin combination, a solvent or a solvent blend, and most often additives and pigments. This mixing is done according to a specific formulation so that when the applied paint

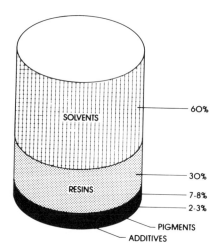

**Figure 1-1. Paint Composition by Volume**

## Resins

The *resins*, or binders, are the organic film-former portion of paints. After application onto a surface, resin enables the paint to cure into a continuous uniform layer that holds all of the paint's other constituents. A cured paint film is actually a layer of plastic, although few people think of paint in that way or recognize that fact. The cured film normally contains all of the ingredients present in the liquid paint except the fluidizers (solvents) portion, which evaporate during the paint application and curing processes. The resin (or resins) must be considered the most important component in any paint for it must be present, although paints need not contain any pigments, additives, or fluidizers.

The resins are usually made up only of polymers, but in some radiation cured paints they are present exclusively as monomers. A *polymer* in organic chemistry is a chain linkage of many repeating individual chemical structures (*monomers*). Many of the polymers used as resins tend to be highly viscous and therefore are generally thinned with a solvent. A paint manufacturer will usually buy resins with some solvents already mixed in for easier material pumping.

Polymers for paints can be categorized into two discrete types, lacquers and enamels, depending upon how they form a relatively hard and dry paint film. Every paint—liquid, powder, waterborne, solventborne, air-dry, heat-curing, radiation-curing, high-solids, low-solids—is either a lacquer or an enamel paint. Urethane, acrylic, epoxy, alkyd, vinyl, or polyester—each of these must either be a lacquer or an enamel.

### Lacquers

*Lacquers* in all types of liquid paint form dry paint films simply by evaporation of the solvent. Loss of solvent from the film permits the polymer molecules to approach each other closely and unite to form a continuous film. The resin molecules in lacquers are always highly polar. When the solvent evaporates, the molecules bond together strongly by the magnet-like polar dipole attractions to create a fully cured film. Solvent evaporation may be hastened by heat, or solvents may be allowed to evaporate at ambient conditions.

Powder coatings that are lacquers contain polar resin molecules. Powder coat films will melt when heated in the bake oven and then flow out to form a liquid film. The molecules bond together by the magnetic attraction of dipolar forces. When the molten powder cools, these become a solid lacquer paint film.

### Enamels

All *enamel* paints are characterized by having resin molecules that undergo additional bonding or linking together by chemical polymerization reactions after the paint has been applied to a substrate. Any solvent evaporates as it does with lacquers, but if reactive cross linking did not also happen the enamel resin film would remain a soft gummy layer. The chemical cross linking of enamel resin molecules can occur by a variety of methods, depending on the chemical natures of the resins used. These methods will be discussed directly.

**Air-drying paints.** A major class of air-dry enamel film formers is composed of molecules that include numbers of carbon-to-carbon double bonds. When exposed to air, the double bonds react chemically with atmospheric oxygen to form a cross-linked paint film. The carbon-to-carbon double bonds are termed *unsaturated bonds*. These are identical to the chemical bonds in unsaturated fats and oils recommended by dietary experts for human consumption in place of the saturated type. In fact, many of the resins that have unsaturated bonds are derived from vegetable products such as linseed, tung, soya, and rapeseed oils. Unsaturated oils are used heavily in alkyd paints and also in some of the epoxy polyester paints.

Air dry paints may also react with moisture in the air to cause cross linking. One component (1K) polyurethane paints are examples of this type. Free isocyanate groups in these coating resin molecules react quickly with atmospheric moisture to form carboxylic acid groups, which then undergo the cross linking reactions that cure the paint film.

**Baking enamels.** Heat can be used to activate molecules to undergo cross linking. Thermal energy is normally supplied by hot air or infrared radiation. Isolated instances of microwave and induction heating can be found, but these remain to be of minor interest. Many bake enamels may contain two polymers: a backbone polymer, such as an alkyd, and a cross linking resin, such as melamine or urea-formaldehyde. In the presence of heat, the backbone polymer and cross linking agent react to form new chemical bonds that link the molecules into a paint film. Some baking acrylics have an acrylic polymer backbone plus an acrylic acid or acrylate ester monomer that cross-links the polymer molecules during the bake cycle. Catalysts can be used to hasten the heat-curing reactions.

**Catalytic cure enamels.** A limited number of enamels are cross-linked without an energy input. Instead, a catalyst vapor chamber into which painted parts are placed is used. The catalyst, which may be mixed into the paint during application rather than using a catalyst vapor chamber, greatly increases the rate of cross-link curing without any heat or radiant energy being required.

**Radiation cure (radcure) enamels.** When free radical-curable film formers are exposed to electron beams (EB) or ultraviolet light (UV), they cross-link into a paint film, often within a few seconds. While some radcure paints contain volatile solvents, others have an advantage in that they produce no volatile emissions. These zero volatile organic compounds (VOC) paints are either 100% solids or in another way are 100% nonvolatile. In 100% solids coatings, small molecule polymers with low viscosities are used to avoid any need for solvents. With higher viscosity polymers, various *reactive diluents* are added that act to reduce viscosity but which will cross-link with a base polymer during curing. Rather than evaporating and producing volatile emissions, the reactive diluents become part of the cured paint film itself. Since UV and EB cause curing without heating the film, the advantage of such coatings for use on temperature-sensitive substrates is obvious.

**Two-part reactive enamels.** If two resins are highly reactive, they can cross-link spontaneously when they are mixed together. Once mixed they will begin to react; thus, the mixture must be used within a time limit or else the cross linking will have proceeded too far to be able to produce an acceptable coating film. Paint that has exceeded its *pot life* should be disposed. Some two-component (2K) enamels have as long as 12–16 hours (h) pot life, while very reactive types may have only a 5 minute (min) pot life. Equipment to use all types is marketed and used widely.

# PAINT COMPONENTS AND THEIR FUNCTIONS

## *Resin Qualities*

The types of polymers used in binders include acrylics, alkyds, aminoplasts, cellulosics, epoxies, chlorofluorocarbons, fluorocarbons, natural plant oils, phenolics, polyesters, polyurethanes, silicones, and vinyls. The chemistry of these polymers and various cross linking resins can become exceedingly intricate and esoteric for nonchemists. Some idea of the complexity is seen in the statement that "a polyfunctional homopolymeric backbone polymer needs hydroxy (-OH), carboxy (—COOH), and monosubstituted or unsubstituted amino (—CH$_2$NH—) functionalities in order to cross-link into polyamide resins." That much scientific detail is beyond the scope of this book but is discussed in detail in any organic chemistry text for those who are interested. A full study of the organic chemistry of these polymers and their cross linking systems is needed by paint makers, but is not necessary for most paint users.

When paint color is important, paint resins must be completely transparent to prevent possible interference with the shade and color of the pigment. The resins must not be susceptible to yellowing or discoloration with age or from exposure to long periods of sunlight, because this too would gradually shift the color of the film. Certain naturally occuring impurities in some resins can contribute to color variation and must be removed before the resin can be used in paints.

In addition to serving as the "heart" or primary building block of any paint, resins need to perform a number of other important functions. These include:

- Bonding
- Encapsulating the pigment particles
- Flowing out uniformly
- Providing the required physical properties

**Bonding.** Resins are also called *binders* because they need to grab hold of the substrate in a bonding action. This is achieved largely by mechanical keying into the slight roughness of the surface of the substrates. It is difficult for paint to adhere to extremely smooth surfaces such as glass and plated chrome. There is little chemical bonding that occurs between the substrate and the paint, only a few percent of the adhesion is due to van der Waals' forces which are those weak attractions that develop between all forms of matter.

**Encapsulating the pigment.** Each pigment particle must be completely wetted out and surrounded by resin. The resins should encapsulate and hold the pigment particles separate, not allowing them to cluster (aggregate). Clustering reduces the effectiveness of the pigment particles in providing hiding and color.

**Flowing out uniformly.** Resins must be able to flow out and form a smooth, continuous film, which provides optimum film properties and allows good gloss by maximizing the reflection of light rays. Poor flowout weakens the film's physical properties and also distorts and dulls the film surface with an orange peel effect.

**Providing the required physical properties.** Resins must be chosen for the most desirable physical and chemical properties for an end-use application. Important properties include:

- Hardness
- Hot/cold resistance
- Weather resistance
- Impact resistance
- Flexibility
- Chemical resistance

# 6   INDUSTRIAL PAINTING

- Initial gloss and gloss retention
- Water resistance
- Heat resistance
- Abrasion resistance
- Stain resistance
- Recoatability
- Corrosion resistance
- Detergent resistance
- Sunlight resistance

Various products require paint films that possess properties specific to that particular item's typical needs. For example, automobiles and trucks require finishes that resist road salt, gasoline, oil, harsh chemicals, detergents, bird droppings, tree saps, and acidic components in rain and air. Home appliances such as washing machines must have finishes that stand up to water, bleach, cloth dyes, and alkaline detergents. Machinery finishes must resist cleaning agents, heat-transfer fluids, cutting fluids, stamping oils, and a variety of organic and inorganic lubricants.

## Pigments

*Pigments* are tiny solid particles that are used to enhance appearance by providing color and/or to improve the physical (functional) properties of the paint film. Pigments used to provide color generally range from 0.2–0.4 microns in diameter; functional pigments are typically 2–4 microns in diameter, but they may range as high as 50 microns. The pigments need to be permanently insoluble in the binder of the coating. If they were to dissolve, they could change the properties of the binder and lose their capability to provide a particular appearance enhancement and functional improvement. Pigment particles must be large enough to block light rays; otherwise, the paint film would be low in hiding power and appear as a tinted clearcoat instead of a solid color film that hides the substrate surface. Tinted clearcoats are used, but the color comes from soluble toners, not pigments.

Paints must be well stirred before use (and in some instances during use) because the pigments are particulate in nature, and thus by gravity they will always tend to sink to the bottom of their containers. Where color is critical, the paint is kept in constant circulation to prevent pigments from settling in the lines and hoses.

### *Appearance Pigments*

Among the types of pigments that enhance appearance are white pigments, colored pigments and pearlescents, metallic powders, and flakes. Examples of white pigments are antimony oxide, leaded zinc oxide, titanium dioxide (anatase and rutile), white lead, and zinc oxide. Included in the category of white pigments are extender pigments, which can include barytes, bentonite, calcium carbonate, China clay, mica, silica, and talc. The relatively low-cost extender pigments can be added in small percentages to reduce the amount of white pigment required.

Colored pigments are available in both inorganic and organic materials; this contrasts with most other kinds of pigments that are mostly inorganic types. Inorganic pigments have a number of important and desirable properties, including low water solubility, negligible organic solvent solubility, high ultraviolet light stability, and very high thermal stability. Most organic pigments are much weaker than inorganic pigments in these attributes; nevertheless, organics are often used because of their superior visual purity, crispness, and brilliance. Organic pigments have great liveliness and excitement when compared with inorganic pigments of the same color.

They display none of the muddy overtones commonly noticed in the inorganic pigments.

Examples of colored pigments are phthalocyanine blue, cadmium yellow and orange, molybdate orange, chrome orange, toluidine red, cadmium red, phthalocyanine green, chrome green, black iron oxide, yellow iron oxide, red iron oxide, and brown iron oxide. Colored pigments are often used in blends of two or more types to yield specific color tints. Pigment materials containing cadmium, chromium, and lead have increasingly been phased out of use because of their toxicity. Without them some shades of color are almost impossible to achieve in stable pigment blends.

**Figure 1-2. Photomicrograph of Metallic Pigments**

Examples of metallic pigments are aluminum and bronze flake (shown greatly magnified in Figure 1-2), which produce a sparkly effect by randomly reflecting light. The same type of effect can be achieved with mica-type nonmetallic pigments. The metallic pigments, due to their conductivity, tend to align themselves differently in a paint film when applied by electrostatic and nonelectrostatic techniques. The mica-type pigments do not have this characteristic because of their nonconductivity. Mica flakes can also be coated with various materials to generate interesting optical effects such as opalescence and pearlescence.

Nonleafing grade aluminum flake is coated with oleic acid or a similar unsaturated fatty acid to enable the flake to be uniformly distributed throughout the paint film. This form is the one utilized for decorative metallic paints of any color shade. The distribution through the film allows it to be clearcoated with excellent intercoat adhesion.

### *Functional Pigments*

Aluminum flake pigments are also available in a leafing form. The leafing type is primarily used in maintenance and roof coating. It is only found in paints with a metallic aluminum color. The flake is coated with a saturated fatty acid such as stearic acid, which causes the flakes to lie flat in the film predominantly on the top portion of the paint layer. It is not readily topcoated because the aluminum on the surface affords poor adhesion. However, after weathering for several years, less aluminum flake is present at the surface and the film becomes more paintable.

Zinc dust is an example of a functional pigment that is added to a binder, often in extremely large amounts, to improve corrosion resistance. The heavy concentration of zinc particles allows them to touch each other, providing continuous electrical conductivity. This conductivity permits the zinc to provide sacrificial protection in the corrosion process, meaning that if the zinc-loaded paint film is scratched through to a steel substrate, the zinc will corrode sacrificially in place of the steel. Some zinc-rich coatings may contain as much as 90% zinc by weight.

The functional properties of a paint film affected by pigments can include adhesion, corrosion resistance, film strength, water resistance, and gloss. The type of functional pigment se-

## 8  INDUSTRIAL PAINTING

lected for a particular paint formulation will depend on the required end use properties of the paint.

### *Pigment Milling and Dispersion*

Paint is prepared by mixing together a particular resin or resin combination, solvent or solvent blend, and most often additives and pigments to produce a specific formulation that when properly cured will possess certain performance characteristics and display specified physical properties. Most dry pigments purchased by paint manufacturers tend to contain numerous tightly adhering clusters of particles. If not broken up, these agglomerates appear in the paint film as unsightly specks. The clusters are broken down into their original separate particles in one of several types of milling and dispersing processes. One kind of mill for this process consists of a container with a motor-driven fan-type blade that is shaped for optimum shearing and circulation of the material being milled. The material fed into the mill includes a precisely measured amount of binder and pigment. Solvents may also be present to abet the blending procedure. Ball or pebble mills, sand mills, bead mills, and roller mills may be used to create pigment dispersions and coatings.

Pigments are often purchased by paint makers already dispersed in slurry form for easier mixing. With dry pigments, in addition to breaking up pigment clusters, the milling/dispersing process has another function: to wet each pigment particle with binder. If pigment particles are not properly wetted and dispersed, clustering will result. Clustering reduces the effectiveness of the pigment particles, as shown in Figure 1-3. The maximum amount of pigment wettable by the binder is known as the critical pigment volume concentration (CPVC).

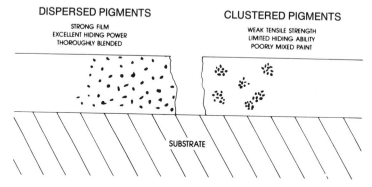

**Figure 1-3. Clustering of Pigment Particles Adversely Affects Paint Film Characteristics**

Decreasing or increasing the amount of pigment in a binder-pigment dispersion away from the CPVC will bring striking changes in the paint film properties. These changes can be graphed and will produce characteristic curves for each property. Each graph of pigment-to-volume concentration versus a particular paint film property will show a distinctive change of direction at the CPVC, as shown in Figure 1-4.

# PAINT COMPONENTS AND THEIR FUNCTIONS 9

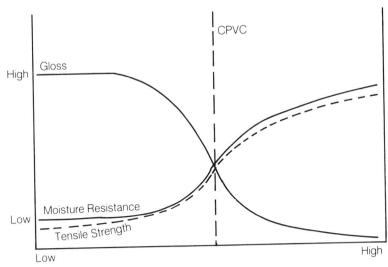

**Figure 1-4. Paint Film Characteristics vs. Pigment Volume Concentration (PVC)**

When the pigment loading in a paint is high, pigment particles tend to protrude from the applied paint film, creating a microscopically rough surface. This roughness scatters light, giving a low gloss appearance to the film. A low pigment concentration and particularly with a fine pigment grind will be more likely to leave a smooth paint film surface, resulting in a high gloss, as shown in Figure 1-5.

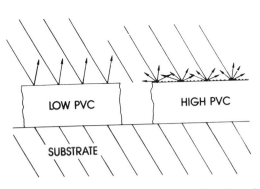

**Figure 1-5. Light Diffusion from a High PVC Paint Reduces Surface Gloss**

## Solvents (Fluidizers)

The fluidizer is usually a blend and often simply termed "solvent" more for convenience than technical accuracy—a practice we will follow here also. In some paints it truly dissolves the resins; in others it merely acts as a carrier without actually dissolving the resin molecules. In either case, the *fluidizer*, or *solvent*, is a liquid that is able to lower the viscosity of a paint formulation sufficiently to allow application onto a product. It then slowly evaporates to permit the formation of a paint film. The fluidizer's function in paint is, therefore, transitory. It is a vital part of the liquid paint but leaves as the paint film solidifies and the resin cross links.

### Solvent Categories

Solvents can be categorized into five types:

- Petroleum hydrocarbon solvents
- Chlorinated hydrocarbon solvents
- Oxygenated solvents
- Terpene solvents
- Aqueous solvents

**Petroleum hydrocarbon solvents.** This category consists of aliphatic (straight-chained) solvents and aromatic (containing benzene rings) solvents. Aliphatic solvents, because of their low polarity, tend to be limited in their ability to dissolve many resins. Examples of aliphatic solvents are mineral spirits and Varnish Makers' & Painters' (VM&P) naphtha in a narrow boiling range fraction of petroleum with boiling points of about 93°–149° F (34°–65° C). Aromatic solvents, however, are higher in solvent power. Only three aromatic solvents are commonly used in coatings: medium-flash aromatic naphtha, high-flash aromatic naphtha, and xylene. Three related xylene compounds are found in petroleum, and blends of these are simply labeled "xylene" since their solvent characteristics are nearly the same.

**Chlorinated hydrocarbon and chlorofluorocarbon (CFC) solvents.** Chlorinated hydrocarbon and CFC solvents used in paints are characterized by high solvent power, very fast evaporation rates, and virtually no flash point (virtually nonflammable). Examples of chlorinated hydrocarbon and CFC solvents that have been used in coatings are trifluoromethyl parachlorobenzene (Oxsol 100), 1,1,1-trichloroethane (methyl chloroform), and methylene chloride (dichloromethane). Initially, in the mid- to late-1980s, chlorinated hydrocarbons and CFCs were of great interest as paint solvents because they do not contribute to air pollution and thus did not need be considered as VOC under many (but not all) state and provincial rules. However, in many countries most chlorinated solvents are, or will be, outlawed for use or manufacture because of their role in the depletion of the earth's ozone shield in the upper atmosphere. Some are also listed as Hazardous Air Pollutants (HAPs). As a result, their importance as paint solvents is now becoming negligible except in certain countries. (See also the section on solvent cleaning in Chapter Seven for more details on which countries may continue to use halogenated solvents for cleaning and in paints.)

**Oxygenated solvents.** Oxygenated solvents have a range of solvent power that can be adjusted to almost any value by blending various types of these solvents. The four types used in coatings include alcohols, esters, glyceryl and glycol ethers (and their acetates), and ketones. In June, 1995, the U.S. Environmental Protection Agency (EPA) declared acetone (dimethyl ketone) to be nearly photochemically inactive and thus not a VOC or HAP. However, individual states still have the power to restrict its use.

**Terpene solvents.** Terpene solvents are derived from pine tree sap and are characterized by high solvent power. Common terpene solvent types are turpentine, dipentene, and pine oil.

**Aqueous solvents.** Since water is a highly polar substance, it can only dissolve resins that are also quite polar, but water alone is rarely able to dissolve paint resins. Polar groups that are used on resins to generate water solubility are formed by acid-base reactions with aqueous hydrogen ion donors (acids) and hydroxide ion donors (bases). In mildly acidic aqueous solutions, tertiary amine groups become highly polar quaternary ammonium cation groups, and in mild aqueous-base, carboxylic groups form polar carboxylate anions. Cathodic and anodic electrodeposition coatings (explained in Chapter Ten) are examples of paints that use (respectively) tertiary amine groups and carboxylic acid groups to achieve water solubility. Many early versions of waterborne solvent paints used ammonia and amines as bases to solubilize resin molecules that had carboxylic acid groups on them.

Even today, however, most so-called waterborne paints also contain an amount of water-soluble organic solvents. For reasons of better film flowout and greater resin solubility, the majority of paints that have water as their principle solvent will also include anywhere from 0.5% to as much as 20% of oxygenated solvent(s). They are termed co-solvents because they assist water in resin solubilization.

### Selecting a Solvent

Chemists have devised many parameters for the properties of solvents, but the two most important for coatings are solvent power (or, more precisely, solubility parameter, an explanation of which is too technical for this book) and solvent evaporation rate. The solvent power is the ability of a solvent to dissolve a given resin. The different types of solvents have a range of solvent powers, and the numerous classes of resins have different capabilities to be dissolved.

The evaporation rate of a solvent from a paint film is of crucial importance because the solvent needs to leave the paint at an ideal rate that is neither too slow nor too fast. This will allow the paint film to form and cross-link without solvent popping, also at an acceptable rapid rate. Solvents with fast evaporation rates include acetone, methyl alcohol, methyl ethyl ketone (MEK), methyl isobutyl ketone (MIBK), and ethyl acetate. Some solvents with slower evaporation rates are ethylene glycol monobutyl ether acetate, di-isobutyl ketone, and amyl alcohol. Solvents with evaporation rates between fast and slow are mineral spirits and xylene. Interestingly, as shown in Figure 1-6, some slow solvents have even slower evaporation rates than water.

Solvents may be further classified as true or active solvents, diluents, reducers, and thinners. A true solvent can dissolve and rapidly reduce the viscosity of a binder; a diluent is unable by itself to dissolve the resin. Within limits, a diluent solvent is able to reduce the viscosity of a concentrated mixture of resin and solvents, but if diluent is added in too large an amount, it will force the resins out of solution. Various solvents can function as true solvents or diluents, depending on the particular type of resins.

The terms "thinner" and "reducer" are vague. They can refer to any volatile liquid used in lowering the viscosity of a paint to the level appropriate for application.

### Coating Solvents and the Environment

In the early 1970s, following California's lead, the EPA began to draft regulations to limit the amount of VOC emissions from coatings. Its main method of doing so was to force paint users to

| Solvent Type | Common Solvents | Lbs./Gal. | Evaporation Rate (BuAc = 1) |
|---|---|---|---|
| Aliphatics | VM&P Naptha | 06.24 | 01.60 |
|  | Mineral Spirits | 06.63 | 00.10 |
| Aromatics | Toluol | 07.25 | 02.00 |
|  | Xylol | 07.25 | 00.60 |
|  | Aromatic 100 | 07.28 | 00.20 |
|  | Aromatic 150 | 07.49 | 00.04 |
| Alcohols | Ethanol | 06.78 | 02.80 |
|  | Isopropanol (99%) | 06.58 | 02.10 |
|  | n-Butanol | 06.78 | 00.45 |
| Esters | Ethyl Acetate (99%) | 07.55 | 05.50 |
|  | Isopropyl Acetate | 07.30 | 04.60 |
|  | Butyl Acetate (BuAc) | 07.38 | 01.00 |
|  | 2-Ethoxyethyl Acetate | 08.15 | 00.21 |
| Ketones | Acetone | 06.63 | 11.60 |
|  | MEK | 06.75 | 05.80 |
|  | MIBK | 06.70 | 01.60 |
|  | Diacetone Alcohol | 07.84 | 00.14 |
| Glycol-Ethers | 2-Methoxyethanol | 08.07 | 00.47 |
|  | 2-Ethoxyethanol | 07.77 | 00.32 |
|  | 2-Butoxyethanol | 07.54 | 00.06 |
|  | 2-Methoxyethoxyethanol | 08.56 | 00.01 |
|  | 2-Ethoxyethoxyethanol | 08.55 | 00.01 |
|  | 2-Butoxyethoxyethanol | 07.94 | 00.01 |
| Misc | Methylene Chloride | 11.11 | 14.50 |
|  | 1,1,1,-trichloroethane | 10.99 | 03.50 |
|  | N-Methyl-2-pyrrolidone | 08.55 | 00.01 |
| Water |  | 08.33 | 00.36 |

Figure 1-6. Properties of Common Solvents

lower the solvent content of the coatings they applied. This trend has continued since then and will not lessen. Both the provincial and federal agencies in Canada have increasingly similar restrictions on VOC emissions. The Clean Air Act Amendments of 1990 heavily restricted HAP use after 1994. The Title V definitions of a plant's "potential to emit" was based upon a 365 day/yr, 24 h/day capability to cause HAP and VOC emissions. This obviously tightened the compliance picture. For some plants that had a history of a 6 day/wk, 2 shift/day painting operation, this was a reasonable calculation method. It was unduly harsh, however, for any plants that historically had never run their paint shops more than 1 shift/day.

The Title V provisions of the 1990 amendments to the Environmental Protection Act are by far the most significant clean air regulations that face painting operations. Federal rules designate a "major source" as one that has the potential to emit annually more than 100 tons VOC, or 10 tons of a single HAP, or more than 25 tons of all HAPs. This category is one most plants strongly

## PAINT COMPONENTS AND THEIR FUNCTIONS 13

wish to avoid. The category of "synthetic minor source" is for plants who would otherwise be major sources, but they either accept limitations on VOC in paint in addition to time restrictions on painting operations, or they add abatement control equipment. A plant that falls below the numbers for a major source is classified as a "minor source."

A typical low-solids paint loses more than half of its volume as it changes from a wet film to a dry film due to evaporation of volatile solvents, as depicted in Figure 1-7. In addition to causing air pollution, this obviously is wasteful. The method of application determines to a large extent at what point in the application and curing processes the solvents are driven out of the paint film. For example, far more solvent is lost during and immediately after application in a spray process than in a dip- or flow-coat operation.

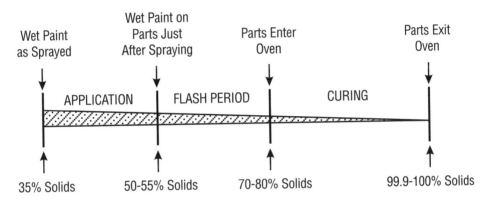

**Figure 1-7. Effects of Solvent Evaporation**

## Additives

Literally hundreds of chemicals can be added to paint formulations (both liquid paints and powder paints) to improve the coating performance in a specific way. These additives are usually considered to be a distinct category of paint ingredients. Sometimes, however, an additive can also be a pigment. A unique type of binder can, in some instances, be considered an additive, such as when some low molecular weight epoxy resin is included to increase the smoothness of the coating film.

### Additive Categories

Some additive types include:

- Antiblock agents
- Antifreeze
- Blending aids
- Curing agents
- Antimicrobial agents
- Defoamers
- Flow control agents
- Gloss modifiers
- Softening agents (plasticizers)
- Storage stabilizers
- Thixotropes
- UV stabilizers
- Antimar agents

**Antiblock agents.** These can be added so that painted parts can be stacked together soon after they exit the curing oven. Antiblock agents reduce the tendency for paint films to stick together when they are warm or not quite fully cured.

**Antifreeze.** Waterborne emulsions can be ruined if they freeze. Antifreeze materials may be added to waterborne paints to reduce the likelihood of this occurring.

**Blending aids.** These are designed to increase the efficiency and speed of paint manufacturing. Mixing aids such as dispersants, emulsifiers, surfactants, antifloats, and related compounds help simplify production of paint by reducing mixing and blending times.

**Curing agents.** This group of additives can improve paint curing properties. These catalysts, driers, or activators are used to increase resin cross linking during enamel curing to speed handling, packaging, and shipping of painted parts.

**Antimicrobial agents.** Bacteria are able to use the organic resin portions of paint films as nutrients for growth and replication. Ocean vessels are subject to barnacles and other marine organisms that foul ship hulls and increase fuel consumption. Painting with coatings resistant to unwanted organisms reduces the problem. House paints in hot humid regions also need fungicide additives to prevent them turning an unsightly green from mildew.

**Defoamers (antifoamers).** Foam eliminators are frequently necessary with waterborne emulsion paints. This is especially likely in dipping and flow-coating operations. The emulsifiers tend to create foam on waterborne paints, just as oil emulsifying detergents will generate foam in dishwashers.

**Flow control agents.** These agents can ease paint application. Sag balancers, bodiers, flow-control additives (leveling agents), and thixotropes will affect the rheology (flow properties) and viscosity of wet paint films. Polyethylene oxide is used in waterborne paints to enhance atomization at low air pressures. Spraying with low air pressures reduces paint overspray waste. Acrylic lacquers often contain cellulose acetate butyrate (CAB), which eases breakup into spray droplets and improves wet paint flowout. CAB also holds the paint film "open" to reduce the tendency for solvent popping.

**Gloss modifiers.** Various additives can reduce gloss levels without affecting film strength if low or nongloss surfaces are desired. Clays are frequently used for this purpose.

**Softening agents.** These plasticizers can give flexibility to brittle resins such as those found in acrylic and vinyl lacquers. They act as internal lubricants between molecules, increasing their freedom of movement under stress.

**Storage stabilizers.** A paint manufacturer may use additives to increase paint storage stability. These additives will reduce the tendency of paints to skin over and can reduce pigment settling during storage. Stabilizers may also allow a uniform viscosity to be maintained for long periods.

**Thixotropes.** Thixotropes change a paint's shear/viscosity relationship. A Newtonian fluid such as water has the same viscosity at rest as it does when sheared (mixed). A thixotrope added to a paint will cause it to be viscous at rest, but significantly less viscous when pumped, sprayed, or stirred, as shown in Figure 1-8. It aids in preventing paint sagging on vertical surfaces.

**UV stabilizers.** This category of additives can increase a paint film's resistance to ultraviolet (UV) light. Nearly all organic polymers are somewhat susceptible to attack and damage by exposure to UV light. Even natural protein polymers such as human skin can be harmed by sunlight. Frequent exposure to sunlight can damage skin due to the breakdown of organic

# PAINT COMPONENTS AND THEIR FUNCTIONS 15

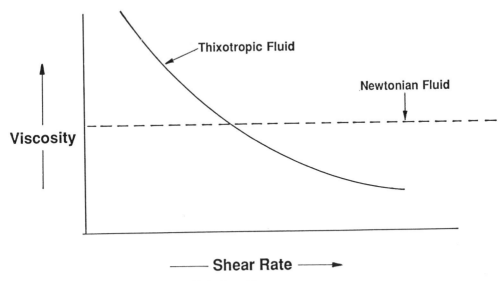

**Figure 1-8. Paint's Shear/Viscosity Relationship**

cellular materials. Skin pigments and sun screening lotions lessen this danger. Nonpigmented paint films lack sun-blocking protection. In nonpigmented coatings, such as spar varnish for marine use and automotive clearcoats, UV light protection is provided by screeners, quenchers, or selective UV light-ray absorbers. UV light is very powerful and tends to degrade exterior finishes on cars, buildings, boats, and homes noticeably more rapidly in sunny regions such as San Antonio and Miami than in the temperate climes of Saskatoon and Minneapolis.

**Antimar agents.** Finely divided teflon powder and various waxes may be utilized to reduce the surface friction on paint surfaces. The potential for scratching and marring is thereby lessened.

**Other additives.** Other additive types include antifloating agents, antioxidants, antiskin agents, antistatic agents, catalysts, coupling agents, dispersants, driers, flame retardants, friction reducers, plasticizers, and thickeners.

# Chapter 2

# Classifying Industrial Paints

## Three Major Groups of Paint

Although paints could be categorized in many ways, it is probably most useful to segregate them into three major categories: (1) trade sale or consumer paints, (2) industrial paints, and (3) maintenance paints. Under each of these major groupings, numerous subcategories are commonly defined.

Trade sale paints are those purchased primarily in relatively small quantities by residents or home and office decorators for residential and commercial painting. The paint is intended primarily for walls, ceilings, and building exteriors. The vast majority are waterborne emulsion paints for easy use and cleanup. Nearly all trade sale paints are bought at a retail outlet, such as WalMart or Sears, or at a paint or hardware store. The shelves of such stores are likely to carry paints that could fall into many classifications including:

- High-gloss
- Semigloss
- Flat
- Deck
- Marine
- Floor
- Oil-based
- Latex
- Varnish
- Shellac
- Lacquer
- House
- Trim
- Enamel
- Wall

Industrial paints are more likely to be purchased directly from the paint manufacturer, often in large lot quantities. Package sizes range from single gallons to 300-gal tote tanks. Barrels and 5-gal pails are also prevalent. Generally, industrial paints are grouped into four distinct main categories:

- End use characteristics
- Types of resin
- Physical makeup
- Cure mechanism

Each of these three categories have been further divided into subcategories. End use characteristics will be discussed later in this chapter; resin types, physical makeup, and cure mechanisms will be discussed in Chapter 3.

Maintenance paints make up an extensive and varied group that is used in exceptionally large volumes each year. Considerable amounts of paint are applied onto objects that are immobile or too large to allow force-curing in an oven. In addition, maintenance coatings are always

# 18 INDUSTRIAL PAINTING

air-dry paints. Examples of their use are traffic-lane markings on roads and parking lots, the interior and exterior of industrial buildings, ships, large construction equipment, and highway bridges and overpasses. Much maintenance paint is purchased directly from the paint manufacturer through large commercial accounts.

## Industrial Paint Types According to End Use Characteristics

This category identifies industrial paints according to either the substrate onto which it is applied or by the way the paint is intended to be used. The following are some examples:

- Primers
- Sealers
- Surfacers
- Basecoats
- Clearcoats
- Topcoats
- Concrete paints
- Wood finishes
- Marine finishes
- Peel coats
- Chemical agent resistant coatings

### *Primers*

A *primer* is a paint formulated to be applied, often directly, to a substrate before any other paint is applied. This is suggested by the name "prime," since it connotes "first." Sometimes only primer is ever applied to an object, but more often it is put on and then another layer of paint is put on over it. A major function of primers is to promote lasting adhesion of subsequent paint layers to the substrate. Often, a topcoat paint, if applied directly to a surface, will not adhere sufficiently. When a primer is applied first and then a topcoat, film durability is substantially enhanced. Another important function of a primer is to isolate the substrate from the effects of weather more completely than is possible by the topcoat alone. This is perhaps of greatest importance for corrosion prone metal substrates, especially ferrous metals. The red color of iron oxide seems to be more unsightly than the white oxide products of zinc and aluminum. Figure 2-1 shows the basics of a metal corrosion process occurring under a coat of paint.

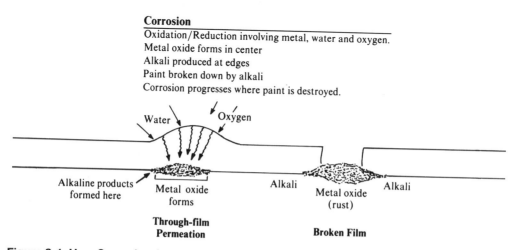

Figure 2-1. How Corrosion Attacks a Paint Film

The most significant types of primer include:

- Wash (etch) primers
- Shopcoat primers
- Shopplate primers
- Flash primers
- Baking and force-dry primers
- Spray primers
- Electrocoat (E-coat) primers

**Wash (etch) primers.** Despite the name, wash (etch) primers do not perform a washing or cleaning function. The name indicates that they are low in viscosity and provide only a very thin or "wash" coating of about 0.25 to 0.5 mils thickness. "Etch" means the primers are formulated to roughen the surface. Wash primers are used to achieve good paint adhesion on clean metal surfaces that have not received a phosphate or chromate conversion coating. Most wash primers are manufactured as two-part systems that are mixed just prior to use.

Wash primers are formulated with an etchant such as phosphoric acid to attack and slightly roughen the surface, giving the wash-primer binder a surface with "teeth" for mechanical bonding. Wash primers most commonly have an acid-resistant vinyl butyrate resin component and, in the past, frequently contained anticorrosion pigments such as zinc chromate or zinc phosphate. The characteristic greenish-yellow color (or dark green) of many wash primers is due to the zinc chromate pigment. Less toxic pigments are now often being used to replace chromium pigments.

Wash primers are intended to be used on clean, bare metal. They are not normally applied over metal that has been phosphated or chromated; however, a conversion coating plus wash primer is required in some military specifications. The acid should be able to react with bare metal. If applied over conversion coatings, the acid in the wash primer may not react and will thus cause blisters when the painted part is exposed to moisture and humidity. Recently, a few military specifications have been updated to omit the wash primer when the substrate is to receive a conversion coating.

**Shopcoat primers.** These primers are usually considered to be temporary coatings of about 1 mil in thickness to protect metal products during outdoor storage. They are especially used on steel to prevent rusting and are frequently totally stripped off before the final painting of a product.

**Shopplate primers.** In shipbuilding, a shopplate primer is applied at the steel mill under carefully controlled conditions. This mill-applied coating is superior to primers that are applied to the steel hull under possible adverse weather conditions. Shopplate primers have excellent durability because the controlled mill environment enhances paint film quality. A shopplate primer is at times also called a shopcoat primer. These primers are often rich in zinc powder for anticorrosion properties, which enables welding of painted parts not possible with most organic coatings. The name "weld-through primers" is also given to such paints.

**Flash primers.** An air-dry primer is called a flash primer because part or all of the solvents evaporate or flash off without the use of added heat before the next coat is applied. The term "flash" distinguishes this class from the baked primers. Sometimes flash primers can be topcoated without first being fully cured; this method of applying the topcoat on the primer is called "wet on wet."

**Bake and forced-dry (forced-curing) primers.** Baking or forced-drying of an applied primer or other paint film can be done if a resin would otherwise cure too slowly at ambient conditions. Many paints will not fully cure unless they are baked because the cross-linking reactions do not begin until bake oven temperatures are reached. This is especially true for powder coats and high-solids paints. These are classified as baking paints.

Other paints will cure slowly at ambient conditions without heat. However, they cure far more quickly at slightly raised temperatures. Forcing can be done with most air-dry paints and two-component (2K) paints. The cure time for such coatings is considerably shortened by "forcing" the process with mildly elevated temperatures. Forced drying or forced curing is usually done so that painted items can be taken off a conveyor line and packaged in less time.

The U.S. EPA classifies bake temperatures below 195° F (90.6° C) as low- or forced-baking conditions and temperatures at or above 195° F as high-baking conditions. This dividing line is arbitrary but reasonable; the government needed it to legally create two distinct classes of coatings and separate regulations for each class.

**Spray primers.** These can be various types of primers that are intended to be spray-applied, as opposed to other types of application. More primers are included in this category than any other. Despite the name, primers of this type can also be applied by rotary atomizers. The name "spray" serves only to show that the application is not performed by immersion methods.

**Electrocoat (E-coat) or electrodeposition primers.** Electrocoating is a technique whereby direct current electrochemically causes paint to deposit on parts that are made the anode or cathode while immersed in special E-coat paints. Depending on the formulation, a deposited E-coat may be considered to be a singlecoat, primer, or topcoat. These are further detailed in Chapter Ten.

### Sealers

The term *sealer* has separate meanings in different industries. In wood finishing, which makes abundant use of sealers, the sealer is applied over a stain. Wood sealers are used to close the pores of the wood, lock in the color of the stain, and raise tiny wood fibers along the edges of the wood grain. Raising these fibers allows them to be sanded readily to produce a smooth surface. This step is crucial for fine-quality furniture finishing.

In metal finishing, sealer function is quite different. The metal sealer may be used between a primer and a topcoat to prevent topcoat solvent migration into the primer. Solvent migration from the topcoat could result in pigment leachout from the primer and discolor the topcoat. This phenomenon is called *bleeding* or *staining*. The leached-out pigment is especially noticeable with dark primers and light topcoats. Solvent migration can also cause swelling of the underlying paint layer, and solvent absorption occurring along fine sanding scratch lines of a paint layer can produce ridges that are visible through the topcoat. The ridges become magnified when they are painted over due to swelling from absorbed solvent. This defect is known as *sand-scratch swelling*.

Sealers are used to bridge large differences in polarity between primer and topcoat, as shown in Figure 2-2. Such large polarity differences reduce bonding adhesion. A low-polarity epoxy E-coat paint can be bridged by a sealer so that a highly polar acrylic lacquer topcoat will adhere. If the acrylic were applied directly over the epoxy, loss of adhesion would readily occur. The situation is similar to the tendency for water to form beads on a waxed surface. Because of their mutual attraction, the highly polar molecules of water tend to pull themselves together into individual drops rather than adhere uniformly to the nonpolar waxed surface. Since lacquers are not much used due to their high solvent content, routine use of paint sealers is seldom required.

# CLASSIFYING INDUSTRIAL PAINTS 21

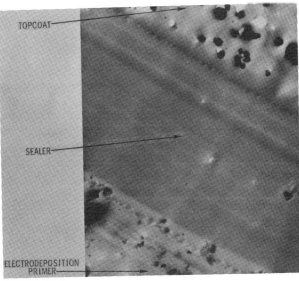

Figure 2-2. Photomicrograph Showing Use of Sealer as Bridge Between Primer and Topcoat

## *Surfacers*

Surfacers are formulated to be readily sanded and are applied to provide a smooth base for a topcoat that must exhibit a high gloss. Smoothness is one of the characteristics of a coating with high gloss. Surfacers are designed to be applied and cured, then sanded as necessary to level out minor roughness. Unless the surface is already smooth, surface sanding is required prior to applying high-gloss finishes. Few paints have enough flow to level out and fill surface irregularities completely. Paints are usually easier to sand than metals, but not all paints can be designed as surfacers. By the proper choice of resin and pigments, paints designed as surfacers are made easy to sand. Because sanding is slow and costly, it tends to be done only where necessary and then only on parts where the cost can be justified.

## *Basecoats*

Any paint applied before the topcoat can be considered a basecoat. The term also can refer to a pigmented (color) coat that will receive a clearcoat over it. A number of industries whose products are enhanced by high gloss use basecoat/clearcoat technology to enhance the brilliance of a finish. The basecoat provides the color; the clearcoat furnishes the shine. These are generally applied over primer coat(s).

European auto manufacturers developed basecoat/clearcoat technology in an attempt to match the gloss of metallic finishes on cars produced in North America. In the 1970s, North American cars were finished mostly with acrylic enamels and lacquers, while European cars were finished mostly with alkyd melamine topcoats. The acrylic enamels and lacquers satisfactorily covered metallic flake added to the paint; the alkyd melamine tended to leave fragments of the flakes extending out of the paint, lowering the gloss.

22  INDUSTRIAL PAINTING

The European car manufacturers, in cooperation with their paint suppliers, developed an unpigmented paint to solve this problem. When applied over the normally pigmented alkyd melamine metallic topcoats, the clearcoat effectively buried the metallic flake, creating a smooth paint surface. The smooth surface exhibited exceptionally high gloss. It was, in fact, noticeably higher in gloss than the North American topcoat acrylic finishes. Japanese car producers immediately adopted this clearcoat technology for their metallic finishes. North American auto manufacturers soon followed suit.

Most automakers worldwide use clearcoats on both nonmetallic and metallic colors because of the high gloss they provide. Acrylic, polyester, and urethane topcoat materials have been used for the base colorcoat and the clearcoat finishes. The film thicknesses on North American vehicles typically are 0.75 mil of metallic color basecoat plus approximately 1.2–1.6 mils of clearcoat applied over it. Nonmetallic shades use more basecoat and slightly less clearcoat.

The latest development in low-VOC basecoat/clearcoat technology uses a thixotropic waterborne color basecoat that is dehydrated in a quick bake to remove nearly 95% of the water. The

Figure 2-3. Brass Bed Coated with Clearcoat

waterborne color basecoat is then topcoated with a high-solids, powder, or waterborne clearcoat. The best metallic appearance is achieved with color basecoats that have a large shrink factor, that is, paints that contain high percentages of volatile materials.

Since VOC restrictions negate the possibility of using organic solvents in large quantities, the method of choice must be to use a safe and nonpolluting solvent to achieve the necessary high film shrinkage. Since water is the only solvent that meets these criteria, waterborne colorbase technology has been perfected. Waterborne clearcoats further reduce VOC emissions.

### *Clearcoats*

Clearcoats are defined as paints without pigments. Figure 2-3 shows a brass bed coated with a clearcoat to prevent tarnish. Figure 2-4 shows drawings of metal flake projecting through the surface, reducing film gloss. A clearcoat applied over the projections would increase gloss. A clearcoat applied to wood is usually termed a varnish, but some BMW car advertisements call their exterior-finish material a varnish. Considerable work has been and is being done to develop powdered paints for use as clearcoats to reduce VOC emissions because current clearcoats have a relatively high solvent content. Special auto paint effects are achieved using a tinted clearcoat plus a plain clearcoat. This technique is just beginning to be used for some pearlescent colors.

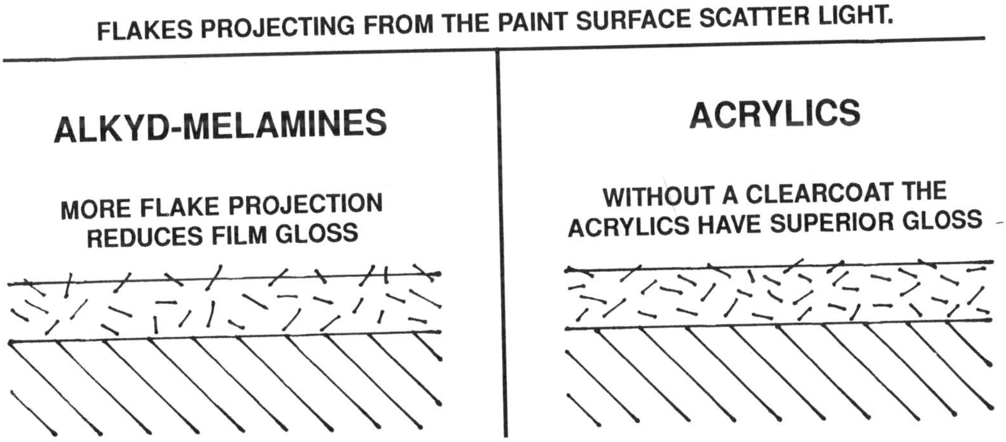

**FLAKES PROJECTING FROM THE PAINT SURFACE SCATTER LIGHT.**

| **ALKYD-MELAMINES** | **ACRYLICS** |
|---|---|
| MORE FLAKE PROJECTION REDUCES FILM GLOSS | WITHOUT A CLEARCOAT THE ACRYLICS HAVE SUPERIOR GLOSS |

Figure 2-4. Typical Metal Flake Orientation Patterns

### *Topcoats*

The topcoat's major function is to provide a pleasing appearance to the surface it covers. The gloss of a topcoat may vary from none for a photographic darkroom paint to the very high gloss of a bicycle paint. Figure 2-5 shows a photomicrograph cross-section of a topcoat and a primer. Finely ground pigments in the topcoat enhance topcoat gloss; coarse pigments reduce gloss.

## 24  INDUSTRIAL PAINTING

Topcoats must also exhibit the proper shade and intensity of color. The topcoat, being the most exposed surface, also needs to protect the surface during exposure to the normal service environment. This might require a particular ability to fight abrasion, withstand sunlight, resist attack by chemicals, or be durable under special exposure conditions. Sometimes a textured or wrinkled surface is achieved with select resins that distort during curing (but textures are also produced by special spray application techniques).

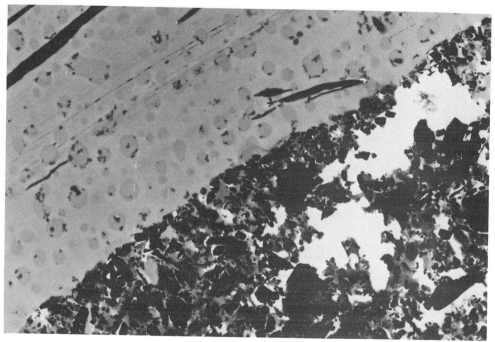

Figure 2-5. Photomicrograph Showing Difference Between Finely Ground Topcoat Pigment (at top) and Coarsely Ground, Heavily Concentrated Primer Pigment

### Concrete Paints

Some paints are formulated especially for concrete roadways. Road life can be significantly extended with these coatings by excluding moisture that can cause severe damage from cracking in freeze-thaw cycles. The paints are extremely durable to withstand constant weather exposure and traffic abrasion.

### Wood Finishes

As the name indicates, these are pigmented and clear coatings formulated for wood. They include fillers, stains, toners, varnishes, sealers, and lacquers. To avoid water loss from the wood, these finishing materials are cured at temperatures at or below 180° F (82° C), rarely at higher temperatures.

## Marine Finishes

Coatings used in a marine (water) environment must have extraordinary resistance to water and humidity. They are applied to boats, docks, harbor buildings, and pilings, as well as other waterside or shore areas.

## Peel Coats

Unlike most paints that are designed to stick, peel coats are designed for easy removal and are typically applied to surfaces that require frequent cleaning. The interiors of paint booths, for example, are often protected with this type of coating. After a period of use, the coating can be readily peeled off, removing all built up overspray and any accumulated dirt along with it. The surface is given a new peel coat immediately after the old one is removed, allowing fast and easy periodic cleaning of the surface.

Surplus military ships, tanks, helicopters, howitzers, and similar items are protected during long outdoor storage with these peelable coatings. "Mothballing" in this fashion combines a sheltered storage condition with easy removal for quick reactivation of the equipment when needed.

## Chemical Agent Resistant Coatings (CARC)

These were developed for military vehicles and equipment to permit ready decontamination in case of chemical, biological, or nuclear incidents. Equipment used in the nuclear power industry is also coated with these finishes. They are formulated to resist strong decontaminant chemicals.

# Other End-use Classifications

Often the protective and decorative functions required of a paint are beyond the capabilities of any single coating formulation. Then two or more different paints might be used for specific aspects of the job. For example, a Canadian locomotive manufacturer uses an alkyd primer for adhesion and corrosion resistance. Next, the manufacturer applies a modified acrylic primer surfacer for smoothness and some additional corrosion protection. The third and final coat is a polyurethane for good gloss and long term weatherability. As a result of applying three different paints, the overall coating system has superior performance compared to what could be achieved by any one paint alone.

## Singlecoat

The term *singlecoat* indicates that only one coat of paint is used. The coating must be able to perform all of the functions expected of the paint on a particular product. The paint might be expected, for example, to have an attractive color and to resist marring and scratching. It is less costly to use just a single application if possible. Some products may be quite suitable for singlecoats, especially if they are low cost or have a short service life expectancy. Singlecoats are appropriate on lawnmower blades or broom handles, for example.

### Multicoats

The term *multicoat* indicates that two or more coatings are applied to a surface. Each paint type has its own special function(s) to perform. By using several different formulations, the overall paint system is stronger and has better performance than any one coating material alone. A wood cabinet may be stained to the proper color, then sealed for sanding smoothness, and then doubly varnished to highlight the finish and protect the wood. Similarly, a bulldozer or farm tractor is epoxy primed for corrosion protection, then topcoated with acrylic enamel for durability in sun and rain and snow. Again, the cost of the item and its expected service life dictate whether or not multicoats can be used cost effectively.

# Chapter 3

# Industrial Paints Categorized by Resin Category, Physical Makeup, and Cure Mechanism

## Resin Categories

It makes sense to name a paint after the category of the resin because the binder, as the film former is also called, constitutes the most important part of a layer of paint. The inherent chemistry of a resin and the extent to which it is cross-linked gives it (and the formulated paint material) its specific properties. For example, the hardness and flexibility of a dry paint film

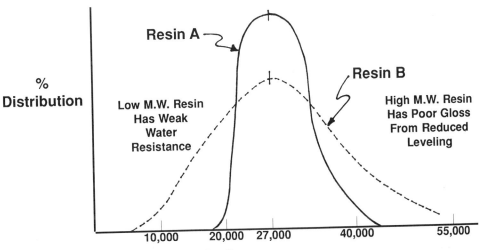

Figure 3-1. Molecular Weight Distribution Curves

vary according to the type of resin in the paint and with the average size and size-distribution range of the resin molecules in a given paint batch. Figure 3-1 shows the molecular weight distribution curves of two resins. Notice that the narrow curve indicates improved paint properties. Figure 3-2 depicts how paint film properties will vary according to the film's softness or hardness.

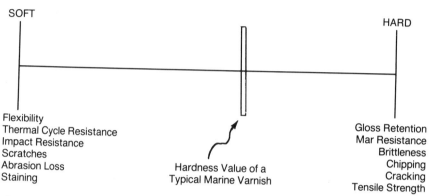

**Figure 3-2. Resin Soft-to-Hard Ranges and Related Properties**

Stating that a paint is, for example, an acrylic, reveals a great deal about the paint. An acrylic paint will have the characteristics of an acrylic resin, modified somewhat by the pigmentation and additives. Within each resin category, several different but chemically related monomeric units may bear the same general name; thus, there are numerous variations among the many possible acrylic types, vinyl types, or alkyd types, for example. Any description of resin properties will be typical for the resin as a category only, and may indeed vary to a degree from one resin of a given molecule type to the next.

Many chemical types of film-forming resins are used as binders, both alone and mixed with other chemical resin types. The following is a list of some of the major resin types and their typical properties based on the chemical nature of the resin molecules:

- Acrylics
- Alkyds
- Cellulosics
- Epoxies
- Halogenateds
- Oleoresins
- Phenolics
- Polyesters
- Silicones
- Urethanes
- Vinyls

## Acrylics

Acrylic resins are able to achieve a high gloss and form hard, highly weatherable surfaces. The hardness of acrylics contributes to good abrasion resistance and increases durability. Acrylic resins are above average in cost. They have very good heat stability and excellent chemical and ultraviolet light resistance. Nearly every truck and car produced in North America from about

1950 through 1990 received either an acrylic lacquer or an acrylic enamel topcoat because of these strong performance properties. Even today the acrylics still make up 90%–95% of topcoats. Paints with hard surfaces have maximum abrasion resistance, but if paint surfaces are too hard, they also tend to be brittle. This can result in a proclivity toward chip formation from stone bruises, bumps, and related impacts. Acrylics are formulated to minimize their tendency to exhibit this defect.

## Alkyds

This resin name, often said incorrectly, is pronounced "Al Kid." The name "alkyd" is derived from the "alcohols" (alk) and "acids" (yd) that react to form them. The alkyd family of resins is widely used because it is so varied; each type of alkyd can be cross-formulated with many different binder chemicals. In this way, it is possible to produce paints with an enormous variety of specific performance properties. Alkyds come close to being "the general-purpose paint" because of their good overall properties plus their moderate cost. Alkyds are chemically modified vegetable (plant) oils, especially soya and linseed oils. Because they incorporate drying (air-drying) oils, they are often used to produce ambient-cure paints. This is an important factor in why alkyds are one of the most heavily used paint resins. The carbon-to-carbon unsaturated intramolecular bonds (double bonds) in alkyds (and in all drying oils) react with oxygen in the air to produce fully cured finishes. Alkyds are the major resins used in combination with others to introduce air-dry capability in a paint formulation. They can also be force-cured and formulated into baking finishes.

Alkyds can be modified with various chemicals to increase specific properties. Styrenated alkyds, for example, have increased chemical resistance compared with pure alkyds. Alkyd melamine resins give durable finishes with excellent weatherability, but they are also somewhat higher in cost than some alkyds. European automobile companies have used alkyd-type paints for primers and topcoats for many decades. The alkyd-melamine finishes are normally used only in baking finishes.

## Cellulosics

These inexpensive materials are produced by the reaction of nitric acid with cellulose, which forms nitrocellulose lacquer resins. They are highly polar and hence limited in solubility, tend to be brittle, and have quite poor heat resistance. They tend to be used because they are relatively low in cost. Although they have poor to fair chemical resistance, cellulosics are attractive for use in low-priced, quick-drying lacquers. They are sometimes used in strippable coatings.

Because of their limited durability and high solvent content, they are now not widely used in industrial painting. Nitrocellulose lacquers were first used for automotive topcoats in 1928 because they air-dried in hours. This property made them more efficient than the oil-based paints that had been used that required a week or longer to fully cure.

## Epoxies

The epoxy resins are produced by the reaction of epichlorohydrin with bisphenols. They have excellent water resistance and one of the highest resistances to attack by alkali of any commercial resin. They form hard, tough, salt- and water-resistant coatings and exhibit superior chemical resistance as well. Their excellent heat and abrasion resistance makes them ideal primers. They are one of the best metal primer resins that can be used in industrial applications. The epoxies are somewhat costly. All one-part enamel epoxies require baking; many formulations require substantial oven times and temperatures.

Two-part epoxies produce equally durable finishes and have an added advantage over one-part epoxy resins. Two-part epoxy paints can be used for ambient-temperature curing when it is impossible to bake the part. This capability is especially valuable in fields, such as the marine industry, where extreme corrosion resistance is required and painting must be done under any weather conditions during all four seasons of the year.

The major weakness of epoxies is their tendency to chalk when exposed to ultraviolet light. This chalking contributes to loss of gloss but is rarely extreme enough to interfere with the structural integrity of the coating. Epoxies should not be used where appearance is critical or on items that could release chalk onto a person's clothing. They are ideal paint resins wherever excellent corrosion resistance and strong chemical resistance are important. Large-scale industrial use of epoxies is directly attributable to their outstanding protective qualities. Because of their strong resistance to degradation by alkali, epoxies are the most frequently used resins for metal primers.

## Halogenated Resins

Fluorocarbons and chlorinated rubber resins are employed in the formulation of superior long-term weather-resistant maintenance coatings. They are expensive, often very difficult to apply, and have relatively low heat resistance. Their chief virtue is their ability to be used in harsh climate environments and to provide up to 30 or 40 years of reliable service. Halogenated resins are used to decorate and protect exterior surfaces of metal panels and similar decorative structural members in architectural construction. Large, modern office buildings use components coated with these resins for striking visual effects.

## Oleoresinous (Oil-Based) Coatings

Oleoresins are some of the oldest resins used to make coatings, but these low-cost binders can still be used to make good paints. Although they dry slowly, drying agents can be added to speed curing. Driers (or catalysts) are frequently salts of lead, cobalt, and manganese. The drying oils (linseed, soya, tung, castor, cottonseed, etc.) form the basis for these binders. They are often modified with synthetic resins to improve drying time or to increase ultimate hardness. Drying oils are sometimes mixed with alkyds to reduce paint costs for noncritical applications.

## Phenolics

Paints and varnishes made with phenol resins tend to be very hard and somewhat brittle, yet they are also extremely resistant to stain, solvent, and acids. Phenol resins are often used for coatings on copper wire and windings for electric motors and transformers. They are not as good for decorative finishes. The aromatic structure of the molecules causes phenolics to turn yellow or brown in sunlight; however, they have been used for dark marine-varnish finishes. Their acid resistance makes them a good choice for many food can linings. Phenolic resins are also used in anticorrosion coatings.

## Polyesters

Polyesters, except for the lack of carbon-to-carbon double bonds, are otherwise much like the alkyds in chemical structure. This is one reason they are used for a wide range of coating needs. They are also moderate in cost. Polyesters do not have unsaturated linkages as do alkyd resins, and for this reason they sometimes are called the "oil-free" alkyds. Without double bonds they cannot react with oxygen, so they do not air-dry. Polyesters are available only as baking finishes.

Polyesters have found extensive application as powder coatings since the early 1980s. They are tough and durable in outdoor exposure, exhibiting good flexibility and excellent hardness in both wet and powder coatings. Two major polyester powder coatings are being used: isocyanate cross-linked and TGIC (triglycidyl isocyanurate) cross-linked. The isocyanate cross-linked powders provide an attractive but slightly less durable finish than the TGIC cross-linked systems. The TGIC powder coatings have outstanding exterior durability. Early reports suggesting that TGIC be outlawed for paint use because in powder paints it initiates health problems have been strongly refuted so TGIC use continues.

## Silicones

Silicone (organo-silicone) resins have superior resistance to water, sunlight, and all other normal exterior surface conditions. They do not chalk. Their most significant attribute, however, is high heat stability, which permits silicone paints to be used wherever conventional organic resin coatings would deteriorate rapidly. Although some are more heat-resistant than others, all organic resins will degrade in sustained heat of 500° F (260° C) or higher. Silicone resin paints are used for applications such as motorcycle mufflers to withstand temperatures of 900° F–1200° F (482° C–649° C). Some can withstand the heat of 2500° F (1371° C) and are used to mark identification codes on the heat-tiles of space shuttles. When these paints reach their use temperatures, the organic portions of the resins oxidize away. This leaves an essentially ceramic coating that is stable at elevated temperatures. The high–heat-resistant silicone paints require long curing times at temperatures of 650° F (343° C) and above.

Silicones are also used at ordinary temperatures when blended with acrylics or alkyds for finishes that have unequalled weatherability and chemical resistance. Alkyd silicones are available in air-dry coatings that are utilized as marine paints to maximize the time between repainting of transoceanic merchant vessels. Silicone maintenance coatings are used on water towers, steeples, and bridges for which their high cost is justified by the complexity and expense of paint application on such structures.

## Urethanes

Among the most outstanding resins in paint usage are the urethanes because of their ability to combine gloss and flexibility with great chemical and stain resistance. Figure 3-3 compares the properties of urethanes with those of acrylics, alkyds, and epoxies. The cost of urethanes is two to five times that of other paints, but they are unique in their amazingly high gloss levels throughout a wide flexibility range. Urethanes make ideal high-solids coatings that require little, if any, additional heat for curing. For coating heat-sensitive substrates with a low–solvent-content paint, urethanes are often the materials of choice. They are physically durable and show excellent water and weather resistance.

| Type | Properties and Characteristics |
|---|---|
| Acrylics | Excellent weather resistance<br>Chemical resistance<br>High heat stability<br>Above average cost<br>Hard, abrasion resistant<br>Slightly brittle |
| Alkyds | Very good overall properties<br>Average cost<br>Air-dry curing<br>Extremely heavy use in paints due to their fine performance<br>Formulation widely variable |
| Epoxies | Outstanding water resistance<br>Tough, heat and abrasion resistant<br>High alkali resistance<br>Above average cost<br>Chalking in sunlight |
| Urethanes | Superior overall properties<br>Excellent mar resistance<br>Flexible, with high gloss possible<br>Outstanding durability<br>Relatively high cost |

Figure 3-3. Comparison of Paint Resin Properties

At least six types of urethane paints are possible, but two-part urethanes (polyol and isocyanate) have the best film properties. Prereacted (one-part) urethanes are sometimes used to form a blend with other resins. The advantage of prereacted urethanes is that no potentially toxic isocyanates are present.

A urethane is formed when an isocyanate chemically reacts with an alcohol. Unreacted isocyanates can cause respiratory problems if an individual breathes the vapors for extended periods. Approved charcoal-cartridge filter masks or head shrouds with a separate air supply are recommended for workers applying two-part urethanes. No filter or cartridge masks are approved by the Occupational Safety and Health Administration (OSHA) or the National Institute of Safety and Health (NIOSH). To gain OSHA or NIOSH approval, the mask wearer must be able to detect any possible malfunction of the device. Since isocyanates have no recognizable odor and since no fail-safe indicator is available to reveal when a cartridge is spent, face masks cannot be approved for use when applying isocyanate-containing paints. Another problem with masks is that facial hair prevents a tight fit; bearded workers may not be adequately protected by masks.

Recently introduced isocyanate-free urethane-type paints and urethane-like coatings, which are for all purposes nearly identical in performance to isocyanate-containing urethane coatings, are considerably safer. A chemical variation of the epoxy linkage is used to cross-link these systems. Because they have outstanding properties, urethanes are used where their flexibility, gloss retention, superior weatherability performance, mar resistance, and general durability can justify the higher coating cost.

## Vinyls

These resins can be rather rigid, or they can be heavily plasticized to achieve great flexibility if desired. Vinyls tend to be low in cost yet have extraordinary acid resistance. They possess outstanding water resistance as well, which makes them a frequent choice for maintenance coatings. Their exterior durability is excellent, although vinyls tend to degrade in heat. For applications in wet environments such as offshore oil rigs, as shown in Figure 3-4, vinyls make excellent coatings. Polyvinyl chloride (PVC) coatings are extensively used as can linings. Solution vinyl coatings also find frequent use in peel coats for spray booth maintenance.

**Figure 3-4. Air-dry Vinyl Resins Are Used in Wet Environments Such as Offshore Oil Rigs**

For structural wood finishes where moisture must be allowed to escape from the building interior, polyvinyl acetate is frequently used because it is porous enough to allow the water vapor to "breathe" through the film. This avoids the blistering that frequently occurs on wood siding when a nonporous paint traps moisture under the film. The water vapor pressure lifts the paint off the surface and either blisters or ruptures the paint film.

## Resin Mixtures

Mixtures of binder resins are possible both within a resin family and in combination with other resin groups. Various acrylic resins are available for the formulation of an acrylic paint, or various vinyl materials can be used in the manufacture of a vinyl paint, for example. But interfamily resin mixtures are also used to produce cost-effective products or to fit a special performance requirement. Thus, a coating formulation may contain an epoxy-polyester or a vinyl-alkyd binder. A paint may have a polymer blend that contains any number of compatible binders.

## Physical Makeup Categories

The third major way to classify a paint is according to its physical makeup. Sometimes the physical makeup differences are obvious, and sometimes they are not. The physical makeup of a two-component paint, for example, is obvious; it comes in two containers that are mixed together before application. Another paint with an obvious physical makeup is a metallic coating.

Sometimes the differences in physical makeup are not as obvious. Consider the terms "solventborne" and "waterborne." A *solventborne* paint uses a traditional solvent to disperse the resins, pigments, and additives; a *waterborne* paint uses mostly water. Note that the terms are "solvent(borne)" and "water(borne)" instead of "solvent(based)" and "water(based)." The term "borne" merely implies "carried or transported by," as in the expression "airborne," which means an object or substance is held or supported in air. The term "based" refers to the paint's film-forming material. A paint can be acrylic-based, epoxy-based, or alkyd-based, for example. Thus, although solventborne or waterborne, a paint can be silicone-based or vinyl-based (etc.), as well.

## Liquid Fluidization Methods

A less obvious physical property of paints is their method of liquid fluidization. The term *liquid fluidization* refers to how the paint solids are allowed to flow by mixing resins with a liquid solvent or liquid carrier. This liquid fluidization should not be confused with the air fluidization of dry powder paints; that subject is covered under powder coatings.

As shown in Figure 3-5, liquid paints can be classified into three major fluidization types: solutions, dispersions, and emulsions. Figure 3-6 compares their viscosity and percent solids.

**Solutions.** In a solution paint, each molecule of resin is dissolved in the solvent. The individual molecules are essentially floating in a large sea of solvent. As increasing amounts of paint resin are added, the viscosity continues to rise proportionally. This occurs because the long thread-like resin molecules physically intertangle with each other. The entwined molecules resist flow and thicken the paint. It is desirable to minimize solvent use, but this would excessively increase the viscosity. The answer to producing a high-solids paint is not simply to withhold solvent; this creates a thick paint that would be difficult or impossible to apply.

**Dispersions.** One way to lower solvent use is with a resin dispersion. A resin uses a blend of solvents with low polarities that force resin molecules into a mild clumping or aggregation. A dispersion paint can force resin molecules to cluster into small nodules containing 10–25 molecules. The viscosity curve for a dispersion paint (see Figure 3-6) has a long plateau region

## INDUSTRIAL PAINTS CATEGORIZED 35

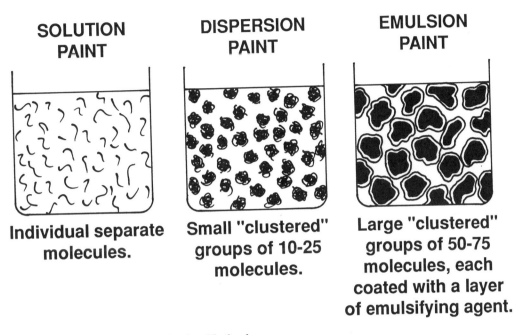

Figure 3-5. Major Resin Fluidization Methods

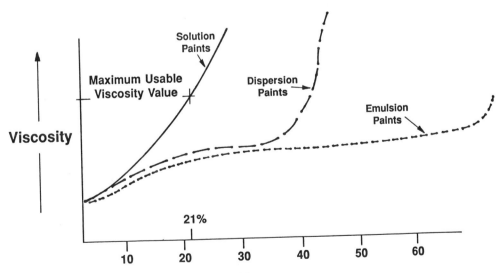

Figure 3-6. Viscosity vs. Percent Resin Solids

into the high-solids range. Viscosity rises rapidly when the concentration gets so high that individual nodules of resin contact each other and begin tangling. The advantage of a dispersion is a reduction in the amount of solvent needed. It produces a paint with a higher solids content at a given viscosity as compared with a solution paint.

**Emulsions.** Producing even larger nodules with perhaps 50–100 molecules in each cluster would allow formulation of paints with even less solvent. The difficulty is that dispersion technology cannot be extended quite that far. Resin nodules of that size rapidly agglomerate and produce a second layer.

The situation is similar to shaking a vegetable dressing with a salad oil and vinegar. Vigorous shaking produces only temporary blending. After standing for just a short time, the oil floats to the top again. Many decades ago condiment manufacturers discovered permanent blending of oil and water could be accomplished if the oil particles are coated with a thin protein layer. The addition of egg to a mixture of salad oil and water (plus a few spices) followed by vigorous blending produces mayonnaise. The separate oil and water phases of this seemingly single-phase product are visible with a microscope.

The egg protein coating on the salad oil droplets is called an *emulsifier*. Although neither protein nor egg is used, emulsion paints are manufactured in a similar fashion. In this case, the emulsifier surrounds the paint resin nodules to prevent forming separation layers. Emulsion technology can produce coatings with higher solids than dispersion technology.

Both dispersions and emulsions can be produced in solventborne or in waterborne formulations. Wall paint for room interiors and house paint for exterior use are nearly all waterborne emulsions, which are often called latex paints. The sap from a rubber tree is a naturally occurring waterborne emulsion and hence was called latex, but the term latex does not imply that the paints contain natural or synthetic rubber-like resins. *Latex* is a chemical term that indicates a waterborne emulsion of organic polymer; for this reason, household paints are often labeled and advertised as being latex paints.

The most significant thing about latexes (latices) is that they use water to fluidize materials that inherently are not mobile fluids and have virtually no water solubility. An example of this is butter solubilized in water. Cow's milk contains on the average about 12% cream. Cream, the major source of dairy butter, behaves as if it were a single-phase liquid. However, when churned, cream separates into clumps of butter and a watery residue called buttermilk. During churning, a protein emulsifier coating is rubbed off the butter particles, permitting the butterfat globules to join together on contact. They will form a large chunk of butter separate from the watery phase. Trying to reverse the emulsifying process is difficult to accomplish, yet that is essentially what is done in the manufacture of latex emulsion paints.

Emulsion paint manufacturing requires an emulsifier to stabilize the mixture of water and resin. Dispersions differ in that the resin clumps are small and do not require added emulsifiers to form a stable fluid mixture. Vigorous physical mixing alone is employed to form the dispersion.

## Cure Mechanism Categories

### Lacquers

Lacquers are paints containing binders with large organic molecules that are fully polymerized before they are applied. After the lacquers are applied and cured, their resins can be softened by heat or solvent. This allows lacquer paint films to be readily spot-repaired. Heat will reflow the resin in the paint film and remove minor scratch lines from repair sanding.

Solvent application to a cured lacquer paint film can dissolve the resin. This facilitates firm bonding between the film being repaired and a newly applied lacquer repair coat. Hot solvent vapor reflow has also been used for lacquers, but like lacquer paints themselves, is now rarely used. This procedure used an oven or a methylene chloride vapor gun to achieve a smooth paint surface with minimal hand polishing and no respraying of the sanded item. Reflow was low in cost and fast compared to manual polishing or respraying.

Lacquer paint films cure through solvent evaporation alone; however, heat may be used to accelerate the evaporation of the solvents. The preparation of lacquers requires a considerable amount of solvent. The large binder molecules need aggressive solvents with strong dissolving properties and oxygenated solvents such as ketones, alcohols, and esters. These solvents cause lacquers to tend to attack enamel paint films, causing blistering and wrinkling.

Since they undergo no cross-linking, lacquer molecules must be highly polar to achieve sufficiently hard and physically durable films. Dipole-to-dipole (magnet-like) bonds attract and hold lacquer molecules together firmly in the dry paint film. Lacquers are often easy to spray without producing runs or sags. Because lacquers form a harder film, they tend to experience chip damage more readily than enamels.

### Enamels

Enamel paints contain resin molecules that are unpolymerized or only partially polymerized. Enamel molecules chemically link together after the paint has been applied. Fully cured enamels are to a degree thermosetting due to the additional chemical cross-linking that occurs among the molecules after the coating has been applied. Sufficient cross-linking must take place to provide the requisite hardness and water resistance to the film; however, the extent of cross-linking in enamels is normally only moderate. If too much cross-linking takes place, the paint film becomes overly hard and brittle with poor chip resistance (low impact strength).

Enamel curing begins to take place rapidly when the solvent evaporates and cross-linking is activated by heat or radiation. The resin links together in many places along the molecule chains by irrevocable chemical reactions. If solvent evaporation alone took place with enamels, a soft, gummy film would result.

Resin molecule cross-linking takes place by a number of mechanisms. Binders may cross-link by reaction with the oxygen in the air; for example, alkyds and oil-based coatings cure this way. Other enamels may react with moisture in the air and undergo subsequent chemical reaction. Most frequently the molecules undergo self-reaction of chemical groups present within the resin. These reactions occur rapidly only after appreciable amounts of the solvents have evaporated and the molecules are in close proximity to each other. Even if molecules are near each other physically, they must be energetically activated or they will not cross-link. Depending upon their chemical properties, resin molecules may be selected that will be activated by means such as heat from a curing oven or by energy absorption from infrared radiation, ultraviolet light, or electron beam rays.

The chemical bonds that are formed thus convert the soft, easily deformed resin into a firm, dry paint film. The paint formulator can chemically modify resins to produce a soft, flexible film; a hard, highly cross-linked film; or something intermediate, depending on which type of film is most appropriate.

Lower cost hydrocarbon solvents can be used in enamels. The short, small molecules of enamel resins dissolve readily with small amounts of solvent. The mild solvents used in enamels rarely adversely affect lacquer paint films.

Enamel paint films are generally not so extensively cross-linked that they become brittle or able to resist being softened by many solvents. But they are sufficiently thermoset so that unlike lacquer paints, they cannot truly be redissolved by solvents or be extensively softened with the application of heat. Not surprisingly then, attempts at producing good results with thermal reflow enamels have not been successful.

A lasting repair on enamels usually must include repainting an entire panel. Once the original paint is cross-linked, it can no longer cross-link with the paint in the repair area. Adhesion between the original and repair paint molecules is therefore due only to mechanical and not chemical hold. Spot repair of enamels is sometimes performed, even on high-priced items such as autos, but it is not recommended. Enamel spot repair is always inferior to repainting an entire panel.

# Chapter 4

# Low- and High-solids Coatings

## The History of Low-solids Coatings

An understanding of low-solids coatings is important to serve as a building block for understanding high-solids coatings. The chemistry of these two coatings systems is considerably different. *Low-solids coatings* are characterized as having a low percentage of solids, both by weight and volume percent (weight percent solids is always higher than volume percent), and therefore a high percentage of volatiles (solvent).

Low-solids coatings have been manufactured and used for a long time. Their formulation and use, however, multiplied greatly after the beginning of the petroleum industry in the 19th century. As the pumping of oil increased, numerous oil-derived solvents and resins began to be developed. Coatings were formulated for thousands of products.

In its early days paintmaking was not a scientific endeavor; the craft evolved purely by trial and error. At the time, almost anyone could make a paint and prosper (provided one had a flair for marketing). One could say with some accuracy that in the early 20th century, the integrity of many paints was in question. Because early paintmakers were not chemical formulators, their products were often mixtures hastily put together by technicians.

However, as time went by, the chemical reliability and batch-to-batch uniformity of the coatings began to improve. The chemical improvements went hand-in-hand with developments in the industrial revolution and with improvements in the mass educational system. As more chemists became available, paint companies, especially those that foresaw the day when paints would have to measure up as a quality product, began to hire them. Chemists were needed not only to develop new formulations but also to devise ways to maintain consistency from one batch of paint to the next. As a consequence, in the 1920s, paint formulation consistency was beginning to be established. This improved quality and performance in low solids formulations increased steadily until 1966. Patents that departed from the typical low-solids technology were being filed during this period, yet petroleum solvents continued to be so cheap that low-solids paints remained as the most practical coatings for most products. Nearly all of the products were variations of what are now called low-solids paints.

Although some low-solids coatings made prior to the mid-20th century were of poor quality, many were excellent. Almost everything that was painted was being finished with these coatings—cars, appliances, airplanes, furniture, and most other products. However, the solids content of these coatings was extremely low, and the solvent content was very high. The solvents were an effective, convenient, and inexpensive means of serving as a fluidization medium for the resins, pigments, and additives. Lacquers were still in wide use with a volume solids content of 10% and less and a solvent content of 90% and more. Most of the industrial coatings by the 1960s probably averaged no more than 20%–30% in solids content.

By the late 1960s, total solvent use for paints had soared, and this large-scale use continued into the early 1970s by sheer momentum. However, some paintmakers, seeing the excessive solvent use as environmentally ominous, were trying to market coatings with higher solids content. But industry forces, principally the low cost of solvent, were not ripe for successful marketing. Solvents were just too cheap to resist using them abundantly. No market force was present to serve as a catalyst to change the technology.

However, in 1966 the first U.S. environmental regulation was adopted in California limiting solvent emissions due to the recognition of the need for solvent emission regulations to curtail the increasingly poor air quality in locations such as the Los Angeles Basin. Then the 1973 Arab oil embargo shook an energy-wasting world into action. These two strong forces precipitated a crisis in paintmaking. Solvent use began to be restricted through national environmental regulations, and at the same time, solvent costs began to rise sharply due to the oil shortage. In short order, solvent content became a major criterion in paintmaking. The industry had arrived at a major turning point onto a path of reduced solvent content in paints, a path the industry is still following today.

The emerging solvent emission regulations caught the industry almost totally off guard. National seminars began to be held annually in the mid-1970s dealing with high-solids and waterborne coatings. Formulators of paints and those who applied coatings flocked to these meetings to learn how to apply or make coatings that would comply with EPA regulations. Marketing projections for low-solids coatings began to turn downward, and those for compliant coatings began to rise.

Most emission regulations now allow low-solids coatings to be used only in limited amounts. Limited exceptions may be allowed if the solvent vapors are destroyed by incineration, bioremediation, or capturing through devices such as carbon adsorpters. The future use of low-solids coatings, if not eliminated entirely, will be limited to companies that can prove the need to use them and are large enough to be able to afford solvent capture or destruction systems. Companies too small in size to be covered by emission regulations and companies in countries not having restrictions may also be able to continue for a time with low-solids paints.

## High-solids Coatings

The EPA's limitations on low-solids coatings paved the way for the emergence of high-solids coatings. The term *high-solids coatings* can broadly include solventbornes, powder coatings, radiation cure coatings, and, under some circumstances, waterborne coatings. In this chapter, however, the term *high-solids* will refer especially to solventborne high-solids coatings. (Waterborne coatings of all types, whether high or low in solids, will be considered in Chapter 5 because they can also be considered a separate and unique class of coatings. Powder coatings, which are certainly "high-solids" because they are practically 100% solids, will be discussed in detail in Chapter 6.)

Some confusion exists among paint formulators and finishers as to what the solids percentage must be to qualify a paint as a high-solids coating. Many in the coatings industry arbitrarily consider that if it is applied at about 50% solids or higher, a coating can be classed as high-solids. That is a convenient way to define them even though it is not a strict rule.

Recognizing the difficulty of defining a high-solids coating, some prefer to call a waterborne or solventborne coating of considerable solids as one having a "higher" solids content. The word "higher" presents another problem: higher than what? It is similarly inexact and its use is not encouraged.

Most high-solids coatings—as applied—typically fall into the range of 50%–70% solids. A few 100% solids coatings are used, especially in radiation-cure finishing, but very few paints are used in general manufacturing that are much above 70% solids as applied.

## High-solids Resins

Easily applied high-solids coatings cannot be produced simply by increasing the solids content in a paint formulation. Remember that fluid viscosity rises rapidly as the percent solids increases in solution-type paints. Only by altering the chemical nature of the resin molecules is it possible to produce a high-solids coating that has a viscosity low enough for normal application. The key difference is that low-molecular-weight resins must be used to produce high-solids coatings. Chemically reactive short resin molecules at a low polymerization level are used in the high-solids formulations. The fluidity of resins will increase as the molecule size is reduced. Short molecule chains do not physically intertwine as extensively, nor do they entangle and ensnare each other as much as the long molecule chains used in low-solids binders. Thus, the relative movement and flow throughout the solution is freer and easier with short molecules.

But merely using short molecule resins is not the complete answer. Small, short molecules would not give the same extent of total system cross-linking during paint curing as would long molecules. Soft and far less water- and abrasion-resistant paint films would result. Since the properties of high-solids coatings must at least be equal to those of the low-solids coatings they are designed to replace, further modification of the small resin molecules is necessary. Increased numbers of reactive sites for cross-linking must be chemically added to these short molecules. This allows the overall film properties to reach the same performance levels achieved with comparable low-solids coatings.

The short resin molecules in high-solids coatings can make VOC calculations difficult. Some of the short resin molecules may volatilize when the paint is baked. The EPA in New York found that several high-solids coatings were being sold by their manufacturers as 70% solids by weight. This was based on the assumption that only the solvent portion of the formulation was volatile. In actual tests, the EPA determined the weight solids to be only 58.4%. This is not a minor discrepancy, nor is it an isolated instance. A number of other paints tested ranged anywhere from 2.0%–9.4% lower in weight solids than the information on the paint manufacturers' product data sheets. If resins are not carefully checked, they may contain low–molecular-weight molecules that will contribute to VOC emissions. To prevent resin volatilization, the molecules should be reactive enough to become part of the film or else be of sufficient size not to evaporate at oven temperatures.

The leading resin types used in the manufacture of high-solids coatings are alkyds, polyesters, epoxies, urethanes, and acrylics. However, nearly any resin types can be made into a high-solids formulation by applying the appropriate chemistry to the molecules.

## Surface Tension

An unfortunate consequence of increasing the number of cross-linkable reactive sites along the length of these short, high-solids resin molecules is that they produce paints with a relatively high polarity and high surface tension. High polarity can cause excessive Faraday cage problems during electrostatic application and poor adhesion on surfaces having low polarity.

*Surface tension* for fluids, often expressed in dynes/cm, is defined as the energy needed to generate an area of interfacial surface between the liquid and the air around it. The surface tension of liquid coatings is critical because it must be lower than the surface tension of the substrate it is to cover. To adhere fully and "wet out" various substrates, the paint needs to be as low as possible in surface tension. This is especially true when coating plastics, which as a group are low in surface tension and therefore somewhat low in wettability.

Some adjustment in the formulation of many high-solids coatings is necessary to compensate for the high surface tension problems. Normally paint formulators strive for a fairly low surface tension for good surface wetting. Paints with high surface tension are characterized by a tendency toward increased internal cohesiveness. This makes it difficult to atomize the coating and increases its tendency to form fisheyes and related defects such as blisters and edge pull.

The sprayability of paints is more closely related to their surface tension and cohesiveness than to their viscosity. Paints are always able to be atomized with less energy when their surface tension and viscosity are low. However, surface tension and cohesiveness are not conveniently measurable, while viscosity is readily determined with cup viscometers. Therefore, coating viscosity is commonly used as an indicator of sprayability. Paint transfer efficiency in air spraying decreases as the energy to atomize a paint (atomizing air pressure) increases due to bounce-off and paint "fog."

Each component in a paint contributes to the overall surface tension and sprayability of the coating. Solvents have the lowest surface tension of all components, ranging from about 17.5–35 dynes/centimeter (dynes/cm); therefore, solvents tend to improve sprayability. However, high-solids coatings contain less solvent than low-solids coatings, which tends to lower the sprayability of high-solids coatings. In addition, high-solids coatings usually contain mostly slow-evaporating solvents, which tend to have relatively high surface tensions, further lowering the sprayability of high-solids coatings.

The resins used in high-solids coatings also contribute to the tendency for high-solids coatings to have poor sprayability. High-solids coating resins are more reactive than those used in low-solids coatings and therefore have higher surface tensions. The surface tension of resins generally vary from about 35–60 dynes/cm.

Because paint viscosity tends to decrease with temperature, high-solids coatings are often heated prior to application to lower viscosity and therefore improve sprayability. Approximately 85% of high solids used industrially are heated for easier spray application.

## High-solids Advantages

The most pronounced advantage of high-solids coatings is the reduced VOC emissions and the resultant improvement in compliance with state and local regulations. Inherent in the lowered VOC emissions is another advantage: solvent usage is reduced and substantial savings can be made, as shown in Figure 4-1. Solvents savings can translate additionally into inventory reductions, plant space savings, and fire hazards reduction. Figure 4-2 shows how increased volume solids decreases overall paint consumption.

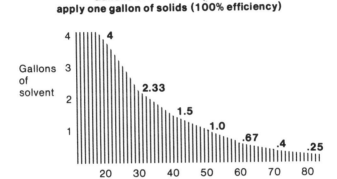

Figure 4-1. Solvent Content of Paint Decreases as Paint Solids Increase

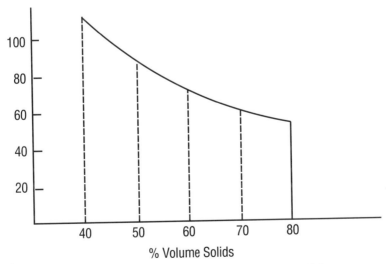

Figure 4-2. The Relation of Paint Sales to Changes in Volume Solids

Another advantage to having increased solids is the associated reduction in the number of spray application strokes to achieve a given film thickness. Theoretically, this can allow increased conveyor speeds. However, in actual practice this is a minor benefit at best.

Although the tendency of high-solids overspray to remain gummy almost invariably causes cleanup problems, it can occasionally be a small advantage because it allows the overspray of some high-solids paints to be reclaimed. One of the most readily reclaimed types of paint is polyester. Many finishers are collecting overspray and either recycling it themselves or having a paint company do it for them. The overspray is collected on vertical baffles and flows down the baffles into a collection trough and finally into a container. All that is needed to reclaim it is to filter, check and adjust the color, and restore the viscosity. Reclaiming the overspray has a bonus: that much less paint needs to be disposed of, which further reduces operating costs. Currently, this is not practical for most plants, yet there are significant savings being realized by some companies who have pioneered these cost saving procedures. In the future we can expect to see considerably more reuse of recovered paint materials.

## High-solids Disadvantages

Although their advantages are pronounced, high-solids do have various disadvantages as well. Low–molecular-weight resins needed in the formulation of these paints generally require high cure temperatures. Figure 4-3 compares the relative cure windows (time and temperature curing cycles) for high- and low-solids coatings. As a result, comparatively few single-component, air-drying coatings are available. A 3 mil high-solids air-dry alkyd may be force-dried in 10 min at 200° F (93° C), but it may need 10 h to air-dry hard at 77° F (25° C). The air-dry or low–temperature-curing high-solids coatings are far more likely to be the two-component types, primarily the two-part epoxies and the two-part urethanes, which cure quite quickly even without heat.

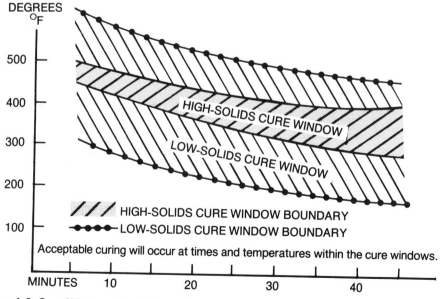

**Figure 4-3. Cure Windows for High- and Low-solids Coatings**

# LOW- AND HIGH-SOLIDS COATINGS   45

Another disadvantage with high solids is their narrow "time-temperature-cure window." This means that oven times and temperatures need to be controlled closely. The cure window profile is not something a paint user would determine; rather, that information is available from the paint supplier and differs from one paint to another.

These low–molecular-weight resins are particularly sensitive to inadequate cleaning of substrates. Minor surface oil contamination can promote cratering, blistering, edge pulling, and picture framing. Such flaws occur with all paints to a degree, but they are more pronounced with high-solids. Blisters and craters can result from oil or grease on the surface because these contaminants have low surface tension. When high-solids coatings are applied over oil or grease, the paint solvents dissolve the oil or grease to some extent and absorb them into the film. This creates a region with low surface tension, causing the material to be pulled into the high–surface-tension paint. What is left forms the blister or crater in the finish. Cleaning must be more thorough with high-solids coatings than with low-solids coatings if these appearance defects are to be avoided. (Defects are discussed in greater detail in Chapter 16.)

The tendency for film thickness to build rapidly can make film-thickness control somewhat difficult with high-solids paints. Extraordinarily fine atomization is often required to overcome this hurdle, and not all spray guns are able to finely atomize high-solids.

Overspray remaining wet becomes a disadvantage during booth cleanup. The tacky overspray can be time-consuming to remove. The sticky nature of the material makes it unpleasant to work around when it accumulates on surfaces, particularly those that people have to contact. Paint accumulations on walking surfaces are fire hazards as well as messy. The coating tends to get tracked over a wide area by people's shoes. The overspray from high-solids can be extremely difficult to detackify in waterwash booths. Newer "kill" agents that are able to reduce paint sludge stickiness have lessened this problem.

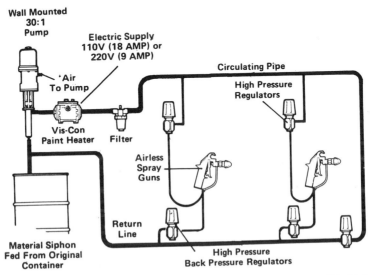

**Figure 4-4. Heating the Paint Supply Reduces the Viscosity of High-solids Paints**

## 46  INDUSTRIAL PAINTING

Because their viscosity tends to be somewhat higher than that of low-solids coatings, high-solids frequently need heaters to produce a low enough viscosity for proper application and good surface appearance, as shown in Figure 4-4. Users do not have much leeway to add more solvents as this would increase the VOC emission levels. The character of high-solids is such that they tend to decrease sharply in viscosity when heated, as shown in Figure 4-5. Thus, mild heating can cause significant viscosity reductions. But this marked response to heat can lead to the formation of sags in the paint film during oven curing. Rheology control agents are utilized since they can help prevent this from occurring, but if the coatings are sprayed too wet, oven sagging will still occur. A spray operator applying high solids coatings soon learns to deposit a "dry-looking" finish to prevent such sagging.

The high-solids coatings are starting to be pushed out by waterborne paints because, in some cases, it is easier to achieve reduced VOC and HAP emission levels that way. Where their use is possible, powder paints offer a method of near total elimination of paint solvent emissions. Although high-solids paints are beginning to be rejected in favor of newer coating technologies, they will not be displaced very quickly. In fact, their utilization in many applications will continue for at least another decade.

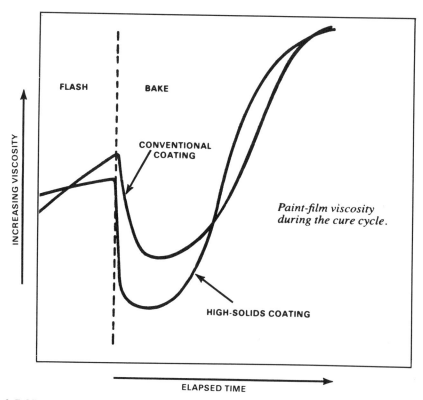

**Figure 4-5. Viscosity Changes During Flash and Bake**

# Chapter 5

# Waterborne Coatings

## Definition of a Waterborne Coating

In solventborne coatings, the fluidizing medium is an organic solvent or a blend of organic solvents. To be classed as a waterborne coating, a paint needs to have water as an important or major fluidizing medium. Most coatings that qualify as waterborne coatings also contain some organic solvent in the fluidizing media. Although a coating may be classified as a waterborne, it often will contain appreciable amounts of organic cosolvents. These tend to be low–molecular-weight ketones, alcohols, and esters because in order to be fully soluble in water they must be somewhat polar. Hydrocarbon solvents such as xylenes and mineral spirits are too low in polarity to be miscible in water.

Strictly speaking, electrocoating (E-coat) formulations are waterborne paints because they contain a great deal of water as the fluidizer. However, E-coat is deposited by electrical current and is a method so individualistic that E-coating will be considered separately in Chapter 10. Waterborne coatings (WBC) as a category considered here are those coatings containing water that are applied by a conventional method such as spray, dip, or roll coating.

Paint binders (resins) of the solventborne type are essentially *hydrophobic* (water fearing literally, indicating not soluble in or not attracted to water). Binders of the WBC variety have been chemically treated to render them *hydrophilic* (water loving)—they have been given a strong affinity for water. The resin manufacturer makes resin molecules hydrophilic by giving WBC resin molecules a distinctly different chemistry than traditional organic solventborne coatings. Almost any paint resin can be chemically modified for use in a WBC formulation. Some of the common WBC resins are acrylics, epoxies, vinyls, alkyds, and polyurethanes.

Like their solventborne "sisters," WBCs can be formulated for air drying, oven baking, or UV curing. As with solventbornes, the air-dry WBCs tend to be somewhat softer, lower in gloss, and markedly lower in water resistance, in comparison to the baked WBCs.

WBCs are categorized into three types according to how the resin is fluidized. These distinct fluidization types are:

- Solutions
- Emulsions
- Dispersions

Solutions, emulsions, and dispersions are not unique to WBCs; they can be used in all liquid coating formulas whether WBCs or solventborne coatings. The categories are not mutually exclusive; some waterborne paint formulations actually contain a combination of both solution and emulsion resins.

As with solventborne coatings, the attributes of each of the three types of WBCs will vary somewhat with the resin used. Alkyds and polyesters, for example, are more susceptible to hydrolysis (reaction with water) compared with acrylic resins, but the latter do not have the hardness of the others. The home- and commercial-decorating business uses acrylic emulsion (latex) paints extensively. Most polyesters require a higher bake temperature or a longer bake time than alkyds, although the polyesters may provide harder, tougher films and exhibit better weathering characteristics.

## Solution Waterborne Coatings

A *solution* is a completely homogeneous mixture of the atoms or molecules of two or more substances. Examples of solutions are sugar molecules dissolved in water molecules, or acrylic resin molecules dissolved in a solvent such as methyl isobutyl ketone. As with sugar molecules in a water solution, each molecule of resin in a solution paint is dissolved in the solvent blend. In the case of a WBC, the solvent blend is primarily water. Solution WBCs are therefore called water-soluble coatings. Many solution WBC resins contain chemically reactive carboxylic acid groups that with the addition of organic amine compounds form ionic groups, thereby making them polar enough to dissolve in water. After the paint has been applied, the solubilizing amine and water are volatilized in the cure process. This then converts the water-soluble polar resins back to their insoluble form. Once it has been fully cured, the WBC film is no longer susceptible to being dissolved by water.

If amines should volatilize from a container of water-soluble paint, the pH can drift down to a neutral value (pH of 7). The loss of solubilizer can make the resin insoluble. When this situation occurs, it is known as *resin kickout*, and unless it can be reversed, it renders the paint unusable. Poor flowout of solution WBCs can in some cases be traced back to partial resin kickout due to low amine solubilizer levels.

For easier solubility, WBC solution resins are typically low in molecular weight compared to emulsion and dispersion resins. Their smaller molecular size also brings the advantages of hydrolytic and mechanical stability, minimal agitation requirements, and long shelf-life. The resins have a high pH, which reduces the tendency of solution WBCs to rust steel application equipment. On the negative side, their appreciable amine content does present a modest health hazard in the form of potential respiratory problems and skin rashes.

Other characteristics of WBC solution resins include:

- Excellent freeze/thaw stability
- Moderate storage limitations
- High gloss capability
- Excellent appearance
- High cost
- Application ease (in most cases)
- High film-build capability
- Minimal gun nozzle clogging
- Minimal water-solvent popping
- High cosolvent requirement (causing combustible flash points in some cases)

## Emulsion Waterborne Coatings

An *emulsion* in physical chemistry terminology is a colloidal suspension of one liquid in another liquid. An emulsion consists of two immiscible liquids, one of which is present as minute globules dispersed in the continuous phase of the other. The globules or micelles of the dispersed phase can be coated by a soap, detergent, or other surface active substance, and thereby form a stable distribution throughout the other liquid. The surface active coating agent is called an *emulsifier*. Its function is to reduce the interfacial tension between the immiscible liquids.

A well-known example of an emulsion is mayonnaise, which consists mostly of a mixture of immiscible water and vegetable oil. The oil particles are coated with a thin emulsifying egg protein layer. The separate oil and water phases can be seen through a microscope even after vigorous blending. In similar fashion, the emulsifier in an emulsion paint surrounds the tiny agglomerates of resin molecules. In a WBC these emulsifier-coated resin micelles are dispersed in a continuous water phase or a phase that is mostly water plus some organic cosolvent.

The manufacture and use of water-emulsion paint involve heterogeneous (having unlike qualities) fluids to create a paint whose behavior is more complex than that of conventional paints and solution WBCs. Some water-emulsion paints, for example, increase in viscosity when temperature rises, which is just the opposite of what happens in a typical paint. As a result, some WBC dip tanks require chilling in summer weather so that the heat does not increase paint viscosity.

Waterborne emulsion paints are also called latex coatings. Contrary to what is frequently believed, latex paints contain no natural or synthetic rubber or rubber-like resins. Scientists use the term *latex* to identify "any emulsion of an organic material in water." Most latex paints air dry rapidly. They are the most common type of paint purchased and used by homeowners for both interior and exterior application. But their value is not limited to trade sales paints (sold in retail stores). Latexes are used extensively for industrial maintenance applications as well.

The chemical nature of emulsion waterborne coatings allows them to be formulated at high volume solids without unduly high paint viscosities. The viscosity of emulsions is basically that of the continuous phase (in this case, water plus cosolvent) over a fairly wide range of paint solids concentrations, which allows the viscosity to be rather constant and not rise significantly at moderately high-solids levels.

If solids levels are increased excessively, the material finally reaches a stage at which viscosity will increase sharply. Note that the plateau in the viscosity-solids curve (see Figure 3-6) indicates the elimination of the possibility of a meaningful viscosity determination as a normal aid in application by the common viscosity cup method and a simultaneous approximation of the percent solids. The correlation between percent solids and viscosity is not linear with aqueous emulsion paints.

The chemical stability of waterborne emulsions is threatened by freezing temperatures. Freezing can permanently separate the paint components, causing insolubilization of the resin—a phenomenon called *kickout* which, as stated earlier, can also be caused by amine loss from solution WBCs.

Baking and air-dry emulsions both cure in the following sequence shown in Figure 5-1: The organic cosolvent evaporates, water evaporates, and the resin particles then coalesce to form a continuous coating. In the case of an enamel, cross linking of the resin occurs after coalescence.

50   INDUSTRIAL PAINTING

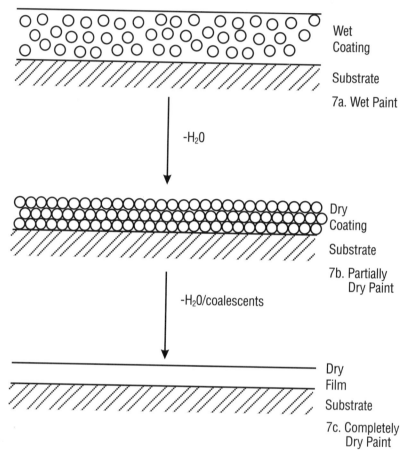

**Figure 5-1. Drying of Emulsion Waterborne Paint**

WBC emulsion resins are high in molecular weight (100,000–3,000,000). The large molecular size brings about the disadvantages of poor hydrolytic and mechanical stability. Over-sized pumps and low paint fluid pressures should be used due to the mechanical shear sensitivity. Shear can result from stirrer blades, gear and reciprocating piston pumps, and flow restriction in paint lines. The low amounts of amine in these resins minimize the health hazard. Resins have a relatively fast air-dry rate, which minimizes recoat times. Unfortunately, their relatively moderate pH tends to encourage flash rusting of steel.

Other properties of emulsion WBCs include:

- Poor freeze/thaw stability
- Rigid storage limitations
- Medium gloss capability (being improved)

- Medium appearance (being improved)
- Moderate in cost
- Ease of application (in most cases)
- High-solids application capability
- Low in odor
- Minimal VOC emissions

Because each of the water-soluble and water-emulsion types has distinct and separate advantages, coatings have been made that are combinations of both resin fluidization types. Some of the combination solution/emulsion WBCs have been utilized in original equipment manufacturer (OEM) automotive paints.

## Dispersion Waterborne Coatings

A liquid dispersion paint is a system of dispersed resin particles suspended in a liquid. A resin might be mechanically dispersed or produced from a solution by careful addition of a solvent (or solvent blend) which has a polarity different from the resin, so that resin molecules are forced into a mild clumping.

A graph of viscosity versus solids for a dispersion paint (see Figure 3-6) also produces a viscosity curve that has a long plateau region into the high-solids range. The curve rises rapidly when the concentration gets so high that individual clusters of resin contact each other and begin tangling and separating out of the paint.

Dispersions differ from emulsions in that the resin clumps (clusters) are small and do not require added emulsifiers to form a stable fluid mixture. Vigorous physical mixture alone is used to form a dispersion. Because many waterborne resins can be thinned using water, they are sometimes referred to by the imprecise terms "water dispersible" or "water reducible."

Waterborne dispersions are slightly different from emulsions but are close enough in properties to be considered as the same general paint type. The properties listed for emulsion WBCs apply practically and almost universally to dispersion WBCs as well.

## Solvent in Waterborne Coatings

During the manufacture of waterborne coatings, organic solvents are added to dissolve the resin. Because paint resins are essentially oil-like and do not readily mix with water, the chosen solvents must be water-miscible compounds such as alcohols, glycols, ketones, and esters. These cosolvents are somewhat polar and contribute dissolving properties to the fluidizer. Suspension or solution of paint resins in water is frequently difficult without these organic cosolvents. The cosolvents also aid considerably in smoothing the film by increasing flowout on the painted surface.

After the solvent and resin are mixed thoroughly, the mixture is made hydrophilic by the addition of an amine or chemically similar material. Then after the addition of pigments and additives, the mixture is reduced to the desired viscosity by the addition of water.

Two component (2K) waterbornes are increasingly popular low-bake paints for plastic substrates, in self-texturing and soft-touch coatings especially. Most 2K waterbornes will cure in 7–12 days at room temperature, but more often a bake at around 165° F (74° C) for 30 min is used to accelerate this cure. Compared to the ordinary melamine cross-linking waterborne resins that need a relatively high bake to cure, the 2K paints provide equally good properties but can be used on many more plastics due to their lower bake requirements.

The organic cosolvent (at application viscosities) of most waterborne coatings is 2%–20% of the fluidizing medium; the remainder is water. Surprisingly, the organic solvent portion in some of these coatings can be quite high, so the coatings do not automatically meet EPA or local VOC emission limits. A number of government and manufacturing representatives have expressed worries that in the early decades of the 21st century, automotive waterborne color basecoat finishes, in particular, may not be able to be applied at low enough pounds of VOC per gallon of applied coating solids (VOC/GACS) in all plants. Best Available Control Technolgy (BACT) is permitted in areas where air purity is satisfactory; in ozone attainment areas, BACT mandates 12.2 lbs VOC/GACS. It is found that on average waterborne basecoats have about 3.0 lbs VOC/gal minus water, so that the normal 50% transfer efficiency of the emission rate overall is in compliance at slightly over 10 lbs VOC/GACS. For ozone nonattainment areas, however, Lowest Available Emission Requirements (LAER) specify no more than 6.2 lbs VOC/GACS, a number far lower than most waterborne basecoats are able to achieve. Much work is being done to develop waterborne coatings with zero VOC, and some success in these ventures has already been achieved but not in paints suitable for all types of products

Why must VOC in WBCs be given "minus water"? The allowable VOC content of waterbornes is given by EPA rules in units such as "lbs VOC/lb applied solids" or "lbs VOC/gal minus water." The "minus water" calculation requires that a theoretical gallon of the nonwater portion of a paint be considered and the VOC reported for that material. People understand the concept but many question the reason behind it. The answer is that by restricting the VOC content reporting that way, no company is tempted to dilute paints with water to lower the VOC per usage gallon and thereby achieve VOC compliance. The usage gallon VOC content is cut in half if a given paint is diluted one-to-one with water, but then twice the normal amount of coating would need to be applied to reach the same dry film thickness. If this happened, no net VOC reduction would result.

Determining the VOC content of a waterborne coating isn't necessarily as easy as simply knowing the amount of cosolvent added by the paint manufacturer. Amine solubilizers also can contribute to VOC emissions. To comply with VOC regulations, the volume percent of organic volatiles must be known. The easiest way to determine total VOC content is by gas-liquid chromatographic analysis, a method that is expensive unless a measuring instrument of this kind is available. Other methods tend to be slow, cumbersome, and inexact. Samples normally can be tested for VOC content by distillation of the volatile organic portions other than water. The volatiles usually will have a lower boiling point than water. Certain mixtures of solvent and water, however, make distillation and subsequent VOC determination impossible because the mixture becomes *azeotropic* (having a single boiling point).

The following case history illustrates how azeotropic solvent and water mixtures can distort VOC measurements based on distillation testing. A plant was cited for a VOC violation with a waterborne coating because the distillation of all volatiles up to 212° F (100° C) showed "an excess of 1.37 lbs/gal VOC over the maximum allowed of 3.00 lbs/gal minus water." The paint manufacturer claimed the paint had a total of only 2.66 lbs/gal minus water. When the regulatory agency was shown that adding water to the paint increased (as measured by distillation) the apparent VOC content, the agency agreed that azeotropic distillation gave an erroneous VOC reading. The fine was promptly canceled.

Whenever WBCs are thinned to achieve the proper paint application viscosity, cosolvents should not be added. One obvious reason is that added solvent will raise the VOC emissions. Because of the viscosity relationship with added solvent, reduction with water must be done in modest increments. Otherwise over-reduction is possible. A small amount of water addition can cause a sharp drop in the viscosity of the paint, as shown in Figure 5-2.

To avoid complications from impurities and dissolved components in water, some plants use only clean deionized or distilled water when making additions to WBCs. Exceptionally pure water is only rarely necessary, but use of deionized water may be advisable depending on the quality of the supply water. The paint manufacturer is the one most capable of determining whether ordinary tap water does or does not reduce film quality.

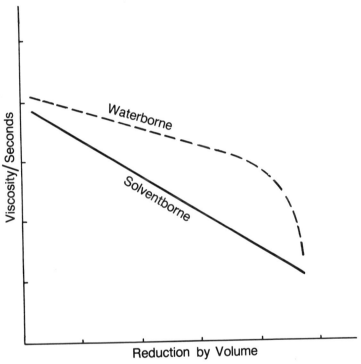

Figure 5-2. Relative Difference Between Waterborne and Solventborne Finish Reduction Rates and Application Viscosities

## Waterborne Wood Finishes

For the past 20 years formulators have attempted to prepare waterborne materials suitable for the enormous wood product markets in furniture, cabinetry, and panelling. The main difficulty is that water tends to raise the grain of most woods. To a lesser degree there is a problem with heat curing, which can extract the natural moisture from the wood and cause warping. As a result, most wood finishes—including fillers, stains, sealers, toners, lacquer varnishes, and enamel varnishes—are still predominantly solventborne types. But EPA regulations are gradually changing this picture. Considerable effort has been expended by paint manufacturers to develop WBCs for all end uses, and they have achieved some success in developing acceptable waterborne wood finishes.

As much as 10%–15% of wood finishes are now waterborne varieties. Currently, roll coating and curtain coating give even better results with WBCs than do atomization application methods. Much of this is due to the greater control in roll and curtain coating of film thickness and to the ability to control humidity and temperature more closely in the application, flash, and curing stages. These variables must be held more tightly with WBCs than with solventbornes. UV curing of WBC on flat wood panels and boards is an area which has had strong growth in just the last few years.

Spraying of WBCs on furniture and cabinets is finding limited use on the less expensive lines of these wood products. In most instances "hybrid systems" are used; only some of the stain, sealer, and topcoats are WBCs, while others are solventbornes. Low cost items may receive only 4–5 separate applications, but fine furniture typically will get 8–12 individual coating applications.

In the future, WBCs can be expected to make inroads into this market due to the continual strengthening of both VOC and HAP regulations. Rule tightening for wood finishing VOC took effect in the summer of 1995, and new HAP standards hit in 1997. Remember, states may not adopt weaker rules than federal guidelines, but they are free to demand more stringent limits if they so choose. When states do this, however, they run the risk of chasing industries to other less restrictive states.

## WBC Considerations

The following are some significant factors that WBC users must consider.

### Operation Permits

The U.S. Clean Air Act (CAA) requires most painting operations to obtain an operating permit. Regulated are VOCs, HAPs, nitrogen oxides ($NO_x$), sulfur oxides, and carbon monoxide, but VOC HAPs are of most concern to painters. Common paint solvents are often both VOCs and HAPs, including toluene, ethyl benzene, xylenes, MEK, and MIBK. (Acetone is neither.)

The CAA provision most opposed was the "potential to emit" condition that classifies plants based on nonstop operation even though the actual maximum operation is eight h/day for five days/week. Major sources are those painting operations whose potential to emit exceeds any of the following:

- 100 tons/yr of air pollutants
- 10 tons/yr of any one HAP or 25 tons/yr total HAPs
- 50 tons/yr VOC or $NO_x$ in "serious" nonattainment areas
- 25 tons/yr VOC or $NO_x$ in "severe" nonattainment areas

- 10 tons/yr VOC or $NO_x$ in "extreme" nonattainment areas
- 50 tons/yr VOC in ozone transport regions not classed "extreme" or "severe"

Since plants must determine their own need to comply and operating officers can be held criminally accountable for CAA rule infractions, some believe this violates their rights against self incrimination. More changes in this law seem likely to be made.

## Pretreatment

Pretreatment when WBCs are to be applied is far more critical than with solventborne paint; in fact, poor or inadequate pretreatment is the major cause of problems associated with the use of waterborne paints. Because their organic solvent content is low, WBCs are not able to dissolve any spots of oil and grease that might remain on surfaces after cleaning. The dirt and oil sensitivity of WBCs tends to create blisters readily. Water has a high surface tension, creating potential edge-pull problems. The choice of cosolvent is crucial in reducing picture framing, craters, blisters, popping, and edge pull.

## Cost

Most WBCs cost more per gallon on an equivalent solids basis than organic solventborne materials. This reflects the higher cost to manufacture them. The cost on an equivalent square-foot-coverage basis for WBCs continues to be slightly higher than for organic solventborne paints, but the gap is narrowing.

## Resin Availability

Not as many resins are available for WBC formulation as with solventborne coatings. Many of those used for waterborne finishes can be cured below 200° F (93.3° C) and thus may be suitable for many wood and plastic substrates. The inherent high viscosity of WBCs restricts their formulation to relatively low-solids levels.

## Modifications

Although conversion from solventborne paints to waterborne paints is still far simpler than switching to powder coating because the handling and application are similar, converting an existing low-solids solventborne coating line to WBC can be complex. Such an existing line may need to have ordinary piping, valves, etc. replaced with stainless steel or other nonrusting materials. This is particularly true when light-colored paints are used in which rust is readily visible. Depending on the local climate, air-conditioning or a heated flash tunnel may have to be installed to stabilize ambient temperature and lower the relative humidity for WBC application.

## Process Commonality

Conventional application processes can be used with WBCs, including all of the various spray methods, disk and turbine application, dip coating, and flow coating. This gives WBCs an advantage over high-solids paints. Dip coating and flow coating are not possible with the majority of high-solids coatings because of paint drain-off troubles, due primarily to their inherent elevated viscosity.

# 56  INDUSTRIAL PAINTING

## Electrostatics

Waterbornes are often applied electrostatically to increase paint transfer efficiency. Because most WBCs are conductive, electrostatic application generally uses a system "isolated" from ground. (Isolated systems should never be used with solventborne paints because the fire danger would be extreme.) Plastic stands and arms are used to isolate all conductive parts of the system from ground. Since the isolated paint equipment components themselves carry a high electrical charge (are "hot"), anything conductive that touches any part of the system will drain off a large quantity of the electrical charge stored on the ungrounded system. Since the human body is electrically conductive, an isolated system must be enclosed by a protective safety cage. The appropriate signs warning people about the shock hazards need to be posted. The added cost for isolation for electrostatic application of waterbornes is moderate and more than offset by the cost saving from reduced paint overspray wastage.

An electrically conductive object located within several feet of an electrostatic paint device, even though it is not actually touching it, can accumulate electrical charges unless the object is fully grounded. The more massive the item, the more electrical charge it can store. The ability of objects to retain electrical charges is termed their *capacitance*. This is more likely to occur with electrostatic application than with nonelectrostatic application, obviously, but static charges can actually be generated with any paint application equipment. Grounding of all objects near painting equipment is a standard safety practice.

*Grounded* WBC systems have had only limited acceptance. Few such systems put on the market have had more than mild success. Remote or secondary charging of the paint particles (see Figure 5-3) can be achieved by locating the charging electrode about 5 inches (in) away from the gun tip or rotary bell head. Secondary charging is generally not very effective. In some

Figure 5-3. Successful Electrostatic Application in Grounded Systems with Remote Electrode Charging

cases, a charging ring has been used with rotary WBC electrostatic application. Another system used for waterborne automotive color basecoats has four secondary charging electrodes located uniformly around the bell head. The amount of electrical potential that can be reached with remote charging is only about 15,000 volts (V), far less than the 75,000–115,000 V possible with normal charging devices.

A number of years ago, grounded WBC systems purportedly would allow electrostatic spraying of WBCs with a fully grounded paint system by electrostatically charging the parts. This required the parts to be hung on insulated hooks. The hooks or hangers had a plastic section that prevented the part from losing its charge to ground. Parts were charged by contact with or proximity to a charging wire that ran along the length of the spray booth. Reports from plants that had installed them stated they were not effective. A similar process used remote "negative" charging of WBCs, plus "positive" charging of large vehicles on isolation stands. The corporation that had installed this system soon abandoned the method of application as ineffective.

Some plants have found that when WBCs are painted electrostatically, frequent "tripouts" of the electrostatic current limiter occur from shorting of the system to ground. This can be a vexing problem if locating the electrical short frequently consumes a considerable amount of time. Plants have switched back to solventborne coatings because of continuous trouble from this problem.

## Atomization

WBCs tend to be high in application viscosity and surface tension (highly cohesive). As a result the paint does not atomize as readily as solventborne paint. This means having to settle for a low finish quality. Also, to get good breakup in air spray, high air pressures are used when applying WBCs which can reduce paint transfer efficiencies below 25%.

## Application Problems

The major disadvantages with WBCs tend to be the difficulty of application without sagging and the problem of solvent popping and boiling when parts are in the oven. Avoiding application problems can be a narrow tightrope to walk. To understand why, let's review what happens when solventborne paint is sprayed.

When solventborne paint is sprayed, the solids content (and viscosity) of the paint increases substantially after exiting the gun tip. This is the result of evaporation of solvents from the very large surface area of the fine atomized droplets moving toward their target. In the short time before the droplets hit the workpiece, the solids can increase 20%–25%. This helps prevent sags on freshly coated parts. Solventborne coatings use a blend of four to eight solvents with varying evaporation rates to avoid a sudden excess of solvent evaporation.

With WBCs, however, much of the volatile portion of the sprayed droplets is water. Water has a high boiling point, so virtually none of it evaporates while the WBC droplets travel from the end of the gun to the workpiece. Thus the "as-sprayed" solids percent is nearly identical to that of the paint on the freshly coated part. Because so little fluidizing media evaporates from the paint droplets between the application device and the workpiece, it is often difficult for the operator to achieve a sufficient film build while at the same time avoiding running or sagging of the material. The applicator must walk a narrow path between inadequate film build and paint runs. Where possible, heating the substrate is effective in reducing the tendency to runs and sags; much more so than heating the paint, which is rarely done.

Paint heaters set at 90° F–105° F (32° C–40.5° C) may, in a few cases, help reduce or prevent such sagging. This generally only has a marginal effect, however. Some WBCs have reverse solubilities, so increased paint temperature can complicate the situation by raising the viscosity instead of helping to alleviate the problem.

High humidity in the spray application area can add to the problem by reducing flashoff to near zero. The application area may need to be air-conditioned to control the humidity. The high cost of air-conditioning may prohibit the use of WBCs. Slow air-dry times will be experienced when humidities are high. WBCs may be unusable under such conditions. WBCs need to be sprayed at somewhat higher viscosities to avoid runs and sags, especially when it is necessary to achieve more than 1–1.5 mils dry-film thickness.

The slow flashoff of water in humid weather can make popping and boiling in the oven a frequent problem, particularly if an appreciable film build is necessary. Heated flash zones are frequently necessary to remove enough of the water so that popping and boiling defects are not experienced in the oven. Zoning of bake ovens may also be necessary to prevent water boiling and popping.

Little of the water evaporates until the temperature of the applied coating reaches 212° F (100° C), then all of the water will tend to evaporate at the same time. Water's high *heat of vaporization* can cause problems in curing WBCs in a bake oven. The "outrush" of water vapor can tend to cause blisters and solvent-popping problems.

## Agitation

Some WBCs are sensitive to being stirred or pumped. This is particularly important since WBCs as a whole tend to settle rapidly. Excessive agitation can have a "butter churn" effect and "break" emulsions. Certain filters can produce the same problem. The kicked-out resin then forms an insoluble mass, which clogs filters and spray guns. Sometimes the resin globules are soft enough to be pushed through filters. Then they can reach the gun and be ejected as small gelatinous globs onto the workpiece. These visible "bumps" must be sanded off after curing before the parts can be resprayed.

## Odor and Cleanup

One of the great advantages of WBCs is their low odor due to the type and quantity of cosolvents in them. Easy cleanup is another advantage for many WBCs, but many others form a tenacious "skin" on the walls of the paint hoses and piping that resists rinsing with water or aqueous solutions. With the former types, paint lines are rinsed clean with water or mild water solutions of cleaning agents. In the case of skinning types, organic solvents must be used to flush out the paint lines to remove the adherent skin not removed with aqueous agents.

However, since the resins are designed to be compatible with water, it is sometimes difficult to get WBC overspray to float or sink in waterwash booth reservoirs. As a consequence, some WBCs are difficult to detackify. Continual improvements are being made in "kill" agents, however.

## Fire Hazard

Another advantage of WBCs is reduced fire hazards (low flash point or no flash point). This can be a significant plus in many plants. If the organic solvent content of a WBC is high, however, these paints can still burn. Some WBCs will not ignite even when an open paint pail is subjected to the direct application of a blowtorch. After all volatiles evaporate, however, the cured WBC film is just as flammable as any cured organic solventborne coating.

## Storage

WBCs should be stored inside to prevent freezing. Expansion that results from freezing can burst paint containers, and ice formation during freezing can damage the paint, especially by kicking out emulsified resins.

## Dip Tanks

In a WBC dip operation the tank size is critical. The tank should be small enough to get fast paint volume turnover. This will help avoid instability problems inherent with WBCs. Complete turnover of the tank volume in one or two months would be considered ideal. With WBC dip coat and flow coat, drain-off continues from the parts for a considerable distance. To recover some of this paint, the drip zone can be rinsed back with a fine water mist or a small amount of cosolvent. One of the most successful recovery methods is to use the paint itself to flush back the coating from the drip zones. This results in a clean and economical operation. Water and cosolvent loss are somewhat reduced when dip tanks are covered during nonuse. This practice is more often used with organic solvent dip systems. Nevertheless, covering the tank is a wise practice because it also helps to keep foreign materials out of the paint.

## Foam

Foam sometimes can be a problem in agitated supply tanks, and especially with flow-coat and dip-coat WBC systems. The emulsifiers and other coating components act as detergents in the system. When the paint material is agitated, it foams like most detergent/water mixtures. The paint foam gets on parts, dries, and creates appearance defects. Foam can stall pumps and distort the readings of monitoring devices.

## WBC Advantages/Disadvantages

WBCs probably have as many advantages as they do disadvantages. One thing is certain, WBC application requires tight process control. It is becoming apparent that solventborne high-solids technology cannot offer a general range of coatings with VOC levels that will meet future projected VOC restrictions, whereas WBCs seem likely to succeed. Aside from a few 85%–100% solids paints, most high-solids paints are not able to go beyond about 70% solids without causing extreme application difficulties.

The most compelling reason for using WBCs is that their low VOC content usually allows compliance with EPA emission limits. This VOC advantage will surely lead to continued WBC growth in the future, especially as VOC limits are tightened. Waterbornes can reduce VOC at the source rather than with additional equipment such as solvent capture and incineration systems.

Yet, waterbornes generally tend to be higher in Hazardous Air Pollutants (HAPs) than solventbornes, and solventborne systems with add-on emission controls, although costly, have set many Best Available Control Technology (BACT), Lowest Available Emission Requirements (LAER), and Maximum Available Control Technology (MACT) standards.

However, WBCs have room for improvement, too. Some plants have been lured to switch systems by some of WBC's advantages but, after using them for a while, have been so discouraged by their disadvantages that they have abandoned waterbornes for other coating systems. Nonetheless, many plants have changed to WBCs and continue to use them with strong satisfaction. The environmental pressures will certainly force many companies to look carefully at WBCs, especially for parts that cannot withstand the high cure temperatures needed for curing powder coat paints.

# Chapter 6

# Powder Coating

## Introduction to Powder Coating

In the 1960s a new painting technology was developed called *powder coating*. Instead of a wet paint, the coating as manufactured and applied is totally dry. Its constituents are practically identical to a wet paint except for the absence of solvent. Like a wet paint, a powder coating contains resin, pigment (if a colorcoat), and additives. The virtual absence of VOCs make powder coatings highly desirable on that basis alone, but their exceptional durability is also a powerful inducement to use them.

Today powders have close to a 20% share of the painting market, and their share is growing. Automotive powder primers and antichip coatings are among the recent large, single markets for powder coats. Flat, precut, or punched metal shapes called "blanks" are also being powder coated ahead of forming to eliminate the need for post-assembly painting in several industries. The precoated blanks, whether painted with liquid or powder paints, have an advantage over using precoated coil stock in that no cut or drilled bare metal edges are present.

The nature of powder coating as a technology can best be understood by examining a particle of powder coating. It is likely to vary in size and shape. If it were a sphere, it would be about 0.5–1.0 mil in diameter. A close examination of the particle will reveal a composite of resin, pigment (if not a clearcoat), and various additives. All components are homogeneous (fused together) because of the way powder coatings are manufactured.

In manufacturing powder coatings, exact amounts of resin, pigment, and additives are dry-blended. The blend is then heated to the resin's melt temperature, turning the dry blend into a fluid-like mass. The hot melt is then extruded into a thin, flat sheet that is quickly cooled and flaked. A mill then pulverizes the flakes into a powder about the consistency of baking flour.

Melting and forming a homogeneous mass of powder coating ingredients prevents any possible component separation that might otherwise occur when the powder coatings are shipped or handled. Potential segregation into pigment-rich and resin-rich portions could result from particle size and density differences among the components if the powder were not prepared in this way.

Returning to the examination of a particle of powder, the understanding of powder coating becomes complete when an analysis is made of how the particle becomes a coating. If the particle is placed on a surface that can withstand heat, and the surface is heated to the resin's melt temperature, the particle will flow until equilibrium sets in. If a hypothetical sphere particle of powder 1 mil in diameter is heated to the melt temperature of its resin, the sphere will collapse and flow to a circular disk about 1.6 mils in diameter and 0.25 mil thick. The actual diameter and thickness would be functions of the rheology of the resin, but this assumption demonstrates the idea of the particle melting into an expanded but shrunken-in-height mass.

If a surface is covered with a single layer of particles of powder, each a hypothetical sphere 1 mil in diameter, and the particles are heated to the resin's melt temperature, the spheres would sag to a continuous coating film about 0.7 mil in thickness. Each melted sphere would merge and become homogeneous with the neighboring sphere, forming a continuous coating.

During an actual powder coating process, however, it would be practically impossible to apply a coating only one powder particle thick. The particles almost certainly would pile up at least several high. Therefore, it is easy to see that the minimum thickness of a powder coating would be in the vicinity of 2 mils or so.

Nearly all the resins used in wet coatings can be used in powder coatings. In practice, however, epoxies, acrylics, polyesters, and polyurethanes are used most. Materials such as nylon, teflon, and polypropylene can be used in powder coatings, although they cannot be dissolved or readily dispersed in liquid systems. Figure 6-1 compares the relative properties of various powder coatings.

Resins used in powder coatings may be either *thermoplastic* (flows when sufficient heat is applied) or *thermosetting* (cross-links when enough heat is applied). Thermoplastic powders are sometimes called lacquers and tend to be used for functional purposes. Thermosetting powders are sometimes called enamels and are usually used for decorative applications. When heated, thermoplastic powder coatings form a paint film by melting and coalescing, just as liquid emulsion coatings coalesce in forming a paint film. Thermosetting coatings, when heated, cross-link to form a paint film, as do enamels. The cross linking occurs between the main resin component and another resin component, designated as a cross-linker.

Powder coatings are packaged in cartons and drums of various sizes, depending on the amount ordered by the coater. Their weights may range from as little as 25 lbs to as much as 300 lbs or more. Each powder container is lined with plastic, which is filled and then sealed to keep out moisture.

Because all powder coatings require heat to flow into a paint film, substrates that can be powder coated must be able to withstand heat ranging from about 250° F–500° F (121° C–260° C). These temperatures rule out the use of powder coatings on wood and most plastics, making powder coating primarily a metal-finishing process.

Surface-preparation requirements for powder coating are generally the same as for WBCs. The degree of pretreatment needed varies with end-use requirements, both for powders and liquid coatings. End uses with extreme requirements would need maximum cleaning, a good conversion coating, and a quality sealer rinse. End uses with low requirements may only require minimal pretreatment so long as the surface is well cleaned. Because powder coating films tend to be thicker than wet coating films, powder coatings can usually get by with less conversion coating than can liquid coatings. Some powder coating end-use requirements permit pretreatment to be limited to blasting with glass beads, aluminum oxide, or steel shot with no conversion coating at all.

## Relative Properties Of Powder Coatings

| | Weather Resistance | Chalk Resistance | Corrosion Resistance | Chemical Resistance | Heat Resistance | Over Bake Resistance | Adhesion | Impact Resistance | Flexibility | Abrasion Resistance | Pencil Hardness Range |
|---|---|---|---|---|---|---|---|---|---|---|---|
| Acrylics | E | E | G | E | E | E | G | G | F | E | H-4H |
| Epoxies | E | P | E | E | VG | VG | E | E | G | E | HB-5H |
| Epoxy Polyester Hybrids | VG | F | VG | VG | G | E | E | VG | VG | E | HB-2H |
| Fluoro-Polymers | E | E | E | E | F | G | May Require Primer | VG | E | G | HB 2H |
| Polyamides | VG | VG | E | VG | F | G | May Require Primer | E | VG | VG | 70-85 Shore D |
| Polyesters | VG | E | E | VG | G | VG | E | VG | VG | VG-E | HB-4H |
| Polyolepins | G | VG | E | Solvents-P Acids And Alkalis-E | F | F | May Require Primer | VG | E | P-F | 30-55 Shore D |
| Vinyls | VG | VG | VG-E | Variable | P | P-F | May Require Primer | E | E | F-G | 30-50 Shore D |

E=Excellent, VG=Very Good, G=Good, F=Fair, P=Poor

**Figure 6-1. Relative Properties of Powder Coatings**

64   INDUSTRIAL PAINTING

## Powder Application Methods

Successful powder coating application involves spreading the powder uniformly over a surface and applying heat to melt and flow the powder into a paint film. Two methods are used to do this:

- Electrostatically applying the powder and then heating the coated part to melt and flow the powder (electrostatic spray and electrostatic fluidized bed).
- Heating the uncoated part to above the powder melt temperature and then applying the powder, which immediately melts and flows (fluidized bed).

### Electrostatic Spray

Electrostatic spraying is the most common method of applying powder coating. Figure 6-2A through D shows various products being powder coated. Prior to coating, the part needs to be grounded. This is usually done by hanging the part onto a properly grounded overhead conveyor, as shown in Figure 6-3.

Powder in the bulk form needs to be fluidized before it can be pumped to a powder spray gun. Fluidizing is accomplished by placing a quantity of powder into a container, or bed, with many perforations in its bottom. Air at a controlled rate is forced up through the bottom, gently fluffing up (fluidizing) the powder.

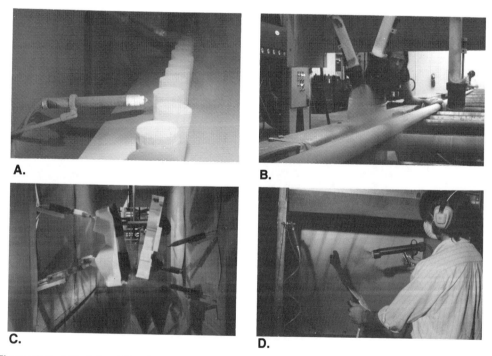

Figure 6-2. A Variety of Products Being Powder Coated

POWDER COATING 65

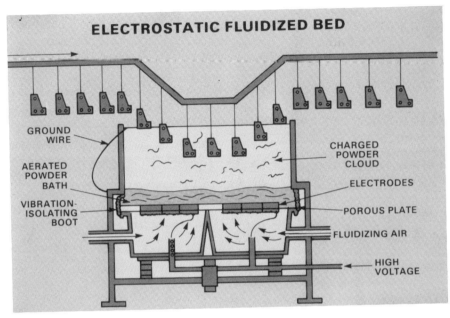

**Figure 6-3. Electrostatic Fluidized Bed**

A venturi pump provides a means of drawing powder from the fluidized bed to the spray gun, which charges the powder electrostatically and expels (sprays) the powder toward the target (part to be painted). Figure 6-4 demonstrates how an air venturi pump moves powder from the fluidized bed to the spray gun, where metered air gently sprays the powder out of the gun barrel. Figure 6-5A through C shows a small powder supply tank and venturi pump, filling the tank, and adjusting the fluidizing air.

The spray pattern is adjusted on some guns with a deflector and on others using air jets. Adjusting the deflector distance from the gun changes the spray pattern. Likewise, on deflector guns, the spray pattern is changed by varying the pressure of air introduced around the end of the gun barrel. Figure 6-6 shows an electrostatic powder gun.

Powder exiting the gun tip is charged electrostatically in ways similar to charging atomized liquid paint. The powder picks up extra electrons from an electrode (charged to 75–90 kV) and becomes negatively charged. The charged particles are attracted to the closest ground, which should be the part to be coated. Some people believe that when the charged particles contact the grounded part, the high resistivity of the powder prevents the charge from discharging suddenly to ground. The charge "hangs on" and holds the particle to the part through electrostatic attraction. Figure 6-7 gives an example of this, showing some small powder coated parts emerging from a spray booth. According to this theory, the attractive forces are sufficient to hold the powder onto the part into the oven, where particles melt into a fused coating. This retained electrostatic charge theory is widely held but scientifically hardly plausible. The author has demonstrated how the powder remains on a part even when it is hung in a steam atmosphere that allows all charges to dissipate.

66  INDUSTRIAL PAINTING

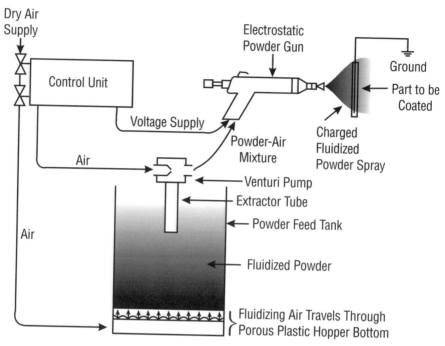

Figure 6-4. Powder Electrostatic Spray System

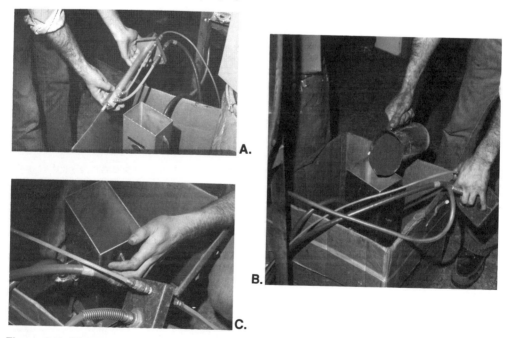

Figure 6-5. (A) A Powder Supply Tank and Venturi Pump; (B) Tank Being Filled; and (C) Adjustment of Fluid Air

POWDER COATING 67

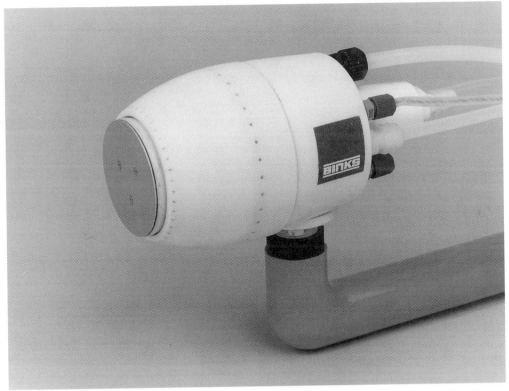

**Figure 6-6. Electrostatic Powder Spray Gun (Drawing & Photo)**

68  INDUSTRIAL PAINTING

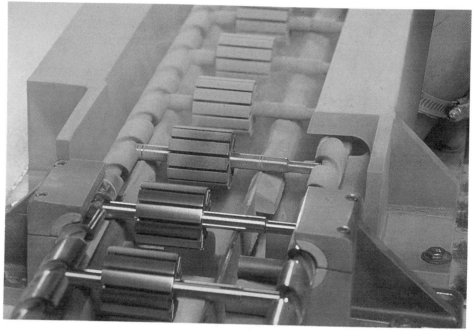

Figure 6-7. Small Powder Coated Products Emerging From Spray Booth

Another explanation, and one preferred by this author, is that a "packing" effect occurs, which holds powder onto the parts firmly. This is why powder particles hold on well even in high humidities that drain away all the electrostatic charges very quickly.

Powder exits a gun very gently, forming a powder cloud through which parts are conveyed. Figure 6-8 shows lawn furniture emerging from a powder booth, the interior of which appears to be filled with a cloud of powder. The grounded parts moving through the cloud attract the charged particles. The amount of powder attracted to a part depends on the charging voltage and the conveyor speed. A part being conveyed rapidly through a weakly charged cloud may draw only a light coating that would flow to a film thickness of about 1–2 mils. A part being conveyed slowly through a highly charged cloud could attract enough powder to build a film thickness of about 3.0–5.0 mils.

Almost all electrostatic spray coating systems apply the powder in confined booths, as shown in Figure 6-9. Powder moving past the part is drawn downward by a gentle downdraft air flow. The air flow must be moderate to prevent blowing the applied powder off the parts. Overspray powder reaching the booth bottom is returned to the fluidized bed to go through the gun again.

The downdraft air creates a negative pressure in the booth, bringing in air from outside the booth. Therefore, the air in the vicinity of the booth must be clean and relatively dry to prevent drawing moisture and contaminants such as dust and lint into the booth.

The air returning the overspray powder to the fluidized bed container is exhausted through powder recovery filters and then through very fine (absolute) filters that trap the remaining ultrafine powder particles, preventing their entering the powder coating room atmosphere. Sometimes fabric bag filters, cartridge filters, or a "cyclone" is used to help recover overspray powder. Figure 6-10 shows how a cyclone recovers the powder.

POWDER COATING 69

Figure 6-8. Lawn Furniture Emerging From a Powder Spray Booth

For small wire baskets, racks, mesh items, and other small-sized parts that require coatings of 2–6 mils, unheated parts can be coated using electrostatic fluidized beds. Electrodes on roughly 4-inch (in) centers are installed in the diffuser bottom of the hopper to add electrons to the

Figure 6-9. Interior of a Powder Application Booth

70  INDUSTRIAL PAINTING

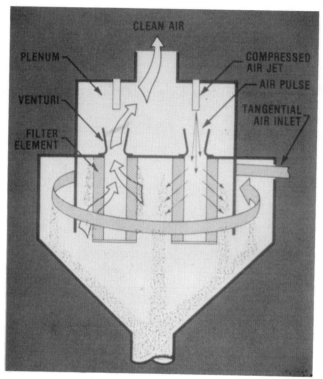

Figure 6-10. Cyclone Cartridge Collector

powder. If the parts were to be immersed fully into the powder cloud, thick coatings would result. However, it is possible to utilize the attraction between the negatively charged powder and the grounded object to be coated. Electrostatic attraction will cause powder particles to be drawn out of the fluidized bed onto the part to be coated.

For small parts the major advantage in using electrostatic fluidized bed application is that thinner coats are possible compared with nonelectrostatic fluidized beds using heated parts. Films of about 4–10 mils are applied on hot parts in nonelectrostatic fluidized beds.

Considerable amounts of fencing, screening, and similar open-mesh coiled stock are coated using electrostatic fluidized powder beds. Screening in coil form is run vertically up between a pair of fluidized beds. The powder that is attracted onto the screening is quickly fused and cured with infrared heat. The screening is then cooled and recoiled. Open-mesh and expanded metal materials can also be coated in coil form by vertical fluidized powder coating. Numerous small parts are powder coated on horizontal conveyors using an electrostatic fluidized bed. The parts must be small because powder cannot be electrostatically attracted out of the fluidized bed for distances of more than 6–8 in. The powder is immediately cured with infrared or convection oven heat.

Just as the downdraft air flow in the powder booth needs to be gentle to avoid blowing deposited powder from the part, so must exhaust and recirculating air in a powder curing oven be gentle, especially at the oven entry. As the parts heat and the powder reaches the melt temperature and begins to flow, the requirement for gentle air ends.

To minimize air circulation, some ovens incorporate infrared heat at the entry. Gas infrared is usually less costly to operate, but temperature control tends to be slow and difficult. Electric infrared lamps or glow-bars are simple to install and to focus if supplementary heating is needed.

It is important that the powder be heated quickly to melt and flow the material before too much cross linking has taken place. In this way, smooth films will be produced. Postcure quenching in water is rarely done, although at one time it was used to create smooth finishes with some vinyl and nylon powders.

### *An Alternative to Electrode Charging*

A method of charging some types of powder without using a charging electrode is termed *friction charging* or *tribo charging*. A tribo gun (see Figure 6-11) is equipped with a grounded teflon insert. Powder particles moving along the insert lose some electrons to ground, becoming positively charged to about 10–20 kV. Small amounts of adsorbed moisture on the particles facilitate the wiping off of electrons. The tribo charge generated in dry environments is often too weak to be effective. Tribo charging will not work with all powder coatings. Acrylic powders, for example, charge very poorly in tribo systems.

Tribo charging is often used as a means of overcoming the Faraday cage tendency for charged particles to avoid depositing in confined areas. Such areas rapidly build up a charge and repel the charged particles. Moderately tribo-charged particles can succeed in overcoming the Faraday cage problem. The same effect can be achieved by turning down the voltage in a conventional electrostatic power supply to 10–20 kV in many instances.

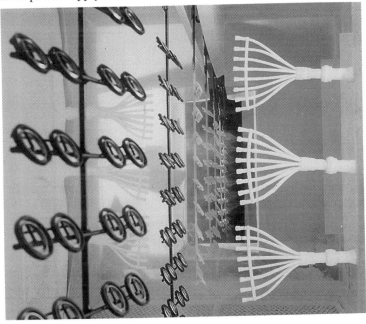

**Figure 6-11. Tribo Guns in Use**

## 72  INDUSTRIAL PAINTING

### Fluidized Bed

In fluidized bed powder coating, heated parts to be coated are first heated to above the powder melt temperature and then dipped into a nonelectrostatic fluidized bed, as diagrammed in Figure 6-12. Figure 6-13 shows a heated part being dipped into a powder fluidized bed. Powder particles contacting the hot part immediately begin to melt and flow into a coating. The coating tends to be thick because all powder contacting the hot part will melt and adhere to the heated surface. Film thickness may vary from 10–50 mils, depending on the temperature of the heated part and the time the part is kept in the powder. Exact film thickness control is impossible with this type of fluidized bed powder coating.

The powder hopper in nonelectrostatic fluidized bed coating is, in essence, like the fluidized bed container used to supply powder to electrostatic powder guns. Air is passed through a porous plastic diffuser plate that functions as the bottom of the hopper. The fluidizing air must be free of oil and particulates and have a dew point below 28° F (-2.2° C). Refrigerated dryers are commonly used. An air flow of 5 ft$^3$/min (at 5–15 psi) per square foot of diffuser surface is typical. The surface of the fluidized powder has the appearance of liquid at a rolling boil. Virtually no powder drifts out of the fluidized bed. An exhaust duct can be built around the top of the hopper to capture possible wandering particles.

The heated parts can be manually or automatically dipped into the fluidized bed and removed, but in practice, except for wire shelving, tend to be manually dipped only. As a coated part is raised from a fluidized bed, powder particles in direct contact with the heated part have flowed to a coating film, but the newest powder on the film surface may not be hot enough to

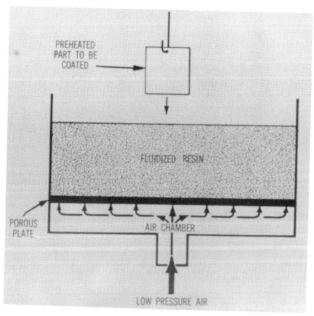

**Figure 6-12. Heated Parts Coated in a Fluidized Bed**

POWDER COATING 73

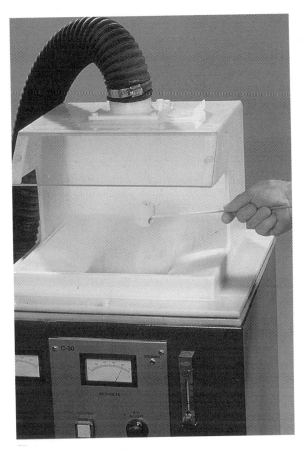

Figure 6-13. Fluidized Bed Powder Coating

have flowed. For this reason, parts removed from a fluidized bed are placed in an oven to complete the flowing of all powder on the part into a film. In some instances, selective heating is done to build film thickness in hard-to-coat locations.

## Specialty Powder Coating Methods

### *Bulk Powder Coating*

Small hard-to-rack parts can be bulk powder coated. Parts preheated on a metal belt fall into and then pass through a vibratory powder coating bowl from which they are fed onto a continuous belt and exposed to additional heat to fuse the powder. Automatic flip-over exposes both sides to curing energy so that no bare spots remain. The minimum film build by this method is 3–5 mils.

### *Non-rotating Powder Coating Disk*

Some vertically reciprocating powder coating disk systems are in operation. As with liquid paint application, the disk is charged negatively at about 100,000 V. The powder coating disk, however, does not rotate as do disks used to apply wet coatings. The high voltage on the disk charges the descending powder negatively. Since like charges repel, the powder particles are electrostatically forced radially outward to the disk edge. Powder particles leaving the disk are attracted to the grounded parts that are conveyed around the disk.

The powder coating disk is claimed to have the output equal to that of six spray guns. It is said to be capable of applying coatings as thin as half a mil with excellent thickness control, at least on parts that are fairly simple in shape. Normally the disk would reciprocate vertically, but it can also operate in a fixed position or be tilted slightly. Exhaust air requirements are identical to those for powder spray booths. Overspray powder is carried down to the bottom of the booth by a stream of air, and powder is recovered for reuse with the usual variety of filtering devices.

### *Rotating Powder Coating Bell*

The design of the powder coating bell is an adaptation of the well-known liquid bell applicator. An air turbine spins a circular head (a so-called "bell," although it is more truncated by far than the normal bell shape), which is electrically charged by a high voltage DC power supply, just as in liquid paint bell application. It is located at the head of the device to create a paint spray pattern; in this case, a cloud-like pattern of powder particles. Fluidized powder is fed through a hose in the turbine housing and then onto the bell via a circular ring of holes in the rotating bell head. The powder particles, as they pass along the bell surface, gain a negative electrostatic charge from the bell. The electrostatic principles of powder application are identical to those of liquid application.

Attempts have been made to use powder slurries for electrodeposition coating, dip coating, and flow coating, and for spraying using ordinary wet paint application systems. To do this involves extremely fine grinding of powder and mixing (slurrying) the material in water. Unless powder is ground extremely fine, it separates too rapidly from the water slurry instead of remaining suspended. The economics of aqueous powder suspensions are rarely favorable.

An advantage of powder slurry coating is that films as low as 0.3 mil (7.5 microns, dry film thickness) can be achieved. Although the reuse of the powder slurry overspray is rather difficult, it should not be completely ruled out. For a short time an E-coat powder slurry was used in an automotive plant, but the finishes were very rough and required excessive sanding.

## Powder Coating Advantages

Powder coating's rapid growth is no doubt due to its many advantages. After 25 years of use, powder coating captured about 16% of the industrial finishing market. This percentage is expected to grow steadily until it probably peaks at about 30%–35% of the industrial finishing market. The following sections discuss several powder coating advantages.

### Cost

The nature of powder coating allows various traditional processes on a finishing line to be omitted or minimized, thereby reducing costs. These include:

- No solvent flash required. This allows shortening the length of the conveyor formerly used for flash time. Parts can enter the bake oven immediately after coating application.

- No coating mix room needed. Powder coating lines have no need for a coating mix room. Powders are completely formulated by the powder coating manufacturer for immediate application.
- Minimal oven length required. Powder coating ovens can be short in length. No gradual heatup is required to drive off solvent slowly to avoid solvent popping. In fact, quick heatup is recommended for powder to avoid film roughness.
- Low ventilation requirement. The absence of solvent greatly reduces requirements for air makeup and exhaust.
- Floor space economy. A properly designed powder coating line requires from two-thirds to three-fourths of the floor space needed for wet painting systems.
- Reduced insurance rates. The absence of solvents and the elimination of solvent fire hazards can reduce insurance rates. Powder in a dust form can explode. However, proper safety precautions can minimize this powder hazard. The fire and explosion hazard with powder is remote in normal operations.

## VOC Compliance

The absence of solvent and the practical elimination of VOCs make powder a certain compliant coating in the eyes of environmental regulatory agencies. The absence of solvents has a number of other advantages:

- No solvent odor. This is highly advantageous inside the plant for employee comfort considerations and outside the plant for eliminating a nuisance factor, which is especially important in residential areas.
- No solvent storage. The elimination of solvents foregoes the need for solvent storage, which saves space and reduces fire hazards.
- No solvent thinning. Since no solvent thinning is needed for viscosity control, fire hazard is reduced.
- No solvent health hazard. Powder reduces the potential health hazards to the sprayers. Solvents can irritate the mucous membranes of the eyes. Powder spilled on the skin is not adsorbed into the body and can be removed with a gentle stream of air or with a vacuum cleaner. Cleanup of skin with soap and water is also recommended and rapidly accomplished. An appropriate particle filter mask is recommended for safety when spraying powders, some of which contain toxic ingredients.

## Powder Coating Quality

Powder coating has acquired a reputation for providing a quality finish. Powder coatings tend to be extremely durable and provide outstanding corrosion resistance. At equal film thicknesses, powder coatings are often superior to their corresponding wet finishes. One reason for this is the virtual absence of shrinkage in the cured powder coating film, which minimizes stress during curing. When wet films cure, the loss of solvent from the coating shrinks the paint film, causing internal stresses. Powder coatings applied onto grit-blasted surfaces that have not been conversion coated are always far superior to wet paints on this type of surface, both in adhesion and in the corrosion protection afforded. Powder coatings can readily be applied thick for endurance coatings. Edge coverage tends to be excellent for powder.

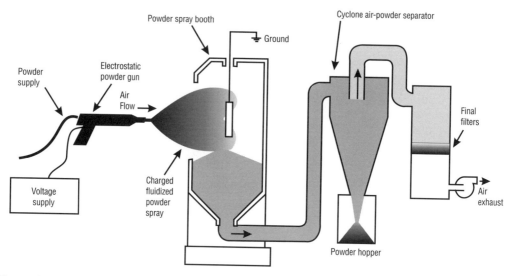

Figure 6-14. Cyclone Powder Recovery System

## Powder Overspray Reuse

Most electrostatic spray powder coating systems recirculate and reuse the overspray powder. For example, a powder application system may include a cyclone air-powder separator, as shown in Figure 6-14. This allows a utilization of about 95% of the powder. The only powder not recovered is the small amount lost during booth cleanup for color changes.

Powder reuse practically eliminates disposal problems. Disposal of spilled or contaminated powder is relatively easy compared with disposal of wet paint sludge. Landfill sites are more likely to refuse to take liquid waste or to charge higher rates for such waste than for dry powders. This is especially true if the paint sludge contains organic solvents. Places for disposal of wet paint wastes containing flammable amounts of solvent are hard to find, which results in high disposal costs. Plants have paid from $300–$1200 to dispose of each 55-gal drum of wet paint sludge! Also, powder coating systems have no contaminated booth water to treat and no used dry filters from spray booths to replace and dispose.

## Ease of Application

Little operator expertise is needed to spray powder. With wet coatings a great deal of spraying practice and finesse is necessary to get uniform coatings and avoid runs and sags. Some plants claim that a person without wet spray experience is often better at powder coating than a skilled wet painter. Manufacturers point out that this allows using unskilled labor for powder coating application, which means a lower labor rate and reduced production costs. Runs and sags, while not impossible to produce, are extremely rare with powder coatings. As long as enough powder is directed into the vicinity of the parts, the electrostatic effect tends to produce a uniform coating (see Figure 6-15). When excess powder is directed at a certain area of a part, the powder simply falls into the recovery system.

# POWDER COATING 77

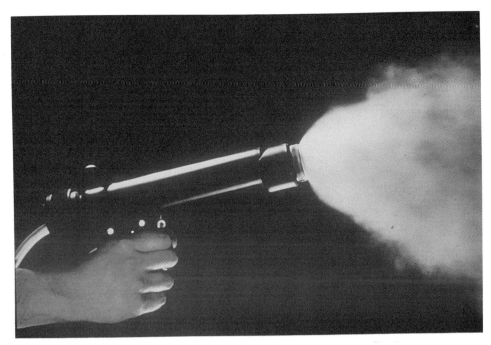

**Figure 6-15. Powder Spray Requires Less Manual Application Skill Than Wet Spray**

Coverage on corners, edges, and projections is particularly good with powder coatings because of the added electrostatic attraction and because powder coatings exhibit very little edge pull. Powder spray patterns and the electrostatic voltage can be adjusted to produce uniform part coverage, even on parts with complex shapes. Voltage adjustment enables careful control of coating thicknesses. Spray patterns can be varied from a thin powder stream for covering deeply recessed and concave areas to a wide cloud to coat broad, flat surfaces. Automatic powder spray guns are readily engineered for high-volume production, and multiple-gun systems are abundant.

Powders can be selected that incorporate topcoat and primer properties in one powder, which can be applied in a single coating operation. In this way the separate primer and topcoat operations can be combined into one step.

## Energy Savings

Although powder coating systems require a dryoff oven after pretreatment and a powder bake oven, they still use less energy than wet systems. Energy for a dryoff oven can often be conserved by using waste heat generated by the bake oven. A powder coating system's greatest energy savings comes from the greatly reduced air-exhaust and air-makeup requirements. Only about 10% as much air is needed for a powder booth as compared with a wet spray booth. Because no solvent vapors form in the powder booth, only enough air is used to recover overspray powder. The absence of solvent allows a powder spray booth to be filtered and exhausted into the plant.

## 78   INDUSTRIAL PAINTING

Powder's requirement for a relatively small oven may save energy, but the bake temperatures tend to be somewhat higher than for wet paints. Air turnover can be low because no solvent vapors need to be exhausted.

### Quick "Packageability"

When a powder-coated part emerges from an oven, it can be packaged when cool enough to handle. Unlike many wet coatings thant continue to cure for days or even weeks, powder coatings develop a full cure during the bake process. As a result, powder-coated parts resist handling abuse immediately out of the oven far better than wet coatings. They are less easily damaged during handling and assembly operations, and they do not require as much care in packaging for shipment. Problems of parts sticking to the packaging because the powder coating is not fully cured do not occur.

### Resin Availability

Resins that are not available in solventborne systems due to limited or total insolubility in solvents suitable for use in painting can be used in powder coating. For example, nylon and teflon powders are virtually unavailable in their pure form except as powder coatings. The same is true for polyethylene and polypropylene.

## Powder Coating Disadvantages

Although powder coating has numerous advantages, it also has a number of disadvantages. Probably its main disadvantage is the requirement for heat (300°–500° F, depending on the type of resin) to flow and fuse the powder into a coating film, which virtually restricts powder to being a finish for metal. In a few limited cases it has been possible to lower the powder's cure temperature to 250° F (121° C). For electrostatic application, powder coating also requires a conductive substrate, which again practically limits it to metal. Powder coating's other disadvantages include:

- Manufacturing limitations
- Application problems
- Repair difficulties

### Manufacturing Limitations

Because of the way powder coatings are made (mix, melt, extrude, flake, and mill), it is often not economically feasible to make small amounts. Powder manufacturers generally like to make a minimum of 1,000–2,000 lbs. A few powder producers specialize in preparing small-size orders, but usually at a cost premium.

One of the problems in manufacturing a new powder formulation relates to color matching. Exact powder color matching usually involves trial-and-error procedures, which adds to the "gearing up" costs. A considerable amount of raw materials may have to be used before the exact color is matched.

Color matching with powder starts at the beginning of the powder manufacturing process. Mixing of powders is not the same as for wet paint. If a can of white and a can of red liquid paints are mixed, two cans of pink paint are produced. If a box of white and a box of red powder are mixed, two boxes of a "white and red" powder will be produced. Even when the powder is applied and cured, the film would be a mixture of red and white. The colors do not mix together with powder as they do with wet systems. Minute amounts of cross-color powder coating contamination are often visible because extraneous color particles fail to blend.

Another powder manufacturing drawback is that wrinkle and texture finishes are limited. In preparing textured powder coatings, control of texture size and distribution is limited, unlike the wide variation possible with wet texturing.

Difficulties also exist in preparing metallic powder coatings. Powder coatings that contain mica, metal powder, or metal flake cannot fully duplicate the attractive look and glamour of wet metallic finishes. This has kept powder out of the large automotive topcoat market. In North America about 70% of all cars produced have metallic paint finishes. The lack of "shrink" of the powder coating gives less metallic brilliance in the appearance of the paint film. Shrinkage of wet films as solvents evaporate forces the metal flakes into a predominantly parallel orientation to the substrate surface, known as metallic "flop" or "travel." A sharp brilliance results from increased reflection because of light striking a greater surface area of metallic flakes. However, recent powder developments have yielded metallic powder coatings that are excellent for many types of parts. One such development encapsulates the metallic particles before they are added to the manufactured powder.

## Application Problems

One of the most basic problems with electrostatic powder spray is that the powder recirculating system creates a negative pressure in the booth (unlike the positive pressure of a wet booth). This allows plant air contaminants to be drawn into the booth and into the powder collection system. To prevent this contamination requires a clean room atmosphere.

Another problem related to air movement is that electrostatic powder spray must use a gentle stream of air to deliver the powder out of the gun. An excessive air flow would tend to blow the electrostatically attracted powder off the parts. Similarly, when electrostatic powder guns are reciprocated, the reciprocating speed must be slow to prevent creating turbulence in the powder cloud.

The required gentle spray from an electrostatic powder gun presents yet another problem: an enhanced Faraday cage effect. The air pressure cannot be stepped up to force the charged powder into a confined area. The enhanced Faraday cage phenomenon increases the tendency for powder to deposit at the confined area entry and not within the area, causing wide differences in powder film thickness in these regions.

Powder coating applications have film thickness problems because of the basic size of the powder particles. This makes thin film below 1.0–1.5 mils difficult to achieve. With wet systems, it is not particularly difficult to achieve films 0.5 mil thick. With the proper powder, careful application, and precise conveyor hanging techniques, it may be possible to apply low film thicknesses satisfactorily. Extremely thick powder coating films (see Figure 6-16) can be produced with one pass, but not everyone wants thick coatings. Powder's tendency to yield thick films can be considered an advantage or disadvantage, depending on whether or not the thick coating is desired.

80  INDUSTRIAL PAINTING

**Figure 6-16. Commercial Food-mixing Paddle Powder Coated with a Sterilizable Non-stick Coating (10-14 mils thick, 250-350 microns)**

Another problem with powder coating application is the formation of clumps of powder that can be ejected from the spray gun, causing paint film blemishes and rejects. One cause of clumps is impact fusion, which is the frictional heating and partial melting of powder moving through the fluidized delivery and recovery systems. Impact fusion tends to occur when fast-moving powder collides with a surface, such as when a powder flow makes an abrupt right angle, sending particles crashing into a wall. Good powder circulation design avoids such sharp turns. Powder clumping can also be caused by wet or moist powder, powder being stored in areas warmer than the recommended 75°–85° F (24° C–30° C), and powder stored beyond its shelf life.

The difficulty in changing powder coating colors is another application disadvantage. If only one booth is used, all surfaces of the booth and powder reclaim system must be purged of

# POWDER COATING 81

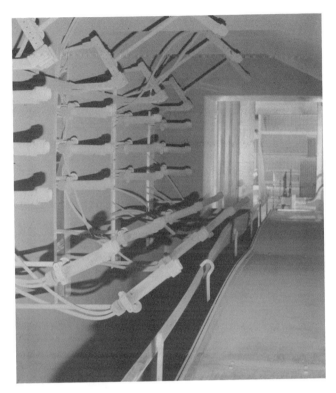

**Figure 6-17. Interior of Booth Showing Accumulation of Overspray**

powder before a new powder is entered into the system. Figure 6-17 shows the amount of powder that collects in the interior of a booth that has been applying powder. All surfaces of the entire powder recirculation system accumulate powder in the same manner. Before a new color can be applied in a booth that has been applying powder, every trace of the previous powder must be removed. Any remaining particles of the previous powder will deposit along with the second color, giving a "salt and pepper" effect.

Various quick-color change systems are used for powder. The most common system uses separate booths for each color. This allows a booth to be moved on and off line as its color is needed. Another system has a booth with plastic walls that are rolled up after each color change. Figure 6-18A and B shows a powder booth equipped with a continuous fabric collection belt floor and vacuum pickup device to somewhat simplify powder color changes.

Some powder coating users make no attempt to recover the sprayed powder, a practice called "spraying to waste." They collect all sprayed powders in a common container and either dispose of it or sell it to someone who can use a mixture of many different colors. Such a mixture can be used on parts where appearance is unimportant.

The *transfer efficiency* of electrostatic powder spray needs to be defined. The percentage of powder attracted to parts moving through a charged powder cloud is about 50% (depending on part size and shape). Although this is true transfer efficiency, it is deceptive because most

82  INDUSTRIAL PAINTING

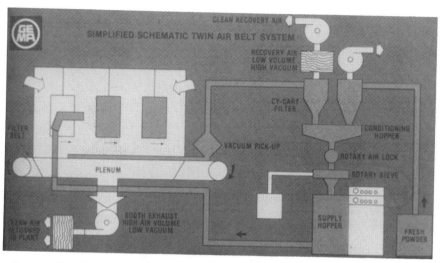

(A) Continuous Belt With Vacuum Collector

(B) Vacuum Pickup Head Collects and Conveys Powder to Small Cyclones (at left)

Figure 6-18. Powder Spray Booth With Continuous Fabric Collection Belt Floor and Vacuum Pickup Device

electrostatic spray powder coating systems recirculate and reuse overspray powder. If powder coating transfer efficiency is defined as the total percent of powder used, taking into consideration the recirculating and reclaiming of the powder, then the figure is around 95% or higher. However, on systems that do not reclaim overspray, the transfer efficiency probably ranges from 20% for thin, spindly parts to perhaps as high as 55% for parts with large surface areas, such as panels.

The absence of solvent makes it difficult for some powder resins to achieve adequate flowout. For this reason the part should be heated rapidly so the powder melts and flows before cross-linking occurs. Additional heating will cause the curing of the coating. This restricted flow tends to cause orange peel and a low distinction of image. Fine grinding of the powder is costly, but in some formulations this can help reduce the visual effects of restricted resin flow rheology.

Powder's requirement to be stored in a cool, dry place can be considered a disadvantage because it adds to capital investment. Powder needs to be kept cool to prevent possible premature cross linking. It must be kept dry to prevent clump formation. Powder storage areas should be air-conditioned to maintain cool temperatures and low relative humidity.

A powder coating's unique properties make it troublesome to strip with conventional caustic or cold solvent materials. The most common way of removing powder coatings from hooks and hangers is by burnoff in high-temperature ovens. This works well providing that the temper of the steel in the hooks and hangers is not adversely affected. Stripping parts for repainting may present similar difficulties.

## Coating Repairs

Repairing blemishes in powder coating films can be a difficult process. If the defective product is put back on the line and powder coated again, the coating can become thick and brittle. If this happens, the part will need to be stripped and recoated. A considerable difference among powders is found in their ability to produce a smooth appearance when recoated with powder. Intercoat adhesion of recoated powder coatings also varies widely.

Defective areas on powder-coated surfaces can often be recoated with liquid coatings. However, formulating a liquid coating to match a deposited powder coating can be difficult. Some companies or consumers have rejected the potential use of powder coating because no liquid repair paint could be found that would exactly match the color, gloss, and texture of the powder coating. For example, the state of New York will not purchase highway lighting poles that are powder coated for this reason.

# Chapter 7

# Cleaning the Surface

## Cleaning a Variety of Materials

For the proper bonding of paint to any surface, the surface of the material to be painted must be free of contamination. No water, dirt, or other impurities should remain between the surface and the paint because foreign matter can reduce bonding integrity and will certainly detract from a paint's appearance. Substrates to be painted fall into five general categories of materials: cloth, paper, wood, plastic, and metal. Composite materials, although comprised of a combination of these, tend to be predominantly one of these materials on the surface.

Cloth and paper are typically manufactured in long sheets that are wound on rolls. They are clean as manufactured and generally need not be cleaned before being painted. Wood, plastic, and metal items are manufactured in all sorts of sizes, styles, and configurations. The paint applicator almost always needs to clean metal items, and often must clean plastic items before painting. In some cases freshly molded or extruded plastic parts can be painted immediately, which is a desirable processing procedure since it eliminates the need to clean the plastic.

Wood surface cleaning often involves a mechanical smoothing such as sanding with fine sandpaper and wiping with a clean, lint-free cloth. Using fluids to clean wood is usually avoided; liquids such as water or solvents can raise the wood's grain, and finish sanding of the surface is then again necessary before a smooth coating can be applied.

Cleaning plastic surfaces most often involves a simple washing with a low-foaming aqueous detergent solution, plus rinsing and drying. The wash solution should usually consist of a low concentration of detergent at a mildly acidic pH. Solvent cleaning of plastics, when required, demands great care; solvents can damage plastics by interfering with their molecular macrostructure. The result can be distortion, swelling, crazing, or even cracking of the item.

## Metal Surface Cleaning

Methods for cleaning metal surfaces before painting vary with the type of metal. Cleaning metal becomes much more involved than cleaning wood or plastic because metal fabricating operations typically use various oils and greases. These lubricate the metal piece as it is pressed, punched, folded and bent, roll formed, and otherwise manipulated. The fabricated part consequently is often oil-soaked and covered with metal powder, filings, and chips. The contaminants might additionally include glues, fabrication shop markings, and weld spatter. In addition, some fabricated metals stored outdoors before painting can collect deposits of oxides. Even when the item remains indoors, iron will form iron oxide (rust), zinc will collect zinc oxide (white rust), and aluminum will develop aluminum oxide. These oxides interfere with paint adhesion and should be removed or treated before painting.

Most manufactured metal surfaces to be painted fall into one of the following categories: ferrous (containing iron), zinc (typically galvanized), and aluminum. Although all of these metals might have different optimum cleaning chemicals, the actual cleaning procedures themselves are practically identical. These procedures can include:

- Mechanical cleaning
- Solvent cleaning
- Aqueous cleaning
- Acid cleaning
- Alkaline detergent cleaning

The optimum cleaning method for a particular metal part depends on the type of contaminants to be removed. Methods for cleaning the different types of metal, for example, steel as compared to zinc, may also vary.

## Mechanical Cleaning

Mechanical cleaning can be something as ordinary as wiping dust from the surface with a clean cloth. However, when firmly attached metal scale or heavy rust must be removed, vigorous cleaning methods are required. Abrasive removal using a sanding belt or disk or a manually applied abrasive pad can be effective. Equally beneficial may be the use of a wire brush or wheel to remove tenaciously held rust. Very high pressure water alone is also effective for surface cleaning.

When very large areas are to be cleaned, it often is efficient to use sand or other abrasive material blasting with the grit propelled in a high-pressure air or water blast. A combination of air and water can also be utilized for this purpose. The grits used in this technique can include sand, steel shot, plastics, glass beads, aluminum oxides, and soft-cutting organic materials such as ground walnut hulls or ground corncobs. Ice, carbon dioxide, and sodium bicarbonate (baking soda) pellets are used as well for their ease in cleanup. Rather than being carried by an air or water blast, grit materials may also be vigorously hurled from a rotating wheel or moving belt at the object to be cleaned. With grit cleaning or blasting, care must be exercised to avoid leaving grit particles embedded in the surface as these can cause poor paint adhesion and possible early corrosion under the paint film. When recirculated grit is used, provision should be made for separating and removing accumulated "fines" and debris.

Mechanical cleaning may be coupled with liquid cleaners using cushioned plastic wool pads, which have the advantage of not loading up and becoming "blinded" and dirt-embedded. As with sanding cloths and papers, the pads are available in a range of physical sizes and grit sizes from coarse to fine. Scoth-Brite, Brade-Ex, and Brushlon are common brand names for these three-dimensional abrasive materials.

## Solvent Cleaning

The restrictions that are increasingly being placed on most solvent emissions in many Asian, European, and American nations have produced a severe decline in solvent use for cleaning purposes. Nonetheless, because oils and greases are not reliably removed by mechanical action, for some manufacturing plants suitable solvents can provide more effective methods of contaminant removal. Despite the high labor and the possibilities for missed spots, hand wipe cleaning is still best for some metal and plastic parts. Solvent cleaning by spray, vapor, or dip may be coupled with wiping with a solvent-saturated sponge or cloth. The wiping material and any solvents used in spraying, dipping, or wiping must be clean or else dirt, oil, or grease will be left on the part being cleaned. Dip tanks of clean solvent soon become dirty from the contaminants they remove. The resulting problem with dip-type solvent cleaning, then, is that parts may be only partially cleaned. Solvent spray is not commonly used for cleaning, but when it is, recontamination of parts with dirty solvent will also occur unless parts are sprayed only with fresh solvent. Continuous distillation can ensure that the used solvent remains clean. The expense of continuous distillation, however, is not usually practical unless warranted by a sufficiently high cleanliness standard or large production volume. Solvent cleaning needs to be done in a confined, well-ventilated space to keep vapors away from people and other processes in the vicinity. Flammable solvents should only be used with extreme care.

Known as *vapor degreasing*, this once widely used method of solvent cleaning enclosed and condensed solvent vapors that were continuously distilled away from contaminants removed from parts. This method could only be used safely with completely nonflammable solvents. Solvent vapors would otherwise present totally unacceptable fire and explosion hazards. In the past, various halogenated hydrocarbons termed chlorofluorocarbons (CFCs), such as freons, methylene chloride, and 1,1,1-trichloroethane (methyl chloroform) were often used in vapor degreasing. But the international Montreal Protocol agreements of the past decade, which phase out both the manufacture and use of CFCs to prevent depletion of the earth's ozone layer, have largely killed this cleaning procedure. A few less developed countries that are part of the Montreal Protocol are still permitted some manufacture of vapor-degreasing compounds. In Mexico and Vietnam some manufacture of Ozone Depleting Substances (ODSs) will be allowed through the year 2006. Still other countries, such as China and India, did not even sign the agreement and so may be assumed to continue the manufacture and use of ODSs for vapor degreasing.

The basic components of the vapor degreaser shown in Figure 7-1 are a tank, a source of heat to vaporize the solvent, condensing coils and accompanying refrigeration components, a water separator, solvent recirculation pump, filter, and a supply of solvent. Vapor-degreasing tanks may range in size from as small as about 5 gal to over 20,000 gal. The heat source maintains the solvent at its boiling temperature, sending vapor upward to the refrigerated condensing coils, where the vapor condenses and flows back to the solvent supply. The water separator catches moisture collecting on the condensing coils, which keeps the water out of the solvent. The condensing coils are located about 10–18 in from the top of the tank, leaving a distance between called the *freeboard*. The freeboard length is calculated to prevent solvent vapor escape from the tank.

A part is immersed into a vapor degreaser only as far as the vapor zone. The hot solvent vapor contacting the ambient (room) temperature part will condense and flow off the part, thus solubilizing and carrying away oils and similar materials. The condensing action will continue until

88   INDUSTRIAL PAINTING

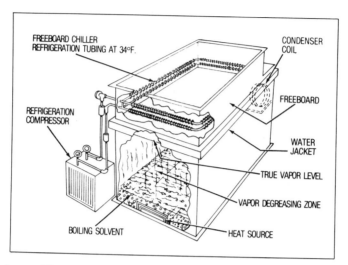

**Figure 7-1. Cutaway View of a Common Type of Vapor Degreaser**

the temperature of the part reaches that of the solvent vapor. Withdrawing the part slowly through the freeboard area allows the solvent to drain off or evaporate and return to the vapor zone. The part entry/exit movement rate must be controlled carefully to minimize vapor dragout.

Whenever halogenated solvents are used, care must be taken to keep the vapors from being drawn into direct-fired ovens. When any vapors of halogenated hydrocarbons come into contact with flames, corrosive gases such as hydrofluoric and hydrochloric acid vapors are formed. These acid vapors can wrinkle the finish on freshly painted parts in an oven; moreover, the acid vapors present a potential health hazard to persons who are sensitive to these irritants.

CFCs do not burn, nor will bacteria break them down so they persist almost forever in the ground. Because it is widely believed that halogenated hydrocarbon vapors are a major cause of the atmospheric ozone layer destruction at the earth's South and North Poles, chemical companies have searched for suitable replacement vapor-degreasing solvents that are less environmentally offensive. Unfortunately, no ideal "drop in" substitute solvent or solvent blend has as yet been identified.

The CFCs are certainly not the only paint and cleaning solvents whose use has been virtually eliminated by a greater awareness of the hazards in using them. Many traditional chemical solvents are no longer acceptable for use, or they are allowed for use only in strictly controlled circumstances. Benzene was eliminated by worries over liver and kidney damage. A number of ketones or glycols pose possible nerve damage, cell damage, or cancer dangers. Carbon tetrachloride, for example, once used as a fire extinguisher material and as a common cleaning agent, was found to be carcinogenic and, thus, is also no longer acceptable for cleaning usage.

Other solvents, in addition to the CFCs, are restricted because they are considered Hazardous Air Pollutants (HAPs) and more destructive to the environment than other solvents. In the U.S., the Environmental Protection Agency issued final National Emission Standards for HAPs (NESHAP) that required conformance by the end of 1997. Typical paint and cleaning solvent

HAPs include toluene, all three xylenes, methyl ethyl ketone (MEK), and methyl isobutyl ketone (MIBK), but heavy metals such as lead and chromium are also on this list of nearly 200 materials.

## Aqueous Cleaning

*Aqueous cleaning* systems—water plus detergent and frequently also small amounts of acid or alkali—are the most popular cleaning systems for industrial finishing. Two main advantages of these systems are the absence of solvent emissions and the ready adaptability to practically every cleaning requirement and every type of substrate.

The basic types of aqueous cleaning are spray and dip (immersion) either with parts conveyed on a hook, carried on a mesh belt, or placed in a wire basket to hold them during the cleaning operation. Parts may be lowered into and out of the cleaning liquid or carried through the cleaning spray or dip zone. Spray cleaning is most commonly used and is faster because of the scrubbing action of the impingement of cleaning solution. But the spray cannot reach into hidden regions of complex part shapes. Dip cleaning is slower, but the solution will reach into areas not accessible to the spray. To increase cleaning rates, the dip cleaner solutions are often higher in detergent concentration than spray cleaner solutions. Low-volume operations commonly use manual spray or dip; high-volume manufacturing generally uses automatic spray or dip. The choices of manual or automatic and whether to use spray, dip, or a combination of spray and dip are often determined by the nature of the soils on the parts and the size or configuration of the products to be cleaned. Parts volume is another important factor.

Dip cleaning can be done with small parts by placing them in a wire basket that can be lowered into the cleaner tank, possibly with some mechanical agitation of the basket for faster cleaning. Basket cleaning utilizing spray cabinets is also possible but is not commonly done.

Large, bulky, relatively low-volume items—such as construction cranes, for example—are often cleaned manually by spraying with aqueous detergent. Most high production items, such as car bodies, are cleaned automatically in huge dip tanks. A manufacturer might choose dip cleaning for a complex-shaped product that requires cleaning in deep, hard-to-reach areas. A few manufacturers have cleaning systems that incorporate both spray and dip cleaning: impingement spray to clean accessible exterior surfaces and dip to enable cleaning solution to penetrate into hard-to-reach areas. Formulating a cleaner to do both spray and dip cleaning thoroughly is not really feasible. Separate cleaning solutions for each method are far more efficient and effective.

Manual spray is usually accomplished with a spray nozzle attached to the end of a long wand as shown in Figure 7-2. Moderately high fluid pressures (75–200 psi) are usually used in manual spray cleaning. The wand permits the operator to reach into sizable cavities and yet stay a safe distance from the nozzle in order to avoid getting splashed excessively by hot detergent solutions.

In contrast, automatic belt or conveyorized spray cleaning systems use lower fluid pressures and incorporate fixed plumbing components that include headers, risers, and nozzles. Headers, constructed of large-diameter (2–6 in) piping, carry the cleaning solution to the risers, which are small-diameter (1–2 in) piping. The risers are usually arrayed somewhat U-shaped, containing straight horizontal and slightly curved or angled vertical members. The parts to be cleaned are

90  INDUSTRIAL PAINTING

**Figure 7-2. Manual Cleaning Using a Wand Extension**

hung on a conveyor that passes through the U-shaped spray. Nozzles attached to the risers—about 12 inches apart—provide a zone of spray through which the part to be cleaned passes. Numerous riser sets can be used; systems have riser sets spaced throughout the length of the cleaner zone, which often ranges from 6–20 ft long. Cleaning time is affected by the line speed and the length of the spray zone. Thus if 90 seconds (s) of cleaning are required and the conveyor speed is 5 ft/min, the risers must extend for 7.5 ft.

The choice of nozzle type is dependent on the stage in which it is used and what chemicals are being sprayed. Whirl-jet nozzles give a flood pattern with low impingement force but lots of fluid volume, but V-jet styles combine high volume and high surface impingement in a flat, wide spray. The K-jet and fan jet nozzles are similarly flat but result in less overspray and allow better containment of chemicals within the spray stage. They are low in both volume and impingement force.

Drainage zones following each of the stages also assist in chemical containment at each stage. The best drainage floor has an inverted off-center "V" shape with the apex located about 60%–65% away from the exit of the spray zone. Solution tank bottoms should not be flat; ceilings can be similar in shape to the drainage floor or totally sloped. For composite and stainless steel enclosures, a 15° ceiling slope is effective, but for mild steel ceilings the slope must be at least 30°. This is because chemical growths form on mild steel and act as drip points. Sloped metal ceilings also have sharply higher construction costs. Slotted top ceilings with the rail outside the oven and pretreat machine offer a possible advantage in chain life but may allow more steam to escape. Opinion on whether they are cleaner or not is sharply divided. Tank bottoms should slope to a drain valve for easier cleaning and to allow complete draining of the tank's contents.

CLEANING THE SURFACE 91

Figure 7-3. Overhead Conveyor Carrying Parts Through an Automatic Spray Washer

Belt washers use a continuous metal mesh belt that travels through the various detergent and rinse zones. A dryoff oven zone may be included afterwards as well. Parts are normally simply placed on the belt and taken off at the other end manually.

Large, heavy, and high-volume parts are usually hung on an overhead conveyor and carried through the cleaning and rinse risers, as shown in Figure 7-3. Various configurations of the conveyor and risers are possible to minimize moisture reaching the conveyor parts; moisture contributes to high conveyor maintenance and may allow drippage of conveyor soils onto the parts being cleaned.

The cleaner section is usually the first stage of an automatic spray washer machine. Cleaning solution is held in a tank and a pump delivers it to the risers. In addition to the riser and nozzle sets, automatic spray washers commonly incorporate external reservoir tanks that contain the chemical solutions (see Figure 7-4) being sprayed onto the parts to be cleaned. The reservoir tanks provide a convenient means of heating the solutions and replenishing the chemical solutions as they become depleted with use. Flame tubes directly inside the main tanks are still used for heating, but separate heat exchangers are better for most solutions because they cause less chemical breakdown.

## Acid Cleaning

Rust, scale, and oxides can be removed rapidly from parts to be painted by using an acid pickling (brightening) solution. Although the oxides react with acid faster than the metal itself, the pickling will also to some degree etch the surface of the metal. The acid may have added inhibitor in it to slow the etching action to a desired rate most suited to the part being cleaned and the type of soil to be removed. Both hydrochloric acid and sulfuric acid can be used for steel; aluminum is frequently cleaned with nitric acid solutions usually referred to as brighteners.

92  INDUSTRIAL PAINTING

Figure 7-4. Automatic Spray Washer Boxlike Protrusions (right) Contain Chemical Solutions and Rinses

Hydrogen embrittlement can occur readily in acid cleaning because hydrogen gas is produced even in inhibited acid pickling solutions. Hydrogen gas embrittlement is a somewhat complex metallurgical phenomenon that occurs when hydrogen is in contact with metals. Briefly, the tiny molecules of hydrogen penetrate into the crystal lattice formed by the much larger atoms of metals, thereby making the metal significantly more rigid and thus more brittle and prone to stress fracture. This embrittlement can reduce the metal's physical properties such as tensile strength and elastic modulus.

Other methods must be used for oxide removal if hydrogen embrittlement of the metal cannot be tolerated. Alkaline derusting of steel is done with very highly caustic (pH = 12–13) aqueous solutions to avoid hydrogen embrittlement. Solutions this high in alkalinity will react so vigorously as to be dangerous with more active metals such as zinc, magnesium, and aluminum. Therefore, alkaline derusting is limited to use on steel alloys only.

## Alkaline Detergent Cleaning

Aqueous alkaline detergent solutions are used to clean all types of plastic and metal products. They are now the preferred cleaning agents since they contain little or no material that is harmful to the environment. In general, the stronger the alkali, the faster the soil is removed and the more thoroughly the soil is dispersed in the cleaning solution. Better soil dispersion results in a longer cleaner life. Just enough alkalinity is introduced into the detergent formulation for good soil removal and dispersion; too much alkalinity results in slow and incomplete rinsing (poor rinsability). This is an important consideration because thorough rinsing of alkaline cleaners is crucial for good painting.

Cleaners may contain sodium hydroxide (caustic soda), sodium carbonate (soda ash), or sodium silicates as builders and may also contain additives such as pyrophosphates, orthophosphates, borates, and chelating agents. Additives are included in detergent formulations to soften water, reduce foaming, increase wetting, aid emulsification of soils, and reduce *nubbing* (the term used for localized over-etching of zinc alloys).

In addition to a caustic builder, the source of the alkalinity, a typical cleaner formulation contains a low-foaming synthetic detergent rather than a soap. Occasionally, alcohols and glycol ether solvents are added to speed cleaning. As a rule, total cleaning effectiveness is increased with nonionic detergents, compared with the more frequently used anionic detergents. The nonionic detergents are more expensive, but they work well even at relatively low temperatures and are particularly low-foaming.

## Emulsion Cleaners

Emulsion cleaners that have kerosene or similar organic solvents dispersed in them have been available for a long time. Some are so high in solvent that they are flammable. They work well for removing some greases but not for the removal of a wide range of soils and contaminants. Emulsion cleaners also tend to cost more than ordinary aqueous cleaners, and so they are not used much. Some of the citrus oil-containing cleaners have found good acceptance, however. The smell of citrus is evident when these are used.

## Cleaners for Aluminum

Aluminum can be cleaned with various aqueous chemical solutions. The configuration of the parts, the alloy composition, and the desired results strongly influence the cleaner selection. Chemicals used on aluminum include:

- Silicated alkaline cleaners. Silicated alkaline cleaners, known also as nonetching cleaners, are used at a pH (acid/alkali rating) of 11–13 (see Figure 7-5) and can remove grease and waxy soils.
- Nonsilicated alkaline cleaners. Nonsilicated alkaline cleaners operate at a pH of 8–10 and are mostly used prior to a caustic etch before chromating or anodizing. Nonsilicated cleaners are excellent products and present little problem of cleaning agents drying on parts. Although they are generally considered to be nonetching, nonsilicated cleaners are actually capable of microetching most aluminum alloys if they are operated too hot or at excessively high concentrations.
- Organic acid cleaners. Organic acid cleaners use a high concentration of organic surface active agents, called *surfactants*, to remove stains, streaks, and related blemishes. They are expensive, but long bath life may justify their use to take off materials not removed by alkaline cleaners.
- Acid etching cleaners. Acid etching cleaners, containing phosphoric acid, hydrofluoric acid, wetting agent, and solvents, such as ethylene glycol or propylene glycol, are used to

94  INDUSTRIAL PAINTING

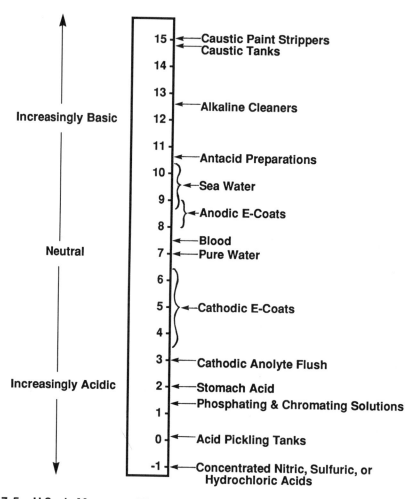

Figure 7-5. pH Scale Measures Alkalinity and Acidity

# CLEANING THE SURFACE

remove light soil and are useful in preparing aluminum for chromating or direct painting. The acid provides a smooth microetched surface that is highly suitable for the formation of a chromate conversion coating.
- Alkaline etching cleaners. Alkaline etching cleaners are used in hot, strong sodium hydroxide (caustic soda) solutions containing chelates and wetting agents. These cleaners produce a heavily etched surface that has a visible pattern after the black oxides (smut) produced by strong alkali are removed by an acid dip.

To ensure an acceptable operating environment, a blanket layer of detergent foam on parts to be cleaned is needed to suppress the alkaline fumes that result from the reaction between aluminum and the hydroxide. If the foam blanket becomes depleted due to oils from the parts, a small amount of high-foaming detergent can be added. Parts should be reasonably clean prior to etching in order to avoid the formation of blotches on the aluminum.

## Other Considerations

Immersion (dip) cleaning has an advantage over spray in that cleaning solution reaches all exterior surfaces, but since it does not clean as fast as a spray process, it tends to require more concentrated cleaners. In some cases, silicated immersion cleaners are used. Silicated cleaners can inhibit excessive alkali attack on aluminum and zinc when concentrated solutions must be used to clean heavily soiled parts. Although these clean fast, they may passivate (deactivate) the surface and thus cause slower chromating. Silicated cleaners can produce unsightly streaking that in some cases is visible after painting.

Most aqueous cleaners are operated at 125° F–160° F (49° C–71° C). Cleaners of all types work better hot than cold because oil and grease are more mobile and easier to remove when hot. Heating the cleaner and rinse stages raises the temperature of the metal to be compatible with the temperature of phosphating solutions that often follow cleaning (see Chapter 8). The first cleaning stage is normally operated at a high temperature to facilitate cleaning and heating of the metal, and the remaining stages are gradually lowered in temperature to that of the subsequent phosphate or chromate stage.

## Rinsing After Cleaning

In addition to the items discussed so far, there are other aspects of the cleaning process that are also important. Good rinsing after cleaning is crucial, for example, because the alkali in the cleaner will neutralize conversion coating chemicals in phosphating and chromating. A portion of the phosphating chemicals is thus wasted because they react with the caustic materials to create insoluble sludge. Additional cost is incurred because the sludge must be treated. (Figure 7-6 shows an itemized list of rinsing cost factors.) Poorly rinsed caustic cleaners can also interfere with subsequent zinc phosphating by coarsening the phosphate crystal size and promoting void areas where no phosphate coating is formed. Surfactant solutions need to be rinsed thoroughly from surfaces to be painted. Alkaline solutions are often not easy to rinse, however, and so special care in rinsing should be taken to avoid a reduction in the quality of chromate and phosphate conversion coatings.

# 96 INDUSTRIAL PAINTING

---

**Total Cost of a Rinsing Operation**

Water
Drag-out Chemical
Waste Treatment Chemical
Waste Disposal
Waste Treatment Labor
Waste Treatment Equipment

---

**Figure 7-6. Rinsing Operation Cost Factors**

Cleaning can be improved using a double-wash system: either a "wash—rinse; wash—rinse," or a "wash—wash; rinse—rinse." In double-washing systems, the first wash removes the brunt of the soil, and the second removes any remaining soil. The first wash tank can be replenished by overflowing the second wash into the first. Because of the necessity for very thorough rinsing after cleaning, the "wash—wash; rinse—rinse" system is a better procedure. Rinsing between cleaning stages is unnecessary; double rinsing after cleaning is more effective.

The water quantity, quality, and temperature all play a part in determining the efficiency of the rinse. For optimal results and efficiency of chemical use, the hardness of the water used in each of the cleaning and pretreatment stages, including rinse zones, should be below 300 parts per million (ppm). Chloride and sulfate anions (negatively charged ions) combined should be no higher than 500–600 ppm; the sum of the ammonium, sodium, and potassium cations (positively charged ions) should be below 250 ppm.

Rinses serve to wash away undesired chemicals from parts that have been cleaned or cleaned and pretreated. These chemicals slowly build up in the recirculating rinse water. Proper overflow of rinse tanks to drain will be needed to keep rinse solutions low in contamination from the chemicals dragged in on parts going through the various process stages. When parts are dry, any visible spots and residues found on the parts are indications of marginal rinsing.

A simple check of the quality of rinse water can be made by measuring the conductivity of the final drain-off rinse. Rinse water as it drips from the parts can be collected and tested for conductivity. A small portion of the drain-off can also be evaporated to see how much residue remains. High conductivity and noticeable residue from evaporated rinse water are signs that the rinse water quality may need to be improved. The supply water may be too high in solids, or the rinse water may require more frequent changing or a faster rate of overflow to drain.

When rinse water is dried, it leaves behind on the parts nearly all of the dissolved materials that may be present in the rinse water. Total dissolved solids (TDS) of rinse water should be minimal. In some paint systems the dissolved solids in the rinse water are held as low as 10 ppm or below 30 micromhos/cm conductivity. For critical applications, several paint systems require deionized water that has less than 2 ppm silica and conductivity below 10 micromhos/cm. For less sensitive systems, the dissolved solids can be as high as several hundred ppm. In order to correct for high conductivity in the rinse water, a brief deionized water (DI) rinse may be required

CLEANING THE SURFACE 97

after the normal supply water rinse. A recirculating DI followed by a halo or "mist" rinse of fresh DI make-up water will provide superb rinsing. The rinsing of chemicals is accomplished by the supply water; then the supply water is rinsed away by the DI water.

In these instances, a recirculating DI rinse of about 5 cc (ml) needs to be supplied for each square foot of work surface. In power washers, 2–4 risers per side are generally adequate. Deionized water is produced by passing water through two regenerable ion exchange beds. The first ion exchange bed removes all cations such as sodium, calcium, magnesium, and iron; the second bed takes out all anions such as chloride, sulfate, nitrate, and phosphate. The cation and anion replacements are hydrogen and hydroxide ions. The hydrogen and hydroxide ions do not exist for long; they immediately react with each other to form water molecules. Thus, pure water without ions results, and hence the name *deionized water*.

## Wastewater Treatment

The strong emphasis on fighting water pollution makes the proper management of cleaner and rinse solution disposal vital. Plants must determine and record the type of contaminants present, their concentration, and water quantities for all their effluent water streams. While a detailed coverage of wastewater treatment is not possible in this book, some typical materials

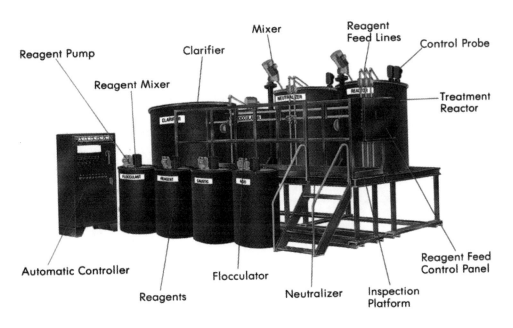

**Figure 7-7. Wastewater Treatment System**

that must be removed before release are heavy metals, suspended solids, soluble and insoluble oils, phosphates, acids, and alkalis. Waste water treatments used include flocculation, settling, skimming, precipitation, filtering, and pH adjustment (see Figure 7-7). The maximum allowable contamination concentrations in wastewater released to a stream or to a municipal treatment plant can vary widely from location to location. In many instances, plants have chosen to reuse the water after treatment rather than release it to drain. This need not be an added expense to manufacturing operations. More and more plants are finding that wastewater reuse and minimization programs can actually be cost effective.

# Chapter 8

# Conversion Coatings

## Conversion Coating Qualifications

The importance of a clean surface prior to painting cannot be overemphasized; however, a clean surface may not be all that is needed. Whether or not additional mechanical or chemical surface treatment is required depends principally on the substrate to be painted, although the properties of the paint also come into play in some instances. In the case of wood, a clean surface is all that is required (excluding surface smoothness considerations). A wood surface is unique because its polar cellulosic composition provides an excellent adhesion base for most paints.

Although a clean wood surface requires no chemical treatment before painting, the same is only sometimes true for clean plastic surfaces. A good bond must be formed between the plastic substrate and the paint film, and that is determined by both the plastic type and the paint type. It is not often possible, but some plants will paint plastic parts almost immediately after the parts are molded. By painting before the plastic has a chance to become soiled, manufacturers save washing and drying steps. If painting parts this quickly is not possible, many plastics can simply be cleaned and then painted. Others, primarily the low-polarity plastics, such as polyethylene and polypropylene, even when clean, do not form good adhesion to paints. These plastics usually require an additional treatment to provide a surface on which paint can find good adhesion.

To improve paint adhesion, plastic surfaces can be oxidized by electrostatic corona discharge, by ultraviolet light-initiated decomposition of sprayed or dipped-on activators such as organic peroxides, by controlled flame treatment, or by special oxidizing paints. Such oxidative treatments introduce carboxyl (COOH), carbonyl (CO), and hydroxyl (OH) groups into polymer molecules on the plastic surface. The presence of these groups, which are oxidized forms of the plastic itself, can be verified by reflective infrared spectrophotometry. The effect of flame treating lasts only about 10–20 min. It is suitable for relatively flat parts or for manual treatment on a limited number of pieces. The corona discharge activation is effective for up to an hour or so, while the UV light plus activator method allows painting for several hours. The latter method may be effective up to as much as a day later, although a shorter interval is obviously advisable.

As with plastics, some metal parts are painted immediately after surface cleaning. These metal items may not have severe requirements for adhesion or corrosion resistance. Low-cost products that are expected to have a short life cycle may fall into this category. In other cases, a pretreatment may not be used even though corrosion resistance requirements are high. Items that are too large or that are not convenient for undergoing chemical pretreatments may be painted after being cleaned only. Marine vessels, construction equipment, or highway bridges, for example, are frequently sand blasted and then painted soon thereafter. While chemical pretreatment would not, strictly speaking, be impossible in these instances, it would be highly inconvenient to administer. By using special high-performance coatings, the lack of pretreatment can be overcome and excellent adhesion plus good finish durability are possible.

## Conversion Coatings for Metal

Most metal products that are to be painted receive an insoluble salt-like conversion coating after cleaning. A *conversion coating* is a chemically developed inorganic surface layer that helps shield the metal from air, moisture, industrial gases, and the like, thus increasing corrosion resistance and simultaneously providing a roughened surface to improve paint adhesion. The conversion coating surface has microporosity and microroughness in the form of pores, fissures, and undercuts that form an excellent stratum into which the paint can flow and create "fingers" for a good hold onto the surface. In addition, the conversion process leaves a slightly acidic residual pH on the surface, which is more receptive to good paint adhesion than is an alkaline pH. Alkaline solutions can degrade paint resins, and thus they find use in paint strippers. Alkalinity on surfaces to be painted is generally undesirable, but there are exceptions for some very mild materials.

Long-term paint adhesion is enhanced by the thermal expansion and contraction buffer provided by the conversion coating between the metal and the paint. The expansion and contraction coefficient of the paint film is many times that of the underlying metal. The conversion coating is intermediate in its expansion and contraction characteristics. As a result the severity of the difference in the expansion rate between the paint film and the underlying metal is lessened when sudden hot and cold changes are experienced.

A conversion coating reduces the lateral creep of corrosion away from gouges and scratches that have gone through the paint down to bare metal. The conversion coating inhibits the undercutting of the paint by loss of the underlying metal, and the spread of rust away from a scratch line is slowed significantly.

Conversion coating thicknesses are given in weight per unit area. (In the United States, the reading is traditionally in mg/sq ft, but multiplying that value by 10.76 gives weights in units of mg/sq meter.) Conversion coatings for metal surfaces to be painted may vary from a few mg/sq ft to as much as 500 mg/sq ft. Conversion coating weights are usually specified within a certain range for optimum adhesion and corrosion resistance for a given product.

Conversion coatings are usually formulated specifically for a particular type of metal and may differ in chemical composition and anticorrosion properties. Metals used most commonly in manufacturing and of primary interest in industrial painting are hot rolled steels, cold-rolled steels, stainless steel, clad and metal-coated steel, hot dipped galvanized steel, electrogalvanized steel, galvannealed steel, galvalume steel, diecast zinc, and aluminum alloys.

Hot-dipped galvanizing is done by passing heated steel through a bath of molten zinc. The adherent zinc layer thickness is controlled by air knives (steam may also be utilized) directed at the sides of the steel. Hot-dipped galvanneal results from a postgalvanizing annealing process that produces an iron-zinc coating instead of more pure zinc only. Galvalume is a hot-dipped steel that has been given an aluminum-zinc coating that is more than 50% aluminum. It has better heat reflectivity and greater ductility than hot-dipped galvanized steel. The smoothest and most uniform zinc layer is deposited by electroplating zinc onto steel, forming electrogalvanized steel.

## Conversion Coatings for Steel

Conversion coatings for steel are usually either iron phosphate or zinc phosphate. Somewhat different chemicals are used in iron phosphating than in zinc phosphating, but the overall processes are similar. They both include use of a dilute aqueous solution of phosphoric acid containing a soluble dihydrogen phosphate salt.

An important prerequisite for applying a phosphate coating is to have a thoroughly clean surface. Cleaning may be done as a preparatory step, but in some iron phosphating operations, cleaning and phosphate coating deposition are accomplished with the same solution. If a separate cleaning stage is used, all traces of the cleaning solution must be rinsed off the metal. Unrinsed alkaline cleaner will reduce the quality of the phosphate coating.

During the phosphating process, phosphoric acid attacks the metal surface, removing soluble ions into the coating bath and simultaneously lowering the acid strength at the interface of the metal and the phosphating solution. In the area of the metal surface where the acid strength has been partially depleted, the type 1 (primary phosphate salt) converts to type 2 (a monohydrogen phosphate salt) and then immediately to a type 3 (tertiary) phosphate. Although it is not necessary to understand the complete chemistry of this process, it is beneficial to know the chemical steps that take place as mentioned in the previous sentences.

In iron phosphating, type 3 iron phosphate is formed; in zinc phosphating, type 3 zinc phosphate is formed. Both type 3 phosphate salts are insoluble. During phosphating they crystallize rapidly out of the phosphate solution, bonding tightly to the metal as they grow onto and into the surface irregularities and etch pits created by the acid attack.

Additional chemicals such as accelerators, additives, and oxidizers are present in the phosphating solution. They include nitrites, nitrobenzene sulphonates, chlorates and bromates, fluorides, manganates, nickel ion, cobalt ion, and more. Accelerators are added to keep the many chemical reactions occurring at an acceptable speed. Additives are utilized for such purposes as reducing sludge formation and preventing "seedy" deposits in the phosphate layer.

When ferrous metals are phosphated, the oxidizers convert the iron ion dissolved by the attack of the phosphoric acid, from the +2 (ferrous iron) to the +3 (ferric iron) oxidation state. At the same time, oxidizers also prevent the formation of large hydrogen bubbles on the surface of the metal. Hydrogen is formed by the reaction of acid with metals. So that hydrogen gas bubbles do not block the access of acid phosphate solution to the metal, oxidizing agents are added to the formulation to quickly convert any hydrogen that forms to water molecules.

Most iron phosphates and zinc phosphates have a clear to a dull gray appearance; however, the addition of molybdic acid or a molybdate salt added as a corrosion inhibitor will give a distinctly purple cast to some iron phosphate coatings. These types of iron phosphate may be used when no postphosphate sealer rinse is included in the system.

Process time, temperature, and chemical concentrations are the three basic factors that affect the chemical reaction in a phosphate process. The process time is dictated by the speed at which the conveyor line must run. Temperature and concentration of the solutions are the conditions that are monitored during phosphating. Some zinc phosphating systems have extremely tight temperature control—far more so than for iron phosphate—in efforts to increase the fineness of zinc phosphate crystal size.

The bath temperature can dramatically affect phosphating deposition. If the temperature of the preceding stage—such as a cleaning stage—is lower than that of the phosphate stage, problems can develop that will eventually diminish phosphate quality and increase solution maintenance. With low cleaner and rinse stage temperatures, the heating burden shifts to the phosphate stage. The increased demand for heat in the phosphate stage results in excessive sludge. The extra sludge will tend to plug risers and spray nozzles. It will also reduce heating capacity by building scale on the heat exchanger. The reduced heating capacity may drop the temperature of the phosphating bath below specifications. For the phosphate to coat properly under these conditions, a high level of accelerator is required, which will generate even more sludge.

Increasingly, we see the use of plastic headers, risers, and nozzles in cleaning and phosphating spray systems. Not only are these lighter in weight and faster to remove and install, they are considerably less prone to collect sludge deposits (insoluble oxide and phosphate compounds), which often plug metal pipe risers and spray tips. Plumbing connections are simpler as well.

## *Iron Phosphating*

Iron phosphate coatings that form on steel alloys are sometimes called amorphous or noncrystalline iron phosphate. (In actual fact, they are finely crystalline and are not amorphous.) These coatings are primarily a mixture of iron oxides and iron phosphate. Some studies reveal that normally about 65% of the iron phosphate coating is actually comprised of iron oxides, while only 35% is mixed ferrous and ferric phosphates. Despite that, the name "iron phosphate" will not change. The typical coating weight of iron phosphate ranges from 25–100 mg/sq ft and, compared to zinc phosphate, provides moderate corrosion resistance. Iron phosphate system characteristics include relatively low operating costs, simple equipment requirements, low sludge formation, and moderate control requirements. The main variables to regulate and monitor are the treatment times, solution temperatures, and solution concentrations. Iron phosphating can be done with ambient (room) temperature solutions; most, however, operate at between 120° F and 140° F (49° C–60° C).

The color may be, and usually is, varied across the parts, but this does not necessarily denote problems, thickness differences, or inadequacies in the phosphate coating. Iron phosphate color varies generally with thickness (most often expressed in terms of coating weights). Typical colors are correlated to phosphate coating weights in Figure 8-1; but formulation and operating parameters, as well as the substrate alloys, will all have an effect on the color of the iron phosphate layer.

Iron phosphating and cleaning are often done in the same bath, for it is possible to clean and iron phosphate simultaneously in a single solution containing a suitable cleaner and phosphating chemicals. Such a combined solution, however, almost always provides inferior cleaning and phosphating than when these procedures are done separately. Part of the reason for this is that if a cleaning agent is combined with the phosphating chemicals, the cleaner must be an acid-type cleaner. Acid cleaners are less able to remove most soils than alkaline cleaners, but soils do not foam nearly as much in acid solutions as in alkaline cleaners. Acid cleaners remove smut and minor traces of metal oxidation (rust) better and rinse off more easily than alkaline

| Iron Phosphate Weight | Approximate Color |
|---|---|
| 10–15 mg/sq ft (100–150 mg/sq m) | Colorless to faint gray or blue-gray |
| 15–30 mg/sq ft (150–300 mg/sq m) | Blue-gray or gray |
| 30–50 mg/sq ft (300–500 mg/sq m) | Blue or gray plus slight gold tone |
| 50–80 mg/sq ft (500–800 mg/sq m) | Blue or gray plus light red-gold tone |
| 80–125 mg/sq ft (800–1250 mg/sq m) | Blue-purple plus red-gold tone |

**Figure 8-1. Phosphate Weight/Color Chart**

types. The combination cleaner + phosphate solution is usually satisfactory only if the parts have fairly light soil levels and no difficult-to-remove contaminants. Such instances abound, though, and for these applications the combined cleaning and iron phosphating is a totally satisfactory reduced cost way to accomplish cleaning and conversion coating in just a single treatment stage.

If one is ever in doubt, it is advisable to install separate cleaning and iron phosphating stages. This may prove to be a prudent measure, for the extra capital cost for two added stages is not excessive if installed simultaneously. Some plants that have installed combination cleaner + phosphate systems later find that paint adhesion loss is reported by angry customers. Combination cleaner + phosphate baths may prove inadequate when new products or new operating steps are added. Soils may be light on the products for which the combination cleaner + phosphate bath is first designed, but subsequent changes in the type or source of metal used or different manufacturing processes may introduce heavy oils or other contaminants that cannot be fully cleaned in a combination bath. Then expensive retrofitting must be done to add a separate cleaning stage and another rinse stage, which can be done easily only if enough floor space is available. Separate cleaning and phosphating stages are almost always more desirable because of the flexibility they afford. It is best to avoid cleaner + phosphate systems when maximum corrosion protection is needed. On the other hand, combination systems are still very common for they can and do work well in the right circumstances.

One application where combination cleaner + phosphate baths are practical is with manual spray wand pretreatment of products that are too large or too few in number for a conveyor— heavy machinery and construction equipment, for example. In such applications, a steam generator provides the necessary heat to the combination cleaner and iron phosphate solution. The chemicals are low in cost and nontoxic, so they are run to drain rather than reused. The manual spray wand is used for water rinses as well, although iron phosphate + cleaner formulations that require no rinsing are also available. In actual practice, postphosphate sealers are almost never used in wand phosphating; yet with the availability of nontoxic sealer rinses, there is no reason other than economics for not using them.

Solutions that form dense iron oxides—some with coloring agents to make iron or steel parts blue, brown, or black—have found only infrequent use as prepaint treatments and are simply not recommended for use under paints. By themselves, the oxides offer extremely limited rust protection. Most often these oxides are coated with drying oils or waxes to provide a low-cost rust preventative, which is a function they do well under mild indoor exposure conditions. They are not intended for severe service or for outdoor situations, nor are they considered to be decorative coatings; however, they do find use in their niche applications.

## Zinc Phosphating

Zinc phosphate comes in a number of formulations. Pure zinc phosphate is called hopeite. Zinc phosphate coating weights are typically 175–450 mg/sq ft on steel that is to be painted. (For other purposes such as for adhesive bonding, the coating weights may be much higher.) Little increase in paint adhesion or corrosion resistance is gained from depositing heavier coating weights. In fact, heavier coatings may cause a decrease in paint adhesion.

The formation of a fine, dense crystal pattern of zinc phosphate (see Figure 8-2) noticeably improves corrosion resistance and paint adhesion compared with coarse zinc phosphate crystals (see Figure 8-3). One of the most common ways to produce a fine, dense crystal pattern is to add crystal grain refining agents, such as titanium phosphate, either to the zinc phosphating solution or preferably to the rinse preceding the zinc phosphating stage. These titanated rinses

Figure 8-2. Fine, Dense Crystal Pattern of Zinc Phosphate

Figure 8-3. Coarse Zinc Phosphate Crystals

neutralize trapped cleaning agents, orient the crystal growth in a configuration more closely parallel to the surface, and most importantly, initiate the growth of finer, denser zinc phosphate crystals. Titanium ions act as nucleation sites for growth of zinc phosphate crystals. This occurs wherever there are minute amounts of titanium which have been adsorbed on the steel surface during the rinsing process.

Incorporating ferrous iron ions ($Fe^{+2}$) into zinc phosphate coatings as they are forming will produce an iron-rich zinc phosphate known as phosphophyllite, $Zn_2Fe(PO_4)_2$, which is more alkali resistant than pure zinc phosphate. Phosphophyllite's strong tensile properties also reduce crystal fracture under stress. Its increased alkali resistance diminishes the rate of creepback from areas where corrosion has already begun from a paint scratch or from damage by a stone chip. Manganese ions and nickel ions also significantly increase the corrosion resistance of zinc phosphate coatings and are often included (individually and jointly) in immersion phosphate formulations for automobile body pretreatment.

Although zinc phosphating baths can be operated at temperatures as low as 85° F (30° C) and as high as 180° F (82° C), most are maintained at about 120° F–140° F (49° C–60° C). At the midrange 120° F–140° F (49° C–60° C) temperatures, chemical concentration control is somewhat less critical than at higher and lower operating temperatures. Thus, the concentration of phosphating solutions must be maintained within a far narrower range when the operating temperatures are lower or higher, necessitating more frequent titrations to maintain proper chemical levels. Often it is easier to operate at the 120° F–135° F (49° C–57° C) range. The criticality of these variables varies, however, from one formulation to another.

In contrast to iron phosphating, combination cleaning + zinc phosphating stages are not available. It is simply not possible to do a good job of both processes by utilizing a single solution. Zinc phosphating is disrupted by the presence of both cleaning agents and soils; separate stages for cleaning and rinsing are always required prior to zinc phosphating. It is rarely done (but it is indeed possible) to use wand phosphatizing to generate zinc phosphate conversion coatings. This formerly was used on field work when repainting large immobile structures, such as bridges and towers or large refinery equipment. The concern over heavy metals in waste water now all but eliminates use of zinc phosphating in such cases. Where the water can be captured for metal ion and phosphate removal prior to release, it is still possible but now economically prohibitive in nearly all cases.

## Conversion Coatings for Zinc

Many modern finishing systems regularly clean, conversion coat, and paint parts with zinc surfaces. As is true for steel, the best conversion coating for zinc is a zinc phosphate. Because galvanized coatings on steel and pure zinc parts react rapidly with zinc phosphate solution, other metal ions should be added to slow the rate of coating deposition and to limit the coating weight. Nickel and ferric iron ($Fe^{+3}$) are widely used for this purpose. Some hot-dipped galvanized surfaces containing aluminum are difficult to phosphate. Fluoride ion activation can be used to initiate reaction. It forms hydrogen fluoride which strips aluminum oxide off the surface and precipitates the dissolved aluminum ions that would otherwise tend to poison the bath. The ions combine to form insoluble aluminum fluoride, which precipitates out as part of the sludge.

A typical zinc phosphate coating formulated for zinc can also be used for steel. Such formulations generally use nitrite ion—typically added as sodium nitrite—to accelerate phosphate coating formation and to control the solution's iron content. As an oxidizing agent, nitrite readily oxidizes ferrous iron ions to ferric ions. With hot-dipped galvanized steels, a nickel- or cobalt-activated bath can cause white pinpoint deposits or "seeds" that can sometimes be seen through the paint. These appearance defects are eliminated if fluoride ion is also present in the bath. Oxalic acid $(COOH)_2$ systems can be used, although in practice they are extremely rare. When parts with different types of metal are run on a phosphate line, it is preferable to mix the various metals randomly on the conveyor. This keeps the bath metal ion concentrations more balanced than would occur when large blocks of an individual metal substrate are run.

The majority of iron phosphating formulations do not produce any noticeable amounts of coating on zinc, whether electrogalvanized, galvannealed, hot-dipped galvanized, or die-cast. Iron phosphating will remove thin layers of oxide from the surface; if maximum corrosion resistance is not essential, it may be satisfactorily used as a prepaint treatment on zinc substrates. There are iron phosphate formulations, however, that do produce deposits on zinc substrates. The optimal coating weights are low, only in the 5–25 mg/sq ft range for such systems.

## Conversion Coatings for Aluminum

An aluminum surface should be conversion coated prior to painting unless it will only be exposed to mild conditions in service. Aluminum, when cut or abraded, very quickly forms a natural oxide layer that provides only poor to fair adhesion for paint. Like other metal surfaces, aluminum needs to be cleaned completely of all soils before being conversion coated. Aluminum should be cleaned carefully and properly to avoid forming a surface smut (black oxide), which is produced when strongly alkaline cleaning solutions are used. When this is unavoidable due to tenacious soils, the smut can be readily removed by dipping or spraying with a mild nitric acid solution. Dipping is preferred for safety reasons; spraying acid is potentially more hazardous.

The choice of the proper conversion coating for a particular aluminum product depends in large measure on the amount of corrosion resistance desired. In the past, the major way to obtain top corrosion resistance was with chromium-containing pretreatments. This situation has changed as plants desire to avoid the toxicity associated with chromium compounds. But far more use is still made of chromium-containing conversion coatings for aluminum than those without when corrosion resistance is a top requirement. This situation is slowly changing and will continue to shift to chromium-free materials, however.

The choice of the proper conversion coating for a particular aluminum product also depends on whether the product is being processed alone or with other metals. When only aluminum is involved, a chromate or chromium phosphate pretreatment is superior and is usually the method of choice. If aluminum is processed along with zinc and/or steel, then a phosphating treatment would be more appropriate, depending on the relative amounts of each metal to be run.

Chromate is the real workhorse of aluminum conversion coatings. It is also known generically as chrome oxide and amorphous chromate. Chromate provides outstanding paint adhesion and corrosion resistance, especially when sealed with a chromic acid rinse. Chromate processes are most frequently used on aluminum but can also be used on cadmium, copper, magnesium, zinc, and titanium to provide corrosion resistance and paint adhesion. Chromate can be used on iron but is less effective than iron and zinc phosphates.

Chromate coating weights normally range from 10–80 mg/sq ft, but most plants apply weights of 25–35 mg/sq ft. Higher coating weights, as expected, increase the corrosion protection. Heavy coating weights in the range of 65–85 mg/sq ft are common for aluminum parts used in water, such as the outboard components of boat motors and drives.

Below 10 mg/sq ft, chromate is colorless. This low coating weight is sometimes used when clearcoats are applied to aluminum, such as on styled aluminum auto wheels. This invisible chromate coating provides some extra adhesion and corrosion resistance without discoloring the aluminum. At 10 mg/sq ft the coating contains a very low concentration of hexavalent chromium ion, which colors the coating at higher weight levels.

At about 15 mg/sq ft, chromate coatings begin to exhibit an iridescent light yellow-golden color. As additional coating weight develops, however, it forms an increasingly golden appearance and then a darker metallic tan or brownish-bronze color. The exact shade is determined by several factors including the pH, chemical formulation and concentration of the bath, duration the bath has been in use, spray pressures, and aluminum surface roughness.

The various chromating processes include those with accelerated and nonaccelerated coating baths. One of the oldest and best (but highly toxic) accelerators is the ferricyanide ion. A coating that uses an iron ferricyanide accelerator has the composition of $CrFe(CN)_6 \cdot 6\,Cr(OH)_3 \cdot H_2CrO_4 \cdot 4Al_2O_3 \cdot 8H_2O$. Safer accelerators, utilizing ions such as molybdate and molybdate/zinc are more common.

In general, chromate conversion coatings produced by nonaccelerated baths are quite durable, but the processes are slow and require frequent bath replenishment. Nonaccelerated baths have fairly short working lives as well. The nonaccelerated chromate coating typically has the representative chemical composition of $Cr(OH)_2 \cdot H_2CrO_4 \cdot Al_2O_3 \cdot 2H_2O$.

Chromating solutions contain chromic acid and hydrofluoric acid at a pH of 1.5–2.0 and a fairly high concentration of hexavalent chromium ion. While amorphous chromates can be applied by immersion or by spraying, immersion is far more frequently selected because of the inherent dangers in spraying toxic chromium solutions.

The newly chromated coating is subject to two types of damage: the fresh coating is soft and easily abraded and the unpainted chromate is heat-sensitive. A maximum skin temperature for chromated parts during dryoff (usually with hot air) prior to painting is 140° F (60° C). Even 10 min at 300° F (149° C) will cause noticeable chromate decomposition. Once the parts are painted, however, heat no longer causes significant deterioration, and paint oven temperatures can be as high as required without degrading the chromate. The difference is presumed to be due to the exclusion of oxygen plus the effective sealing of the water of hydration in the crystals by the paint layer. Heat destroys chromate effectiveness by altering the chemical composition of the film by expelling the water of hydration in the crystal structure. A color change to a bluish-purple with a powdery look is seen when overheating occurs. Loss of the water of hydration shrinks the crystals and results in bare metal areas devoid of conversion coating. Note in the chemical formulas given previously that the nonaccelerated coatings contain only two "waters of hydration" versus eight for the accelerated versions. The lower water content makes the nonaccelerated coatings somewhat less susceptible to heat deterioration.

## Phosphates for Aluminum

Iron, zinc, and chromium phosphate coatings can be used on aluminum with various levels of effectiveness. Iron phosphating solutions usually deposit little or no coating on aluminum, but they rather effectively clean the surface. Many aluminum alloys exhibit good paint adhesion with only this weak acid cleaning. Some iron phosphate systems leave a small amount of coating on the aluminum, but this is probably a mixture of iron and aluminum oxides rather than a true phosphate. The coating composition will vary widely with the dissolved iron concentration in the bath. When production involves mixed metals with steel plus a small amount of aluminum, the preferred pretreatment is likely to be an iron phosphate.

Zinc phosphating generally would be used on aluminum only if steel and/or zinc were also being surface-converted. A fluoroborate ($BF_4^{-1}$) or fluoride ($F^{-1}$) additive in the bath is helpful to etch the aluminum. The hydrogen fluoride that is formed will remove the aluminum oxide on the surface and allow acid attack on the metal. The fluoride ion also acts to precipitate aluminum ions as aluminum fluoride ($AlF_3$) to avoid poisoning the bath by aluminum ions.

Zinc phosphate produces a clear-to-light-gray coating on aluminum, much the same in appearance as the zinc phosphate coatings on steel and zinc. The immersion-produced zinc phosphate on aluminum, as is true on steel and zinc, gives finer crystal sizes and better corrosion resistance than spray-applied zinc phosphate.

If only small amounts of aluminum are to be processed, a chromium phosphate pretreatment may be used, but chromium phosphate is not restricted to small volume users. Most architectural aluminum siding and panels are pretreated with chromium phosphate. The aluminum is first cleaned in a separate stage, usually by spray. The chromium phosphate bath contains a mixture of hydrofluoric, phosphoric, and chromic acids. The weights of coatings produced range from about 25–200 mg/sq ft; the normal amount is approximately 40–60 mg/sq ft. At around 20 mg/sq ft, chromium phosphate is a rather pale green with a slight iridescence. The green color becomes more evident as the coating weight increases. It has a medium to deep green color with coating weights between 150–200 mg/sq ft.

A typical chromium phosphate coating is $Al_2O_3 \cdot 2CrPO_4 \cdot 8H_2O$. It contains no toxic hexavalent chromium ion ($Cr^{+6}$), and so has the approval of the U.S. Food and Drug Administration for use as a conversion coating inside food and beverage cans. A chromic acid sealer will increase corrosion resistance, but the coated metal cannot be used for food container applications because the chromic acid contains hexavalent chromium. A number of $Cr^{+6}$ compounds are known carcinogens.

At least two chemical suppliers have dried-in-place, no-rinse aluminum conversion coatings. The coating produced is based on tannic acid, tannates, zirconates, or related anions, but full detail is not available on their chemistry. Another company has a similar dried-in-place, no-rinse material for aluminum, based on either titanium or zirconium chemistry. The principal advantage of no-rinse systems is the absence of toxic waste water; nothing goes down the drain. The no-rinse coatings are low in coating weight and chemically low in resistance to corrosive attack. Thus, their usage is limited to parts where corrosion resistance is not a major consideration.

Some aluminum parts are clearcoated so their attractive metal color and mechanical finish will be visible. For outdoor exposures clearcoats must be used for protection, but clearcoat adhesion is weak without a conversion coat. A colorless form of conversion coating is obviously needed for such applications. Low weight (<10 mg/sq ft) chromates or heavier weight chromates with the color from the hexavalent chromium leached out with water have been used for this purpose. New proprietary chromium-free colorless conversion processes are being used that provide outstanding corrosion resistance and paint adhesion. Little information has thus far been revealed by suppliers on these novel systems, but it is believed that transition metal chemistry is the basis for them.

All conversion coatings, even iron and zinc phosphates, are harmed by excessive heat, but the amorphous chromate is the most heat-sensitive of them all. The phosphates are the least susceptible to thermal damage. Chromium phosphate coatings are somewhat more heat tolerant than chromate coatings. Where heating of conversion-coated aluminum parts cannot be avoided because of a particular manufacturing process, one of the chrome phosphate conversion coatings would be a wise choice.

### *Anodizing of Aluminum*

Another type of conversion coating for aluminum is called *anodizing*. This gets its name from the process whereby an aluminum oxide coating forms on the surface when aluminum is made the anode in an electrochemical cell. It is not an exceptionally effective prepaint coating in most cases, however.

Two of the major virtues of some anodized coatings are their corrosion resistance and hardness, which contributes abrasion resistance to anodized parts. Both factors are important in the aerospace and aircraft industries; airplanes may encounter considerable abrasion from sand, airborne particles, raindrops, cold-weather icing, sleet, hail, and snow. For protection of the thin aluminum skin, the anodizing is far more important than a layer of paint. It has been preferred in the aerospace industry because it causes the least fatigue stress to be induced in the aluminum of spacecraft.

Anodizing is usually done in a dilute acid, such as chromic, oxalic, or sulfuric acids. The recent trend is to use sulfuric acid processes, however. The aluminum part is connected to an electric power source so that it becomes an anode in a direct current electrical cell, while immersed in the acid medium. The electrical flow forces the reaction of the aluminum with the solution. The resulting oxide forms a tight uniform layer about 1/4 mil thick that weighs 500–2,000 mg/sq ft. The process is very slow, and to form 1,000 mg of oxide may require 15 or even 20 min. Initially, the oxide film is soft and somewhat gelatinous, but on drying it becomes very durable.

Pores in the newly formed oxide coating can be sealed effectively by immersion in hot water or in aqueous solutions of corrosion-inhibiting substances. A unique process simultaneously hydrates and impregnates the anodized coating with a polyurethane resin.

Color anodizing of aluminum is not a painting operation; it involves placing the freshly anodized part in a dye solution that fills the pores. When the pores are sealed, the color remains trapped in the oxide layer, which gives such parts their distinctive though often slightly garish appearance.

In contrast to the 1,000 mg/sq ft coating weights for aircraft, heavy anodized coatings of up to 5,000 mg/sq ft are mandated by military specifications for aluminum used in underwater applications. Anodizing done in cold solutions produces a much harder oxide layer than in warm solutions. Anodizing bath temperatures as low as 20° F (-6.7° C) have been used.

## Specialty and Mixed Metal Phosphates

For years the appliance industry has used a microcrystalline calcium-modified zinc phosphate bath that gives good corrosion resistance, but which needs a relatively high process temperature close to 160° F (71° C) for consistent coating. This material is an extremely heavy sludge producer. If the phosphate stage of a washer is to use this high-temperature solution, the equipment may have to be modified. This is because of thermally inverse solubility problems with some of the chemicals, in which the compounds become less soluble rather than more soluble as the solution temperature increases. Heaters for the solution should have a large surface area and only a moderate temperature differential of less than 12° F (2° C), rather than running full bore at high temperatures with a small heat tube surface area. Use of a heat exchanger is indicated rather than a less expensive flame tube. This will help to prevent localized hot spots from forcing substances with thermally inverse solubilities out of the solution in the form of sludge.

A lower temperature zinc phosphate system can work fairly well, but it will require extra cleanliness for good operation. Temperature control often has to be within ±2° F for good operation; plus, the concentration of the grain-refiner chemicals is extremely crucial with this process. Stainless steel is rarely phosphated; however, when it is, it is treated somewhat like zinc. Also, the phosphate system should contain fluoride.

Many closely related process formulations fall under the headings of iron phosphating and zinc phosphating. Some industrial magazines and directories may list dozens of manufacturers who formulate and sell the basic phosphatizing chemicals and the numerous proprietary ingredients used. Particularly common in recent years are the zinc phosphates that also contain manganese, which is added to increase corrosion resistance.

## Spray and Dip Processes

Phosphating solutions can be applied to metals either by spray, immersion, or combinations of the two. Spray application (see Figures 8-4 and 8-5) promotes fast crystal growth because fresh solution constantly impinges on the surface at pressures of roughly 15–25 psi. However, recessed areas do not receive much coating because of the difficulty in spraying solution into these areas.

The immersion (dip) process (see Figure 8-6) has a number of distinct advantages, although it is slow and requires a large body of solution, especially in conveyorized lines. The advantages include:

- Higher coating weights can be achieved.
- An immersion zinc phosphate coating on steel is iron-rich.
- A high quality and small size crystalline structure is produced.
- It is relatively easy to maintain because no nozzles have to be cleaned. The problem of clogged nozzles that interfere with the overall spray phosphate quality is avoided.
- Neither heat nor water is lost from spraying hot aqueous solutions into the air.
- Sludge formation is minimal.

CONVERSION COATINGS 111

Figure 8-4. Automatic Spray Line (left) and Dryoff Oven (right)

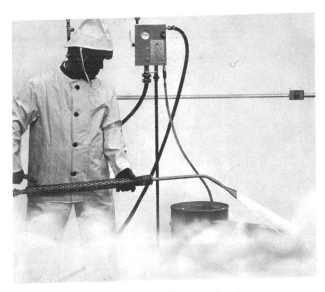

Figure 8-5. Applying Phosphate with Manual Spray Wand

112   INDUSTRIAL PAINTING

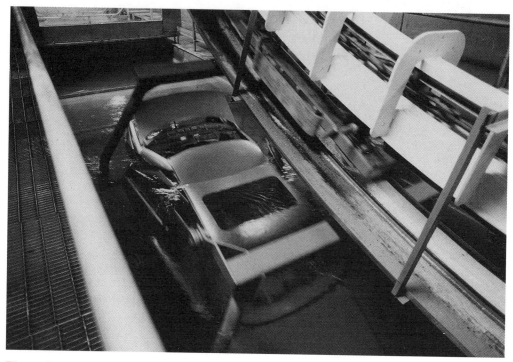

**Figure 8-6. Car Body Immersed in Phosphate Dip Tank**

New automotive painting facilities and retrofit systems installed in recent years use full immersion cleaning and phosphating (and rinsing) to increase corrosion resistance. For years Chrysler used a combination spray + dip process known as the "slipper dip" method. The lower half of the bodies were immersed in an attempt to increase rust protection in these more corrosion-prone areas. But this was not as effective as desired because phosphate baths work best formulated for either spray or dip operation. Trying to do both spray and dip phosphating with the same solution is not recommended.

A typical system for phosphating either iron or zinc parts may have five or often six stages:

1. Aqueous alkaline clean
2. Water rinse (single or double)
3. Phosphate
4. Water rinse
5. Aqueous postphosphate sealer rinse
6. Water rinse, often brief, with deionized water

Zinc phosphating sludge from pretreatment of steel includes salts of zinc and iron (principally iron oxides and phosphates), spent accelerator, and oxidizer. Some of whatever metal is pretreated will invariably be found in the sludge because acid dissolution of the metal is a necessary step in the conversion coating process. Phosphate sludges may be allowed to settle and then be removed, or they can be continuously filtered out. Excessive sludge may be produced when very hard water is used; deionized or softened water may be economical for this season.

## Phosphate Troubleshooting

Troubleshooting problems in phosphating prior to painting requires a considerable knowledge of cleaning, rinsing, and phosphating. The troubleshooting process generally involves three steps:

- Determine the problem.
- List the probable causes of the problem.
- Review the possible remedies to correct the problem.

An examination of four categories of phosphating problems follows. Each problem is first stated and discussed. Then, possible causes are listed, and the possible remedies are reviewed.

## PROBLEM 1
## Streaky, Blotchy, Thin, or Discontinuous Phosphate Coatings

Streaky phosphate coatings might not reduce corrosion resistance, but blotchy, thin, or discontinuous coatings will. It is good practice to take corrective action to obtain phosphate with a uniform, light- to medium-gray color and tight, fine crystal structure.

### Possible Causes
— Excessively dirty metal entering the cleaning stage
— Failure to remove contaminants such as sealants, dried drawing compound, or heavy oil deposits
— Inadequate solvent preclean
— Dirty solvent, rags, or sponges
— Poor rinsing after phosphate or a contaminated rinse

### Possible Remedies
☑ Ensure that equipment is properly maintained and that chemical concentrations, cleanliness, and temperature requirements are met.
☑ Check if sealant is applied only where specified. Remove excess sealant prior to a part's cleaning stages. Excess sealant will contaminate chemicals, plug nozzles, and alter the correct solution spray pattern on parts.
☑ Review the prewash process.

## 114 INDUSTRIAL PAINTING

**Possible Causes**
— Weak cleaner concentration
— Low temperature and pressure in cleaner stage
— Inadequate rinse after the cleaner stage
— Drying of parts between stages due to the cleaner bath being too hot

**Possible Remedies**
☑ Review phosphate process procedures.
☑ Make necessary changes to temperature, spray pressure, and chemical levels.

**Possible Causes**
— Contaminated cleaner
— Plugged or misaligned nozzles

**Possible Remedies**
☑ Make daily checks of nozzles to correct misalignment and replace plugged nozzles. High spray pressures can indicate that some nozzles are plugged.
☑ Remove accumulated dirt from the cleaner.
☑ Ensure that the filtering system works properly.
☑ Drain, clean, and recharge the tank at necessary intervals to keep it operating effectively.

**Possible Causes**
— Low accelerator or high free-acid concentrations
— Other chemical imbalances
— Poor bath agitation

**Possible Remedy**
☑ Make chemical additions as needed. Low accelerator or high free-acid concentrations will lower the ratio of total acid (TA) to free acid (FA).

**Possible Cause**
— Stained, discolored, or oxidized metal surface

**Possible Remedy**
☑ Remove defect by wiping with acid and/or abrasive before parts enter the cleaning stage. Do not allow acid to dry on parts. Rinse thoroughly with water when acid is used.

**Possible Cause**
— Drips of chemicals from washer roof or conveyor

**Possible Remedies**
☑ Adjust nozzles to avoid spraying the roof or the conveyor.
☑ Install or adjust drip pans to catch drips.
☑ Reconstruct the enclosure to have a domed roof and/or move the conveyor outside the washer enclosure.

# CONVERSION COATINGS  115

## PROBLEM 2
### Loose, Powdery, and Nonadherent Phosphate Coatings

Loose, powdery, and nonadherent phosphate will not provide adequate adhesion to metal and will reduce corrosion resistance. The crystalline structure will not be tight or dense, and crystals will be larger than normal. This phosphate coating can be easily wiped off.

**Possible Cause**
— High total-acid/free-acid (TA/FA) ratio in the phosphate stage

**Possible Remedy**
☑ Correct the TA/FA ratio by adding phosphate makeup material per instructions.

**Possible Cause**
— High accelerator concentration

**Possible Remedy**
☑ Add fresh water to reduce the accelerator content. Recheck the total-acid and free-acid concentrations and add phosphate makeup as required. Excess accelerator will neutralize the free-acid content and will cause a high TA/FA ratio. A high TA/FA ratio will yield powdery, nonadherent coatings.

**Possible Cause**
— Poor cleaning

**Possible Remedy**
☑ Check to ensure that the cleaner concentration, pressure, and spray impingement are adequate. If the cleaners are excessively dirty, dump the cleaner stage, clean the tank, and recharge. Inspect all solvent-wipe operations to ensure properly cleaned exterior surfaces. If cleaning is inadequate, the quality of the phosphate coating will be poor.

**Possible Cause**
— Slow line speeds or frequent line stops

**Possible Remedy**
☑ If line speeds are operating below design capacity and if line stoppages are frequent, it may be necessary to reduce the phosphate chemical concentrations. Check with the phosphate supplier for recommended concentration reductions.

## PROBLEM 3
### Salt Deposits, Acid Spots, and Water Spots

If phosphate salt deposits, acid spots, or water spots remain on the metal surface, they will act as contaminants. They may have adequate adhesion to the metal but will usually be the starting point for humidity blistering or scab corrosion. Since they degrade the overall corrosion protection, they must be corrected immediately.

# 116  INDUSTRIAL PAINTING

**Possible Cause**
— Overhead conveyor dripping onto parts due to chemical buildup and condensation

**Possible Remedies**
- ☑ Adjust nozzles and/or pump pressure to avoid spraying the conveyor.
- ☑ Adjust overhead drip pans to catch dripping.

**Possible Cause**
— Inadequate water rinsing after phosphate stage or after sealer rinse stage

**Possible Remedies**
- ☑ Check nozzles in the water rinse stages for inoperative or plugged conditions.
- ☑ Determine if stages provide sufficient rinse water for thorough removal of residual chemicals.
- ☑ Large-capacity rinse tanks and/or pumps, additional risers, and/or extension of rinse risers may be required.

**Possible Cause**
— Parts drying due to line stops

**Possible Remedy**
- ☑ Ensure that water rinses are activated during all line stops.

**Possible Cause**
— Deionized rinse water too high in solids content and/or conductivity

**Possible Remedy**
- ☑ Check the deionized water. If necessary, recharge the deionizer. When recharging, flush all chemicals from water lines prior to rinsing.

## PROBLEM 4
## Flash Rust

If the metal is oxidized or rusted, corrosion resistance may or may not be affected. The rust or oxidation severity will vary, depending on the cause. Corrective action is required in all cases so that flash rust is no longer observed.

**Possible Cause**
— Low accelerator concentration

**Possible Remedy**
- ☑ Adjust the accelerator concentration. A low accelerator concentration will imbalance the TA/FA ratio, resulting in a thin coating susceptible to flash rusting.

**Possible Cause**
— Incorrect TA/FA ratio

**Possible Remedy**
- ☑ Correct the ratio. A low TA value or excessive FA value will result in thin coatings that may flash rust.

## CONVERSION COATINGS

**Possible Cause**
— Drying between stages

**Possible Remedy**
☑ Avoid overheating any stages. Install misting nozzles between stages to keep parts wet. Due to high temperatures, line stoppages, or insufficient rinsing, the drying of parts between stages may occur.

**Possible Cause**
— A weak or insufficient spray pattern in the phosphate stage

**Possible Remedy**
☑ Check for plugged or misaligned nozzles and for low pressures—these may alter the phosphate spray pattern, resulting in discontinuous or thin coatings. These conditions will usually produce rust. Correct as required.

**Possible Cause**
— Rust on metal prior to phosphating

**Possible Remedy**
☑ Remove all rust prior to phosphate by mechanical abrasion or acid strippers. When acid is used, thoroughly rinse with water. Acid must not be allowed to dry on the parts. Rusted parts entering the phosphate systems will result in rusted parts exiting the phosphate system.

**Possible Cause**
— Stripping of phosphate coating in the chromic acid concentration

**Possible Remedy**
☑ Minimize line stoppages. Check the chromic acid concentration. If a phosphated part stops in a chromic acid sealer rinse or the acidity of the chromic acid is too strong, stripping of the phosphate coating may occur and allow flash rusting.

## Sealing Phosphate Coatings

Even when phosphating is done perfectly, one or two percent to as much as five percent of a surface will not be phosphate-coated. This results from hydrogen gas bubbles forming pin holes and crevices that prevent crystalline growth due to interference with the phosphoric acid attack on the metal.

Improved corrosion resistance can be attained when these open areas are sealed with an inhibiting material. In the past a chromic acid rinse or a chrome-free sealer rinse was most often used. Even more effective was an acidic mixture of trivalent ($Cr^{+3}$) and hexavalent ($Cr^{+6}$) chromium compounds known as "reduced chromic acid."

Typically, the chromic acid rinse was used to accomplish these purposes:

- Remove possible traces of unrinsed phosphate solution. Phosphate chemicals are hygroscopic and are able to pull water out of the air; if they are not thoroughly removed, the result may be early blistering of the paint film.

# 118 INDUSTRIAL PAINTING

result may be early blistering of the paint film.
- Remove loosely attached conversion coating deposits.
- Greatly enhance corrosion resistance by sealing open areas or pores in the phosphate coating and by leaving behind residual amounts of the chromate ion antirust agent.

Today few plants use chromium in the sealer rinse because it is a toxic heavy metal. If an extremely dilute aqueous chromate or similar solution is used as a postphosphate sealer, the parts can proceed directly to a dryoff oven and then be painted. Better corrosion resistance is frequently obtained when a more concentrated sealer rinse is used, but this requires that the sealer be rinsed afterwards. Unfortunately, the materials deposited by the sealer are considerably more soluble than the phosphate coatings. If a thorough rinse were to be used, all the newly applied sealer material would be rinsed away. For this reason, a brief "compromise" rinse is used of about 20- to 30-second duration. This removes the excess solution without removing all the deposited sealer compounds. A deionized water spray rinse is usually used to minimize the rinse volume needed and to avoid residual deposits of materials possibly present in plant water.

Because of the extreme toxicity of chromium compounds, a wide variety of nonchrome sealer materials have become available. Silanes work very well on iron and zinc phosphate. Others are phosphates, tannins, titanates, molybdates, zirconates, thioglyconates, and aluminates. Depending largely on the paint materials that are applied, nonchrome sealers may rival chromium-containing materials in anticorrosion performance. Virtually no new installations use chromic rinses unless tests show they are absolutely needed to get the best corrosion resistance on metal parts.

As water (and solids disposal) restrictions are being tightened, nonchrome, nontoxic materials that can run directly to drain after possible pH adjustment and dilution are increasingly being used in industry. As Figure 8-7 indicates, the cost of dumping a toxic solution includes many items beyond the solution cost.

Most of the nontoxic sealers are acidic, but a few are slightly alkaline. They are in many cases as effective or almost as effective as reduced chromic acid sealer rinses, but their performance

**Total Cost Of Dumping a Process Solution**

Wasted Chemicals
Labor
Waste Treatment of Chemicals
Waste Disposal
Waste Treatment Equipment
Lost Production
Quality of Product

**Figure 8-7. Process Solution Disposal Costs**

cannot be expected to exceed that of chromium.

## Other Conversion Coatings

In addition to those previously discussed, other conversion coatings are available. They are generally intended for special appearance properties, corrosion resistance, or to hold lubricants but not intended as a base for paint. Various blueing, browning, and blackening oxide coatings provide some aesthetic value as well as minimal corrosion resistance for iron and steel. Hydrochloric acid and table salt are employed in artificially "aging" or oxidizing copper, producing the attractive variegated green ("verdigris") finish associated with long-term weathering. Heavy manganese phosphate coatings (as much as 2,500 mg/sq ft) are applied on metal to hold die lubricants in place during deep-drawing operations and are applied on friction surfaces of gears and bearings to stop scoring during break-in periods. It also may be used as a base for the glue in adhesive bonding of metals. In rare instances, lead phosphate is used as a conversion coating. Neither of these is used under paint.

## Dryoff Ovens

Parts cleaned or pretreated—whether they are plastic, wood, composite, or metal—still need to be dried before being painted (unless they are metal and are to be immediately E-coated). As much water as possible should be removed before parts enter the dryoff oven. Compressed air can be costly but may be necessary for some items. Manual compressed air vacuums can be used as well. If parts are hung to facilitate drainage and are designed with drain holes, the cost of oven drying can be reduced. Good air turbulence in the oven speeds drying, and air seals or bottom entry and exit will optimize energy consumption.

## Conversion Coating Waste Treatment

Conversion coating solutions and rinses need to be treated before discharge to drain if they do not conform to local, state, and federal discharge codes. Discharge regulations almost always severely limit the concentrations of hexavalent chromium and nearly all heavy metals. Harsh fines can be expected to be levied against companies for allowing release of wastewater containing amounts of regulated substances beyond specified concentrations. Limits on phosphate ion concentrations in wastewater are also established for many locations even though phosphate is not strictly a toxic material.

It is likely that in the future social and legislative pressures may force many manufacturing companies into some form of both voluntary and mandatory material recycling, perhaps of paint sludges or of other chemicals. Societal pressures for such legislation are increasing rapidly since many communities have long established voluntary recycling of paper, metals, and plastics. Landfill restrictions also continue to increase, but for now it is still cheaper simply to buy new paints and chemicals than to recycle them. The cost of the conversion coating chemicals is far too low to consider trying to directly recover from spent pretreatment baths and recycle them, at least at the present time.

# Chapter 9

# Paint Application I—Traditional Methods

## Historical Applications

Although no known records exist about early paint application, the most probable method of application was by some form of spreading. The methods used by cavemen in painting their caves almost certainly included the use of their fingers, sticks, and perhaps straw-like spreading devices. The application of paint by means of a device held in the hand evolved over the years; even now a brush is one of the most common methods of applying paint, either for art or industry.

The essential feature of paint brush construction is primitive—merely tying or binding straw-like fibers together. The quality differentiation of brushes is linked to three parameters: the nature of the bristles, the quantity of bristles, and also how well they are bound together.

That painting with a brush is slow, labor-intensive, and messy is apparent to anyone who has ever done much of it. If the paint is too thick, it won't spread easily; if it is too thin, it will spread easily, but the paint will run and the application will be extremely messy. It may be hazardous in the sense that the paint fumes generally envelop the painter. In the past, many paints commonly contained toxic pigments made of substances such as lead, chromium, and other heavy metals.

## Modern Application Methods

A number of application techniques have been developed in more recent times to improve the effectiveness of paint application and the appearance of the cured coating.

### Rolling

Painting with a roller is faster than painting with a brush, but it, too, has drawbacks. A roller cannot get into hard-to-reach areas, and the application often requires touchup with a brush. In addition, rollers generally absorb—or "load up"—a substantial amount of paint, most of which cannot be saved when the job is done. Devices that pump paint through a hose to the roller are somewhat helpful to speed the rate at which paint can be applied. While not often used, there are still some legitimate uses in manufacturing for paint application with rollers. An example is the black paint applied to plate glass for opaque glazing sections in high-rise buildings. Opaque

## 122  INDUSTRIAL PAINTING

black paints are also rolled onto the thin silver or aluminum layer deposited on glass in mirror making.

## Dipping

Painting by dipping, as shown in Figure 9-1, is much faster than brushing or rolling and much less labor-intensive. Dipping, however, is extremely dependent on the viscosity of the paint, can be somewhat messy, and is potentially hazardous. The viscosity of the paint in a dip tank must remain practically constant if the deposited film quality is to remain at all consistent over a period of time. Even with constant viscosity, this method is not conducive to high quality paint appearance due to variable drainage of the paint from the parts.

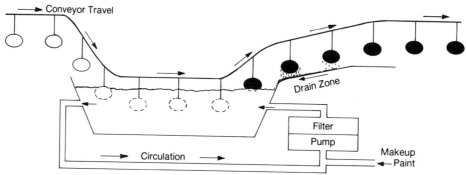

**Figure 9-1. Side-view of Painting by Dipping**

The film thickness in dipping can be controlled primarily by varying the viscosity and to a smaller extent by the rate at which parts are withdrawn from the dip tank. Thinner paint films are produced by a slow withdrawal rate; an increased film thickness results from an abrupt withdrawal.

Dipping is not suitable for many hollow-shaped items. They would, of course, tend to float as they contact the paint and could easily become disengaged from their hangers or hooks.

Color change is very slow because of the volume of paint to be replaced. Thus, color change is generally not feasible for most dip operations, unless multiple tanks mounted on rollers are used. As needed, the various color tanks are rolled into position under the conveyor line. This technique is not done by many plants, but some high-volume paint users find this a convenient way to get multicolor parts. A few companies also pump out one color, clean the tank, and refill with the next color of paint to be used.

An open tank of flammable paint can constitute a major fire hazard, so sometimes safety dump tanks are required by local fire codes. For all such tanks, an efficient fire-extinguishing system must be installed as a safety measure. Paint that drips from parts exiting the dip tank creates an additional mess that adds to the fire hazard.

## Flow Coating

In a flow coat system, as shown in Figures 9-2 and 9-3, a number of individual streams of paint are directed at the parts. The number of streams needed varies according to the parts being painted, but commonly there may be 20–100 of them directed to cover all surfaces of the parts since the excess paint is caught and recirculated. The streams may flow out of holes drilled in pipes or through short, crimped pipe outlets. The paint headers are arranged so that all surfaces of parts carried through the flow coater on the conveyor are painted. Paint is liberally applied to all surfaces of the parts. The paint flowing off the parts drains into a sump, from where it is pumped through filters and recirculated to the flow coat paint reservoir.

**Figure 9-2. The Coating Chamber in Flow Coating**

Both dipping and flow coating tend to be used to coat items whose appearance is not vitally important and to apply primers. Topcoats are not commonly applied by dipping or flow coating. Coatings applied by these two methods have only poor to fair appearance unless parts are rotated during the drippage period to avoid run lines. Currently, only a couple of manufacturers are using flow coating and rotating coated parts. They use flow coating to apply clearcoat to metal handles or to apply basecoats for items to be vacuum metallized.

Both dipping and flow coating have the disadvantage that practically the only way to control dry-film thickness is by the viscosity of the paint. The amount of paint that will drain from the parts depends on viscosity. If the viscosity is allowed to go too low, insufficient paint build will

124　INDUSTRIAL PAINTING

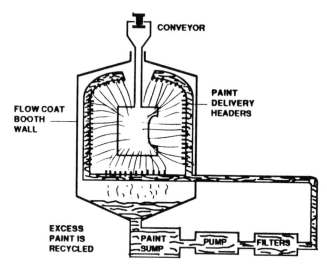

**Figure 9-3. End-view of Flow Coating**

result. This could cause a large number of parts to be rejected and need repainting. On the other hand, if the paint viscosity should rise too high, then extra paint will be applied. This will not only increase paint costs, but extra paint can also plug small holes in the part. Automatic viscosity controllers that will add solvent to the paint supply as required are frequently used to maintain tight control of paint viscosity.

When items are dipped or flow coated, the drain-off pattern may contribute unsightly runs or heavy flow lines on the parts. This can usually be seen around through-holes that are large enough to avoid being filled with paint. Sometimes extra paint draining down builds up as "fatty edges" along the bottoms of parts. Often the excess paint dries as it drips, forming elongated dried paint accumulations that resemble icicles (paint-cicles?).

Another significant disadvantage of dipping and flow coating is that the appearance of the finish is not as attractive as those produced by more sophisticated methods. Despite the limited flexibility, the advantages make them highly desirable methods for painting certain kinds of parts that need paint coverage but for which appearance is not important. Nonappearance (nonvisible) items such as springs inside upholstered furniture and vehicle underbody parts are examples.

A strong plus is that both methods achieve very high paint transfer efficiencies, often 90% and higher. These techniques are well suited for automation with conveyorized paint lines. Manual flow coating is nonexistent, and hand dipping of parts is rare. The latter is normally suitable only when a very limited number of parts is to be painted. An exception is golf club shafts, a high-volume product frequently coated by manually drawing each shaft through a reservoir of paint and then wiping to a uniform thickness through an opening in an elastomeric squeegee.

Painting by dipping or flow coating is fast and easy, involves a relatively low installation cost, requires little maintenance, and has a low labor requirement. In addition to a paint attendant, the only other labor requirement is a worker to hang parts onto the conveyor line and remove them after they have been cured. Thus, in addition to the low labor requirement, the skill required is also low as compared to manual spraying.

## Continuous Coaters

A variation of the flow coat process is the continuous coater, as exemplified in Figure 9-4. An enclosed tunnel-like painting machine captures and reuses the excess paint that runs off coated parts. It uses directed atomized sprays rather than streams of paint. Tennis racquets, metal panels, and engines have been coated by this technique. Reuse of the paint within a confined space contributes to high paint transfer efficiency and reduced VOC emissions because of the solvent vapor capture capability of continuous coaters.

One drawback to the method is that all parts being coated must be roughly the same size and shape. This is because the gun-to-part distance (target distance) has to be kept uniform for best part appearance.

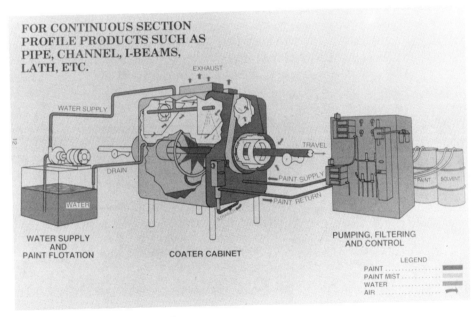

Figure 9-4. Nordson Continuous Coater

## Dip-Spin Coaters

Dip-spin (centrifugal) coaters are designed to paint large-quantity batch loads of small parts that are not amenable to being hung on hooks for painting, such as hairpins, screws, various clips, and related fasteners. There is often no convenient way to hang them; in addition, the huge numbers involved would make hanging them cost prohibitive. A wire basket, which might contain anywhere from 0.5 lb–50 lbs of parts, is fully immersed in a reservoir of paint. The basket is then raised, the parts are allowed to drain, and the basket is spun rapidly to remove excess

## 126  INDUSTRIAL PAINTING

paint. Next, the parts are shaken out onto a mesh conveyor belt and thus carried through a paint bake oven. After a brief cool down, parts are dropped off the belt into containers. The entire process is done automatically.

The main advantage of dip spinning is the extremely high productivity rate. A disadvantage is that some of the painted parts shaken out onto the conveyor may stick together when baked in the oven, leaving paint void spots when they are separated. If this is not acceptable, parts are run through the process a second time to eliminate the defects.

### Curtain Coating

Curtain coating is designed for coating relatively flat and low-profile stock only. In *curtain coating*, paint flows at a controlled rate from a reservoir through a slot whose opening size can be varied onto work being conveyed horizontally through the waterfall-like flow of paint, as shown in Figures 9-5 and 9-6. Curtain coating is suitable only for items that are flat or at least relatively so. They should also be at least 6 inches long and wide. The fixed width of the coating curtain can be constructed to be any dimension but is usually about 18–60 inches wide, depending totally on the width of the material to be coated. The volume of paint released and the speed of the conveyed sheet or rolled work determine the thickness of the applied coating.

Gaps between the sheets of material being fed through the curtain coater do not cause a problem. The paint that drops between the parts from the continuously falling curtain of paint is captured and recirculated; thus, virtually no coating material is wasted. Extremely uniform

**Figure 9-5. Curtain Coating Line**

# PAINT APPLICATION I—TRADITIONAL METHODS    127

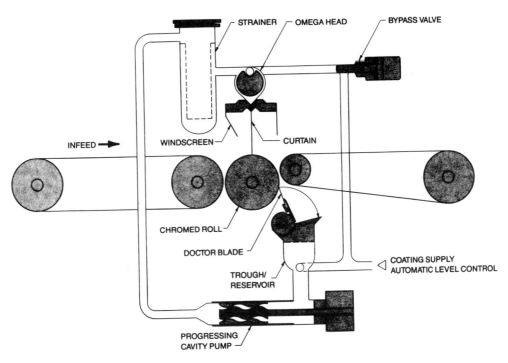

**Figure 9-6. Curtain Coater**

coating thicknesses are possible. However, only the top side of a sheet or roll can be painted at one time.

Curtain coating also is highly viscosity-dependent. Because it is simpler and uses less complex equipment, curtain coating may be preferred over roll coating for lower production runs.

## Roll Coating, Coil Coating

The roll coater is employed as the means of paint application in coil coating. The term *coil coating* refers to the painting of a metal (steel or aluminum) strip up to 72 inches wide. A coil coating line, as shown in Figure 9-7, automatically unwinds a roll of metal, cleans it, applies a conversion coating, paints it, bakes it, and rewinds the painted metal into another coil. Some coil-coating lines operate at speeds in excess of 600 feet (183 m) per minute (fpm). The painted metal is sent to a fabricating plant for shearing, forming, and assembly into various products. The main advantage in fabricating from prepainted metal is that no paint line is required for the completed parts.

The problems associated with VOC emissions have made it attractive for some manufacturers to stop painting in their own plants and to manufacture their products with parts fabricated from prepainted metal. Because the metal is cleaned, conversion coated, and painted while flat, parts made from this material are therefore finished uniformly.

128   INDUSTRIAL PAINTING

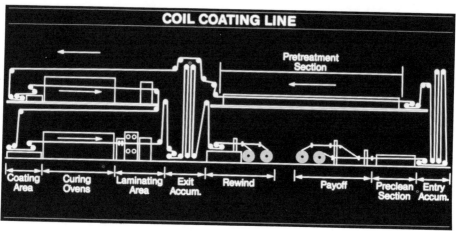

**Figure 9-7. Coil Coating Line**

When fabricating parts from precoated coil, precautions must be taken to hide the uncoated cut edges. This must be done to avoid visible corrosion that can readily form on unprotected metal. Edges can often be hidden by folded seams, but if they cannot be hidden, they need to be painted. Should painting of exposed edges be required, much of the advantage in using prepainted metal is lost.

Prepainted metal is used on autos, appliances, cans and related containers, architectural panels, and on numerous other products. The large volume output that is possible per line is an important factor in the economics of the coil-coating industry. Coil coating was developed around 1930 to paint steel in a thin, narrow, continuous strip for venetian blind manufacturing. In 1997, about 180 coil-coating lines were operating in North America.

Because coil-coating lines may be over 1,000 ft in length, threading a strip of metal through a coil-coating line is a fairly time-consuming process. If every coil had to be threaded, the process would be very inefficient. To avoid this, a scrap coil is always the first and last coil in a run. As the line is started, the leading edge of the new coil is fastened to the trailing edge of the scrap coil. When the scrap coil is wound, its trailing edge is sheared, and the leading edge of the production coil is threaded into a new take-up roll.

An accumulator at the line entry and exit allows new coils to be started and painted coils to be moved out without stopping the lines. The entry accumulator stores some 300 ft of strip to feed the line while a new coil is being loaded and fastened to the end of the in-process coil. The exit accumulator runs empty until time to change coils at the painted end. That accumulator then collects the painted stock until the painted coil is moved out and a new take-up roll is started.

The process begins with a spray application of hot alkaline cleaner to both sides of the strip to remove oil and grease. Next is a rinse with warm water, followed by abrasive brushing (cleaning) on both sides of the strip (if needed). Then the alkaline cleaning is repeated. A final immersion rinse in hot water dissolves any alkaline residues and completes the cleaning operations.

The strip is then ready for dip, spray, or roll-on of the chemical conversion solutions. The conversion coat improves the adhesion of the paint to the metal. The type of pretreatment that is applied depends largely on the metal substrate and the kinds of parts to be manufactured from the prepainted stock. For cold-rolled steel, conversion coatings are microcrystalline iron phosphate or zinc phosphate. Iron phosphate is thinner and more flexible, but zinc phosphate offers superior corrosion protection.

Two conversion coatings used on hot-dipped or electrogalvanized steel are zinc phosphate and chromate-sealed oxide. The oxide coating is formed in three steps. First, the zinc surface is treated in an alkaline solution, which results in the formation of the layer of oxide. A cold-water rinse follows. Unless this rinse is thorough, blisters and paint peeling can result. Any salts left from the first step can react with the metal under high humidity conditions. The last step is a dilute chromic acid rinse that yields a complex oxide to promote paint adhesion and flexibility. Its corrosion resistance is not as good as zinc phosphate. Hot-dipped aluminum-zinc coated steel requires a chromate conversion treatment for the best corrosion protection. Chromates can be used as well on galvanized steels. "React-in-place" and "nonrinsed" conversion coating processes avoid the costs and problems associated with waste water treatment for removal of environmentally objectionable substances. Pretreatments that eliminate rinsing have not equalled the anticorrosion properties of the phosphate or chromate coatings. (See Chapter 8 for more information on conversion coatings.)

The metal strip then passes through an air knife drier (high-intensity air blowoff) before entering the coating section. A roll coater applies paint by forward or reverse coating to one or both sides of the strip, depending on the specific product requirements. Figure 9-8 shows a coil-

**Figure 9-8. Coil Coating Line Roll Coater**

**Figure 9-9. Multihead Roll Coater System**

coating line roll coater applying a single color; Figure 9-9 shows a coil-coating line multihead roll coater system capable of applying several colors. Epoxies, vinyl plastisols, fluorocarbons, polyesters, and silicone-modified polyesters are often used. From the coating rolls, the strip goes to the bake oven to drive out solvents and to cross-link (cure) the paint. To shorten oven dwell time and reduce the length of the oven, a very high temperature (about 750° F or 400° C) is maintained. A 15-min bake for a strip traveling at 10–12 ft/s would require an oven roughly two-thirds of a mile long. For this reason bake times are normally held to about 5 minutes. This still will normally require an oven over 1,000 feet in length. After baking, the strip is air-cooled and water-quenched. Should a second coat be necessary, the coating and curing are repeated.

The coating is inspected visually (and sometimes instrumentally) as the painted strip winds onto the take-up roll. The finished coil is taken off the machine and moved from the area for shipment, storage, embossing, or slitting. Samples are cut from the completed coil at periodic intervals for quality control tests. If required, samples are sent to customers for preshipment approval.

The coil can be slit to any width, cut into flat sheets of various size, or transported in coil form as desired. Lines may also be equipped to print patterns such as wood graining. Thick-film coatings can be embossed into various textures. Many coil lines are able to laminate decorative or functional films onto the metal as well. Acrylic film can be laminated onto galvanized steel or onto aluminum for building applications; solid-color and decoratively printed vinyls are being used on appliances, cabinets, lighting fixtures, and many related items. Even cork, paper, and rubber are being laminated onto specialty goods.

# PAINT APPLICATION I—TRADITIONAL METHODS

Two problems for the coil coating industry are fastening methods for parts made from prepainted coil and protecting the bare metal edges where the strip is cut. Many potential users of prepainted stock depend on welding for joining parts. Although some paints (weldable zinc-rich primers, for example) do not interfere with welding, most organic paints prevent effective welding. The weldable zinc-rich primers, although they allow welding, emit zinc fumes that constitute a health hazard during welding. Joining methods that have been used for coil-coated parts are adhesives, self-tapping screws, rivets, and spring clips. Bare edges can sometimes be located in nonvisible areas with proper product design. Hem folds, lock seams, plastic caps or plugs, and plastic beading are common ways to hide cut edges from view.

## Safety Procedures in the Paint Shop

Safety is of critical importance in any painting operation. Paint facilities are purchased with operator safety as the foremost consideration. All materials, interlocks, and equipment must meet or exceed every local and national safety standard. Make sure you thoroughly understand these procedures.

### Fire Safety Precautions

- No electrical device, smoking material, or any other spark or flame producing device is permitted (a) in or within 3 ft of the spray booth, and (b) within 10 ft of where flammables are mixed or stored.
- Nonessential combustible materials should not be brought into the booth. For example, if parts are wrapped in paper, they should be unwrapped before being brought into the spray booth.

Fire precautions are particularly important while spraying is in progress. Nearly all solventborne coatings and thinners are flammable, as are the dried overspray deposits adhering to booth surfaces and on filters. When atomized for spraying, solventborne materials burn very rapidly and can be considered to be "potentially explosive" (National Fire Prevention Association Regulation 33). However, if the safety procedures described here are followed, the likelihood of injury is no more than with operating most other types of machine shop equipment.

### Cleanliness

- In a spray-coating facility, safety depends on maintaining a neat, well-organized area. Nothing should be in the booth that is not required for spray application.
- Overspray deposits on booth and filter surfaces must be monitored.
- Filters must be replaced when airflow falls below specification.
- Collected overspray deposits and strippable coating may be hazardous waste and must be disposed of properly in approved containers.
- Spills must be cleaned up immediately. Dried deposits should not be permitted to collect on paint containers and hoses or on measuring and mixing equipment.
- No flammables can be stored within three feet of the exterior of the booth with the exception of materials stored in fireproof cabinets.

## Safe Spraying Techniques

- Mixing and solvent cleaning must always be performed in well-ventilated areas. All flammable mixtures must be maintained in tightly covered safety containers (which will not release any of their content if accidentally overturned) when not in use.
- When spray coating, keep as much of the coating as possible on the parts and as little as possible on the walls or floor. Sounds simple enough, but in a typical spray-painting operation, only half of the paint or less ends up on the product.
- Workers should operate near the center of the spray booth and away from the walls so that the cloud of atomized overspray is carried away as quickly as possible.
- A worker should spray with the booth airflow toward the water wall or particulate filters as much as possible so that the paint cloud moves away from him or her.
- Workers should wear an approved, properly fitted respirator or fresh air mask.

## Waste Disposal

- Any paper waste containing dried overspray deposits should be removed from the booth area on a daily basis for local disposal. Empty paint, solvent, and catalyst containers can also be sent to local domestic disposal areas if free of all flammable materials.
- Combustible liquids such as unused paint or lacquer thinner should be removed from the booth area immediately after use. Liquid wastes need to be transferred to a properly identified waste container. The waste container should be stored in a fireproof cabinet located in a secure area. The fireproof cabinet must provide sufficient secondary containment. All flammable liquids temporarily maintained in the booth must be kept in sealed, spill-proof, electrically grounded containers.

# Chapter 10

# Paint Application II—Electrocoating and Autodeposition Coating

## Coating by Immersion

As in ordinary dipping, both electrocoating and autodeposition coating involve immersing the part to be painted into a container of paint. Although dipping can be used with solventborne as well as waterborne paint, electrocoat and autodeposition paints are always waterborne. Immersion of a metallic part in waterborne paint is the only thing these techniques have in common, however. As may be deduced from its name, electrocoating involves the use of electricity to get paint onto the parts; autodeposition involves the use of integrated chemical reactions to achieve this.

## Electrocoating

Electrocoating is also called electrodeposition coating, electrophoretic coating, electropainting, the electrodeposition of polymers, or most commonly, E-coat. For some reason the "el" and the "po" from the ELectrodeposition of POlymers was combined at General Motors where, for over 35 years, this process has continued to be known as ELPO rather than as E-coat. But no matter the term used, *E-coat* is the electrical deposition of a paint film from a waterborne organic solution onto a part—a process in which the part is one of two electrodes forming an electrical cell. The E-coat electrical cell is comprised of a source of electricity, the two electrodes, and the solution. Thus, the part to be painted (which is one of the electrodes), the other electrode, and the solution all must be electrically conductive to complete the cell requirements.

E-coat application is very similar to the electroplating of metals—a process which also involves a source of electricity, two electrodes, and an inorganic electrolyte (instead of an organic electrolyte solution). But plating deposits a film of metal instead of paint. Unlike the electrodeposition of metal in electroplating, the electrodeposited E-coat paint must be cured. Although there are air-dry E-coatings, the baked E-coat paints are far more durable and thus much more likely to be used.

The E-coat process is extremely efficient, depositing a highly uniform paint film on all surfaces that can be reached by electricity. This includes all surfaces except those in confining Faraday cage areas, such as inside a long, narrow-diameter pipe. In such a pipe configuration, the

134  INDUSTRIAL PAINTING

Figure 10-1. Parts Conveyed into E-Coat Bath by Overhead Conveyor

electrodeposition occurs preferentially at the high–current-density areas—that is, the outermost conductive surfaces, especially those closest to the oppositely charged electrode. A rule-of-thumb is that E-coat will deposit inside the end of a pipe only to a distance equal to the internal radius.

Parts to be painted are usually conveyed in and out of an E-coat solution paint bath with an overhead conveyor, as shown in Figure 10-1. Sometimes, however, parts are lowered into an E-coat bath with a hoist "elevator style," as pictured in Figure 10-2. A few systems even raise the entire E-coat tank with a scissors lift up to the parts, as shown in Figure 10-3. This is an advantage in that the process allows changing colors by switching tanks. The basic components most commonly found in an E-coat system are illustrated in Figure 10-4.

PAINT APPLICATION II    135

Figure 10-2. Parts Lowered into E-Coat Bath by Elevator-Style Conveyor

## Electrical Considerations

E-coat power supply requirements vary in proportion to the total surface area of parts to be immersed in the solution at any given moment. The power supply must be able to provide a current density of from 1 to 4 amperes (amps or A) per square foot of surface of the part being coated. This current density will produce a normal film build of approximately 1 mil (25 microns) in thickness within about 90 s. A typical high-capacity power supply can provide 50–900 amps of direct current at 75–500 V.

The part being coated and the conveyor carrying the part are always electrically grounded, which greatly simplifies safety and eliminates the need to insulate the conveyor. The other electrode, however, is charged either above (cathode) or below (anode) the "ground" potential, which creates a difference in electrical potential that results in current flow and the deposition of the paint film.

136  INDUSTRIAL PAINTING

Figure 10-3. E-Coat Tank Hydraulically Raised to Parts

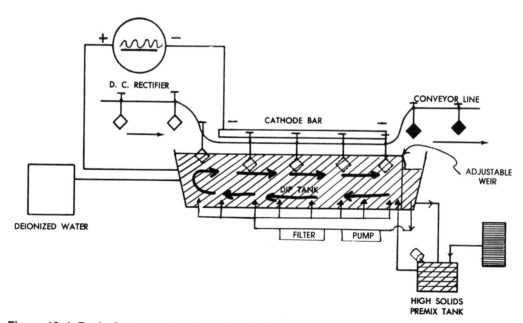

Figure 10-4. Basic Components of an E-Coat System

In a *cathodic*, or *cationic*, E-coat system the part being coated is the grounded cathode, and the other terminal, the anode, is charged positively. In an *anodic* system the part being coated is the grounded anode, and the cathode is charged negatively. The anodic E-coatings were developed first and made commercially before cathodic types. The essential shortcoming of anodic E-coat relates to the nature of a fundamental electroplating cell. In such a cell the anode dissolves, and the dissolved ions enter the deposition solution, where they drift over to and deposit on the cathode. Thus, in anodic E-coat the part (anode) is simultaneously dissolving to a small degree while it is being coated. The dissolved ions of the anode tend to discolor light-colored E-coats and to lower their corrosion resistance properties.

When cathodic E-coats became commercial, the basic defect of anodic E-coats was eliminated. In cathodic E-coating, the cathode—the part being coated—does not tend to dissolve. In addition to eliminating discoloration, cathodic E-coat also improves corrosion resistance. This is due to the inherent chemical superiority of the cathodic resins and to a lesser extent because the absence of deposited substrate metal improves the integrity of the paint film.

Most of the discussions in this chapter will relate to cathodic rather than anodic E-coating. Anodic systems are still in use, and some new anodic systems are being installed, but the trend is toward cathodic E-coat because of the superior films. A disadvantage is that cathodic deposition solutions are more costly than anodic types. For some color shades, however, anodic paints must be used because those pigment colors are only stable in the alkaline (anodic) paint baths and not in the acidic (cathodic) baths.

## E-Coat Paint Constituents

Paints for use in E-coating must be specifically manufactured solely for that purpose and made either to be an anodic or a cathodic paint. The E-coat paints are always used as solution paints. An E-coat paint solution or bath is somewhat similar in content to conventional waterborne paint, but the resin molecules are chemically modified to enhance their water solubility. For this reason E-coat paint cannot be used in ordinary dipping; it must be electrically deposited, or film appearance and durability suffer dramatically.

Like conventional paint, an E-coat bath contains resin, pigments (unless it's a clearcoat), solvent (mostly water plus a cosolvent), and additives. The water content puts E-coat in the family of waterborne coatings. A typical E-coat bath will contain from 2%-6% of a high-boiling water-soluble solvent, such as butyl cellosolve or hexyl cellosolve (ethylene glycol monobutyl ether or ethylene glycol monohexyl ether) to give smoother films. Moderately higher cosolvent levels do not affect the coating process, but if excess cosolvent is added, VOC emissions may become unacceptably high.

Not just any waterborne coating can be used as an E-coat bath; the resins must first be modified to contain certain chemical groups, which makes E-coat paint different from other waterborne paints. The resins are treated with appropriate solubilizer chemicals and form positively charged (cathodic) or negatively charged (anodic) molecules called polymer ions. In their ionic form these resins are very polar, and thus are soluble in water because water is also highly polar. The anodic and cathodic resins, however, are not interchangeable.

For anodic resins, carboxyl groups are chemically attached to resin molecules. The resin is mixed with potassium hydroxide, sodium hydroxide, triethyl amine, or other amine compounds to solubilize the resin molecules. Organic salts are formed. These are polar substances.

Cathodic E-coat resin molecules are chemically modified so that tertiary amine groups are located along the backbone carbon chains of the molecules. Without ionizable groups such as amines, the resin could not be used in an E-coat formulation. The tertiary amine groups, when treated with dilute acid solubilizers, such as acetic acid or formic acid, will form the necessary positively charged ions needed for cathodic E-coat.

In theory any type of resin can be modified for use with E-coat. The most commonly used resins, however, are epoxies and acrylics. Epoxy E-coat paint films, like any other epoxy, tend to chalk when exposed to ultraviolet (UV) light. This is generally not a problem if the epoxy E-coat film is to be a primer. If the epoxy is to function as a topcoat, then it shouldn't be exposed to UV light unless chalking does not present a problem. E-coat paint films that require outdoor exposure and relatively high appearance standards use acrylic resins. These are resistant to attack by UV light and have outstanding weatherability.

In formulating E-coat resins, chemists select cross-linking agents, such as aromatic and aliphatic isocyanates, that are appropriate to the end use of the paint film. The aromatic cross linkers are less costly. However, they tend to cause yellowing in the paint film upon outdoor exposure, whereas the aliphatic cross linkers do not. To avoid darkening problems, acrylic E-coat resins are cured with the aliphatic cross linkers.

Since isocyanates react with water and E-coat baths contain water, chemists "block" the isocyanates to prevent water reaction. The blocking is done chemically by "tying up" the water-sensitive ends of the isocyanate molecule with a blocking agent. The blocked isocyanates deposit into the E-coat film, but the blocking agent is driven off later by the high E-coat bake oven temperatures, thereby allowing the isocyanate group to cross-link with the E-coat resin molecules.

## Bath Chemical Reaction

In both anodic and cathodic E-coat systems, the bath chemical reactions are essentially similar except for different chemical groups and the opposite electrical charge on the polymer ions. For the initial fill of the E-coat tank, insoluble neutral resin is premixed with a solubilizer (most often acetic acid) to produce soluble positive resin ions. Formed simultaneously are negative solubilizer (acetate) ions according to the reaction:

$$\text{neutral resin} + \text{acetic acid} \rightarrow \text{polymer ions} + \text{acetate ions}$$
$$\text{(positive)} \quad \text{(negative)}$$

When the part to be E-coated is immersed in the paint, direct current is passed through the solution. The passage of current causes an electrochemical reaction that converts the water-soluble polar ionic resin polymer into a neutral nonpolar (and therefore insoluble) molecule that can no longer remain dissolved in the paint bath. As a result, resin molecules are deposited directly onto the surface of the parts to be painted.

Such large amounts of direct electrical current pass through the bath that some of the water simultaneously decomposes into oxygen gas and hydrogen gas at the anode, and, more importantly, hydrogen ions ($H^+$) and hydroxide ions ($OH^-$) at the cathode. Normal plant air circulation sweeps away the flammable hydrogen gas to prevent the possibility of an explosion or fire. The amount of hydrogen generated is generally of low volume.

PAINT APPLICATION II 139

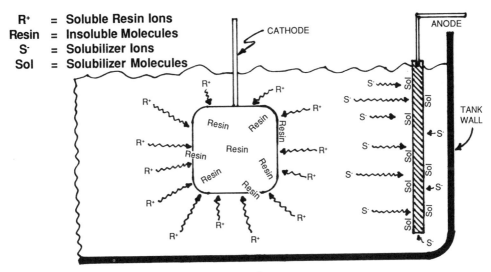

Figure 10-5. Cathodic E-Coat Electrode Reactions

The positively charged (cationic) soluble resin ions interact with the negatively charged hydroxide ions generated at the cathode (the part being painted). Combination of the two oppositely charged species forms water and neutral insoluble paint molecules. In this way items can quickly be coated. As direct current continues to pass through the E-coat, increasing amounts of resin become insolubilized. Figure 10-5 depicts what happens at the electrodes in a cathodic E-coat system.

As additional neutral paint molecules deposit on the metal surface, the part becomes increasingly electrically insulated. This gradually reduces the current flow, slowing the rate at which the paint film is deposited, and finally stops it almost completely. In most cases about 0.5–1.8 mils (13–45 microns) are deposited in 1–4 min. After about 4 min, virtually no additional coating is deposited. It is not possible to produce single E-coat films over 2.0 mils (50 microns) thick.

Only relatively good conducting substrates can be E-coated, so it tends to be done on metal parts only. Just a single coat of paint can be applied by E-coat unless the existing layer of paint is strongly conductive. If an object has a layer of any ordinary paint film, the surface probably will be too limited in conductivity for coating to deposit. Double E-coating is now possible if highly conductive paint is used for the first coat. The first coat is baked before recoating.

The current flow, typically 1–4 A/sq ft (0.1–0.4 A/sq m), is regulated by the voltage setting; greater voltages are used to obtain higher film thicknesses. If the voltage is set too high, however, excess current will flow between electrodes, and gases formed by electrolysis of water can be generated under the newly formed paint film. This causes the paint to lift off in spots as large as an inch in diameter. The voltage at which this occurs is called the *rupture voltage*. Rupture tends to occur on projections or other areas of parts that are closest to the electrodes because current density will be greater in those areas.

At the other electrode, the anode, a process similar to resin neutralization occurs during E-coating with the acetate ions from the solubilizer. Oxygen gas and hydrogen ions are generated at the anode, and the attraction of opposite charges causes negative solubilizer ions to be drawn to the positive anode. Positively charged hydrogen ions are generated by electrolysis of water in the vicinity of the anode. The negative solubilizer ions (acetate ions) combine with positive hydrogen ions to regenerate neutral (in the sense of electrical charge) acetic acid (solubilizer) molecules. The reaction is:

$$CH_3 COO^- + H^+ \rightarrow CH_3 COOH \text{ (acetic acid)}$$

Although neutral, the acetic acid molecules are small and mildly polar so they are water soluble. Resin, solubilizer, and pigment must be added periodically to replenish the paint bath, but as the parts enter and then leave the bath, they take out with them only resin and pigment. If nothing were done, the solubilizer concentration would continually rise and the electrical efficiency would drop. Three methods can be used to prevent excess solubilizer from accumulating in the bath. Any single one, or all of the following, methods may be used in a given system to keep solubilizer levels at an acceptable limit:

- Anode semipermeable membrane (flushable) cells
- Solubilizer-deficient makeup resin
- Ultrafiltration/reverse osmosis

**Anode semipermeable membrane cells.** If the anode is encased in a semipermeable membrane, small molecules and ions (including acetate ions) will be able to go through the membrane. Small negative ions pass through the membrane pores rapidly because they are electrically attracted to the positive anode. The paint resin ions and the pigment particles will be excluded because they are too large to pass through the membrane pores. When acetate solubilizer ions ($CH_3 COO^-$) are neutralized by combining with $H^+$ ions, they become concentrated to a large extent within the anode enclosure. An aqueous flush called the *anolyte solution* is introduced into the bottom of the anode compartment. The anolyte is essentially water that slowly flushes excess acetic acid out of the cell directly to drain or to a holding tank for recirculating the anolyte, as shown in Figure 10-6.

As this flush water recirculates, its acetic acid content slowly increases. Normally, the pH of the anolyte flush ranges from 2.5–5.0, and the electrical conductivity ranges from approximately 250–950 micromhos. Periodically, some of the dilute acetic acid solution in the tank must be released to drain. (Acetic acid is so mild in dilute form that it is used in a large number of foods; vinegar, for example, is a 5% water solution of acetic acid.) Deionized water is added to replace the anolyte solution released to drain. Upon neutralization with caustic soda in the waste-treatment holding tanks, the dilute acetic acid forms sodium acetate, a nontoxic soluble substance that in dilute solution can be drained directly to the sewer.

PAINT APPLICATION II 141

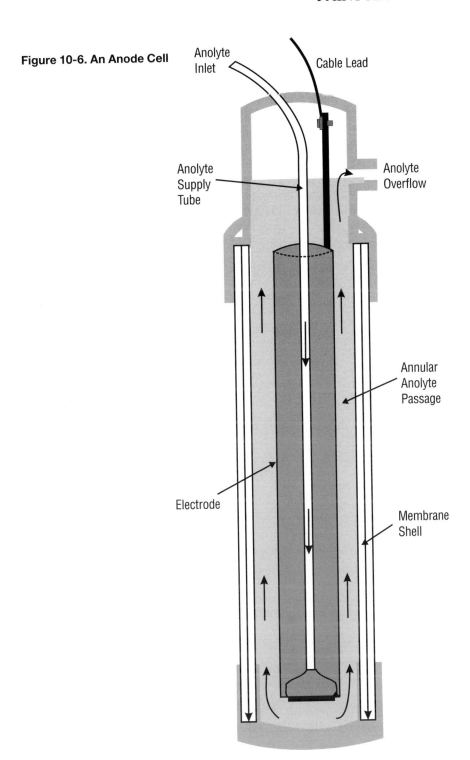

Figure 10-6. An Anode Cell

**Solubilizer-deficient makeup resin.** Another way to reduce the solubilizer content is by using solubilizer-deficient makeup resin. If replenishment resin is added that is not fully neutralized with solubilizer, it is possible to force the excess solubilizer in the bath to become the source of the needed solubilizing acetic acid. If added slowly and with very thorough mixing, the solubilizer-deficient makeup resin can be stirred and blended into a large quantity of the tank's contents. There it reacts with the excess solubilizer in the bath to form fully solubilized resin.

**Ultrafiltration/reverse osmosis.** *Ultrafiltration* and *reverse osmosis* are terms that describe forcing molecules of a liquid through a very fine filter membrane. Both techniques are essentially the same, differing only in the pressures and membranes used. Reverse osmosis uses the higher pressure. Ultrafiltration is used to separate resin and pigment from water (and solubilizer) for rinsing parts. A few systems use reverse osmosis to further purify the ultrafiltration-generated rinse water by removing solubilizer and ionic contaminants. These removed materials are released to the wastewater drain.

## Rinsing

A part emerging from an E-coat bath has a deposited paint film that is covered by a wet solution consisting of bath paint solids (dragout). An elaborate rinse system removes the dragout before the coated parts are baked. Extremely large quantities of rinse water are required to remove all dragout materials that may cling to the part. If dragout were not thoroughly rinsed away, the E-coat film quality would suffer. For example, poor rinsing causes pronounced E-coat roughness that can require sanding of the film after curing. Sanding E-coat films is possible, but it is not easy to do because they are extremely hard. The slow, tedious sanding increases labor costs and slows production. Thorough E-coat rinsing is, therefore, absolutely essential.

The need for such thorough rinsing, however, causes a major dilemma. The rinse water volume must be large, and the amount of rinsed dragout is very low. Discharging all of the rinsed-off bath solids to drain can be cost-prohibitive and would cause an excessively high biological oxygen demand for the sewer system. The problem is complicated because the small amount of paint solids and the large quantity of rinse water cannot economically be separated. Filtration, distillation, and related methods are much too costly for taking paint solids out of the rinse. If this problem could not be resolved, nearly all E-coat processes would be stymied because most plants could not afford to discard the important volume of solids along with the rinse water. Continually adding fresh deionized water to rinses to remove the dragout and then discarding all the ultrafiltration solution would likewise be cost-prohibitive.

A possible answer might be to allow the rinse water and the small amount of dragout to run directly back into the E-coat tank. But if continuous fresh water were used for rinsing, the E-coat tank would rapidly be diluted and accumulate such a large amount of extra water that the tank would quickly overflow.

The remedy to the problem is found by using ultrafiltration to withdraw (temporarily) a small portion of the mostly-water permeate from the E-coat bath. The *permeate* is water plus a small amount of solubilizer and bath salts that is "squeezed" out of the E-coat bath by ultrafiltration. This borrowed permeate is then counterflowed through the rinse system. All of the rinse permeate is collected and reused in several (usually three) separate counterflowing rinse stages. The rinse permeate, having been used and collected for reuse several times, is finally returned to the

PAINT APPLICATION II   143

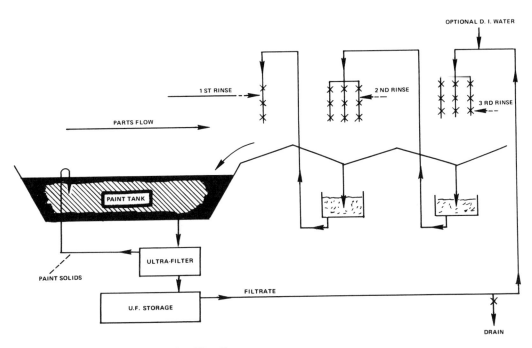

**Figure 10-7. Closed-Loop Ultrafiltration**

bath with an initial rinse located directly over the exit end of the E-coat tank. Figure 10-7 shows how ultrafiltration is tied into an E-coat rinse system.

Thus, the ultrafilter system temporarily appropriates a small portion of the water in the bath, uses it to rinse off the parts several times, and finally allows the permeate plus all the rinsed-off dragout paint to flow back into the E-coat tank. This closed-loop filtration system continually provides enough permeate for rinsing without diluting the bath. The permeate that is extracted from the bath by ultrafiltration is also referred to as *ultrafiltrate* and *flux*.

Factors that can affect the amount of permeate that is produced for E-coat rinsing include:

1. Concentration of paint resin and pigment
2. Pressure of the bath flowing through the ultrafilters
3. Flow rates of bath through the ultrafilters
4. Bath temperature
5. pH
6. Cosolvent concentration
7. Pigment-to-binder ratio of the bath solids
8. Presence of foulants, such as chromates or phosphates, from pretreatment carryover and dissolved solids from the use of tap water rather than deionized water

Under optimum conditions the permeate production rate is about 1.0% of the total bath flow through the ultrafilters. As the filters gradually become fouled, this percentage slowly decreases until cleaning of the filter membranes becomes necessary. Figure 10-8 shows a bank of ultrafilter cartridges in a fairly large E-coat installation.

Figure 10-8. Bank of Ultrafilter Cartridges in a Large E-Coat Installation

Types of ultrafilter construction used in E-coat systems include spiral-wound, shell-and-tube, plate-and-frame, and hollow-fiber multiple-tube. Each has advantages and disadvantages that should be examined by those considering E-coat systems. Individual preferences and preventive maintenance patterns seem to be as significant as operational and performance factors.

Ultrafilters must be properly maintained because good ultrafiltration requires a sufficient flow rate through the filter elements. In many instances manufacturers suggest flow rates of at least 35 gal/min (132 L/min) to prevent fouling of the thin filter membranes. A thorough rinse of the parts after pretreatment will help avoid bringing contaminants into the E-coat bath because the contaminants reduce permeate generation. It is inevitable that some insolubilized resin molecules will not deposit but remain in the bath. These will gradually clog the pores of the ultrafilter membranes.

Depending on throughput rates, normal cleaning of cathodic E-coat ultrafilters may be required at intervals of 6–20 weeks. In normal production, the ultrafilter permeate output slowly decreases. When permeate generation falls to 80% (approximately) of initial output levels, the ultrafilters must be cleaned with a mixture of solvents and concentrated solubilizer solution. Cleaning the ultrafilters frequently will minimize rinsing problems. Extending the interval between cleaning is an invitation for E-coat trouble. If cleaning is postponed too long, permanent impairment of ultrafilter function can result, especially with hollow-fiber and multiple-tube ultrafilters. Flat-plate and spiral-wound types may survive this neglect. However, even these ultrafilters should have regular cleaning as soon as their output diminishes to an ultrafilter manufacturer's designated level.

After ultrafilter rinse, the parts may undergo further rinsing with deionized water to ensure that all contaminants are removed prior to baking. The deionized water rinsing is usually in two stages: a recirculating rinse and a final virgin rinse. The final fresh deionized water rinse of about 1–2 gal/min (3.8–7.6 L/min) should flow into the recirculating deionized rinse tank, which can overflow either to drain or into a permeate rinse tank.

A large amount of deionized water is needed to prepare the E-coat bath and provide the final rinses. Many companies insist that the drain-off deionized rinse in the pretreatment section have maximum conductance of 30 micromhos to prevent contamination of the E-coat bath and the system's ultrafilters.

Solubilizer-deficient makeup resin is added to the E-coat tank through bath recirculation entry injection ports called *eductors* located along the sloped exit end of the tank. Figure 10-9 shows how replenishment materials are added to an E-coat system, and Figure 10-10 shows banks of eductors at the bottom of an E-coat tank. Adding the makeup resin in this way helps prevent fouling the ultrafilters, compared to systems where makeup paint is introduced into the intake of the circulation pump. The eductor entry method ensures thorough mixing of the makeup material.

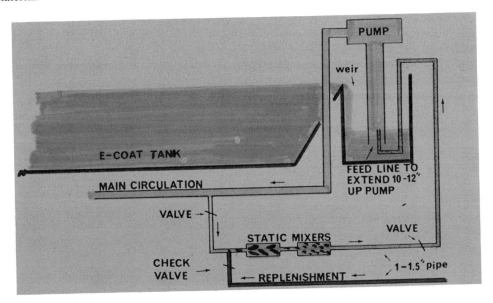

Figure 10-9. E-Coat Single Component Feed

Before the solubilizer-deficient makeup enters the bath, it must already have been fully solubilized. This is essential because even tiny amounts of unsolubilized resin will quickly plug the ultrafilters and sharply curtail the production of permeate rinse.

The bath content must be monitored closely for the buildup of excessive chemicals that can interfere with efficient E-coat operation. For example, an excess concentration of ions in the bath can raise conductivity and interfere with electrical efficiency. To prevent such buildups, a small amount of permeate is normally released to drain continuously and is replaced with an

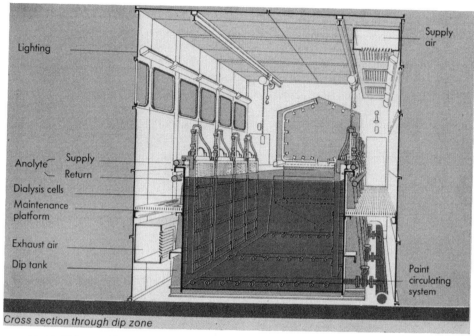

**Figure 10-10. Eductor Entry Method Using Eductors at Bottom of Tank to Ensure Thorough Mixing of Makeup Material**

equal amount of deionized water. Excess solubilizer, contaminants from pretreatment baths, and water hardness salts—all of which are present in small amounts in the permeate—consequently are also released to drain with the discharged permeate. Excess solubilizer in the permeate rinse must be avoided because it can resolubilize the freshly deposited paint film, in effect stripping off the film immediately after application.

## Pigments

The pigment particles are mixed uniformly throughout the bath by vigorous circulation. During the E-coat process the pigment particles become entrapped among resin molecules as the resin is attracted toward and being electrodeposited onto the parts. E-coat pigments never undergo any change in electrical charge; they remain uncharged throughout the process. Only the resins are converted from a soluble ionic state to an insoluble neutral form.

## Bath Parameters

An E-coat bath needs to be monitored closely to ensure quality coating deposition. Meticulous records should be kept of the following parameters:

- Percent total solids
- pH

- Solubilizer concentration
- Solution conductivity
- Temperature
- Relative amounts of pigment and binder
- Film build levels
- Voltage and amperage

The pH for many cathodic systems is held at a slightly acidic level of 6.0–6.5. Some other systems, especially those that eliminate excess solubilizer primarily by flushing ultrafiltrate to drain, may have a pH as low as 3.0–3.5.

A bath sample must be titrated periodically to check the concentration of acetic acid solubilizer. The concentration is usually expressed in units of milliequivalents per liter (meq) of bath material. A typical operating solubilizer concentration is 90 meq.

Overall conductivity affects the electrical efficiency. If it is too high, efficiency drops. Normally, the maximum of roughly 900 micromhos bath conductivity is appropriate.

The substantial amount of direct current used tends to raise the temperature of the paint bath because of the conversion of electrical energy to thermal energy. Chillers and heat exchangers are used in the circulation system to maintain bath temperature in a normal range of 70° F–95° F (24° C–35° C). Excessively high temperatures can cause difficulty with the ultrafiltration process and deteriorate the bath.

As resin and pigment are removed from the bath and deposited as a paint film onto the parts being coated, additional resin (presolubilized with acetic acid) and pigment must be added. Sometimes the resin and pigment are replenished in a single concentrate. However, since the resin and pigment may be used up at different rates as they deposit on the parts, it may be necessary to add each separately. At times a system may use up resin faster than pigment, while at other times the reverse may be true. Paint supplier assistance in monitoring the resin and pigment levels and in establishing the correct pigment-to-binder ratio is usually available. Improper ratios can quickly and easily be readjusted by the appropriate makeup composition.

Continuous circulation and filtering of the bath are necessary to prevent pigment settling and to keep the bath clean to avoid the deposition of foreign contaminants along with the resin and pigment. Pumps circulate the bath at a rate of four to six turnovers per hour. The bath is taken from the area behind the *weir* (an adjustable dam to control the level of the paint in the tank) and is then filtered. The paint is next fed back into the tank along the bottom through a series of pipes fitted with eductor nozzles. The jet force out of the nozzle orifices and the venturi action of the eductors on the surrounding bath help prevent pigments from settling and maintain good pigment dispersal in the bath. Eductors produce a mild scouring action that also reduces the amount of residue that can accumulate on the bottom of the tank. If adequate circulation is not maintained, extremely hard resin/pigment aggregates can build up on the tank bottom. As much as 14 inches of deposit have been found in E-coat tanks due to grossly neglected circulation problems. Parts that fall off the conveyor should be removed immediately from the bottom of the tank so that dead circulation spots are not formed. These would contribute to residue deposits and small bits of resin/pigment aggregates floating in the paint tank. These bits deposit in the E-coat and contribute unwanted visible roughness to the paint film.

## Tank Details

The E-coat tank must be large enough to hold the parts to be coated and the necessary quantity of bath (see Figure 10-11). Some tanks have a capacity as high as 120,000 gal (450,000 L). To paint continuously moving parts on a conveyor line, the tank must be considerably longer than for parts held stationary. A conveyor line operating at 14 ft/min requires a total tank length of roughly 50 ft, including the portions used for gradual part immersion and withdrawal.

Anodes are located vertically along both sides of the E-coat tank and, in some tanks, are located horizontally near the bottom as well. In most cathodic systems, each anode is enclosed within a flushable cell. These cells enable continual flushing to remove excess solubilizer produced by electrochemical reaction at the anodes. Tubular (5 in or 13 cm in diameter) or rectangular (6 in by 24 in or 15 cm by 61 cm) membrane cells of the needed length can be used, but the ease of handling has made the tubular type cells increasingly more popular.

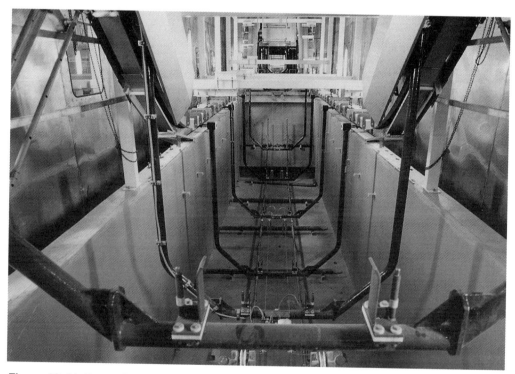

**Figure 10-11. Empty E-coat Tank**

### Voltage at Tank Entry

Several options are available concerning the application of direct current E-coat voltage as parts are being immersed and after full immersion. Two options are as follows.

- The voltage can be turned off during parts immersion. This is called *cold* or *dead entry*. After the parts are fully immersed, the voltage can either be raised gradually to the specified level or be applied to that level all at once. Raising the voltage gradually after immersion will bring a slow, even, and optimum deposition of paint film. Applying full voltage suddenly after full immersion may cause current spiking and the formation of excessive hydrogen gas at the part surface, which can interfere with even paint film buildup, causing film rupture.
- The voltage can remain fully energized before parts begin to be immersed, during immersion, and after immersion. This is termed *hot* or *live entry*. As a part is being immersed during hot entry, a paint film buildup begins on the wetted portion of the part's surfaces. Part surfaces under immersion the longest will have the greatest film buildup. Sometimes the progressive film buildup can leave "hashmarks" or noticeable lines of distinct film buildup around the part, which can become pronounced if the conveyor movement is jerky during immersion.

Two different kinds of systems may be used to lower parts into the E-coat bath: a hoist and an overhead conveyor. In the hoist type, a single part or rack is lowered into the E-coat tank, which is not much bigger in length and width than the dimensions of the part or rack. After a specified bath deposition time, the part or rack is raised and advanced to the next stage, and the cycle repeats with another part or rack. With this type of system, it is relatively easy to have cold entry and raise the voltage gradually for optimum paint film buildup.

With the overhead conveyor system, the tank will have a narrow width but an extended length, which may permit a relatively large number of parts to be immersed at any given moment. To eliminate mechanical complexity, this type of system usually has hot entry, and the bath electrodes are generally spaced equidistant along each side of the tank. Some overhead conveyor systems wire the power supply so that the first few electrodes near the tank entry have reduced voltages. In this way, the parts do not come near an electrode with full specified voltage until a predetermined time after full immersion. Other overhead conveyor systems omit electrodes near the bath entry to permit a gradual paint film deposition both during and immediately after parts immersion. All these methods reduce the current spiking that otherwise would occur with hot (or live) entry of parts.

Because E-coat direct current electrical power may exceed several hundred volts and hundreds of amps, an E-coat tank during production is extremely hazardous. A person falling into an energized E-coat bath could be instantly electrocuted. A locked protective safety enclosure around the tank is necessary to prevent entrance by unauthorized personnel.

## E-coat Curing Cycle

The final E-coat step is oven curing of the deposited coating. Although low-temperature curing and even air-curing are possible, the most durable films require a moderately high-temperature bake to cross-link the resins in the binder. This is often in the order of 300° F–350° F (150° C–177° C) for 12–35 min, depending on the design, configuration, and heat capacity of the part being coated. Convection ovens are used almost exclusively for curing E-coat paints, although infrared ovens can be used as well.

## E-coat Advantages

Some of the advantages of E-coat include:

- Coating thickness uniformity
- High production rates
- Primer/one coat/two coat options
- Low environmental pollution potential

**Coating thickness uniformity.** In electrical deposition systems such as electroplating and E-coat, the consistency of film thickness varies with the separation distance of the part being coated and the other electrode (or electrodes). Part surfaces closest to the charged electrode will receive the greatest film buildup because these are high–current-density areas, where film thickness will always be the greatest. Properly positioned charged electrodes will minimize differences in deposited film thicknesses. The deposition power (throwing power) is usually sufficient to deposit at least some coating thickness into many hidden areas. Auxiliary electrodes can be positioned to improve throwing power into these areas. In many instances the edges of parts become the high–current-density areas and receive extra film deposition.

**High production rates.** E-coat is adaptable to very high production rates because of the fast film deposition that is possible. E-coat wheel lines have operated at a conveyor speed of 48 ft/min.

**Primer/one-coat/two-coat options.** E-coat can be deposited as a primer to add corrosion resistance and to serve as a base for a topcoat. The primers tend to be mainly epoxies; others include alkyds and polyesters. E-coat formulations—generally acrylics—also can be deposited as one-coat systems. By using highly conductive E-coats, it is possible to double E-coat parts. The first coat must be fully cured before the second coat is applied.

**Low environmental pollution potential.** E-coat paints are waterborne types with little co-solvents (in only a few cases with zero co-solvents) so VOC emissions are low. New lead-free epoxies have eliminated the heavy metal problems of earlier epoxy E-coats.

## E-coat Disadvantages

E-coat disadvantages include:

- Substrate limitation
- Color change difficulty
- High cost to install
- Masking problems
- Sophisticated maintenance requirements
- Sanding/stripping difficulty
- Second coat restriction
- Air-entrapment pockets
- Bulk small-part coating difficulty
- Corrosion-resistant equipment requirement
- Conveyor stoppage problems

- Deionized-water quality requirements
- Deionized-water volume requirements
- High energy demand
- Restriction to large volume finishing
- Coating thickness limitation

**Substrate limitation.** E-coat cannot be used for plastics, wood, or other low-conducting and nonconducting materials. Even some high–silicon-containing aluminum alloys cannot be used due to lack of electrical conductivity.

**Color change difficulty.** Color change for most installations is very slow. This is why more than 95% of E-coat installations paint only a single color or use a separate tank for each color. Quick color change requires a separate tank for each different shade when multiple colors are used. This is not generally the method of choice for economic reasons. The plants that change E-coat colors usually pump out the old bath into an agitated holding tank, clean the coating tank and ancillary equipment, and then fill the tank with the new color material.

**High costs.** The initial capital cost and the continuing operating costs are extremely high when compared with simple systems such as dipping and spraying.

**Masking problems.** Since E-coat is an immersion process and requires a subsequent high-temperature bake, the masking requirements are unique. Masking tape, plugs, and caps must be resistant to the E-coat bath and to the oven temperatures. In many instances, they need to be watertight as well.

**Sophisticated maintenance requirements.** Because of the complexity of an E-coat system, sophisticated maintenance is required. Careful maintenance records must be kept; testing must be done on schedule. A paint sample of the E-coat bath is usually sent to the paint supplier biweekly or monthly for detailed testing. Color corrections may be needed on a daily basis.

**Sanding/stripping difficulty.** Because of the hardness of the deposited coating, it is extremely difficult to sand E-coat rework. It also can be difficult to strip E-coat paint from hooks, hangers, and parts baskets.

**Second coat restriction.** Most coated parts cannot be run through an E-coat tank for another coating. The paint in the prior coat will act as an insulator to prevent deposition of another coat of paint over it unless a special highly conductive first coat is used.

**Air-entrapment pockets.** Since E-coat is a dip process, it can be subject to coating voids due to air entrapment in pockets of the parts. A part that tends to form air pockets when immersed may need to be redesigned. Sometimes a conveyor oscillation system is employed to eliminate the air or to move the air back and forth.

**Bulk small-part coating difficulty.** Bulk E-coating of small parts can be done, but it tends to require a complex system. Only a few bulk E-coat systems have been built due to the inherent problems of operation.

**Corrosion-resistant materials requirement.** Because of potential rust problems with all water solutions, it is often necessary to use corrosion-resistant materials for E-coat tanks, piping, valves, heat exchangers, and other components that contact the bath. Also, the acidic character of cathodic E-coat paint can slowly dissolve metal parts.

**Conveyor stoppage problems.** The nature of the deposited E-coat paint film and the ultrafiltrate rinse requires exacting conditions in the E-coat postrinse. If the conveyor stops, coated parts in the ultrafiltrate rinse may be adversely affected. The unrinsed bath solids will dry within 5 min and then cannot be rinsed off. Manual rinsing must be initiated if the conveyor stops for more than a minute or two. Similarly, E-coated parts that remain submerged in the E-coat tank during a line stoppage of more than 5 min can undergo paint film degradation.

**High deionized-water purity requirements.** The deionized water quality must be high. Its conductivity should not exceed 25 micromhos, the resistivity minimum should be about 40,000 ohms/centimeter, and the soluble salt content must be no greater than 10 ppm.

**High deionized-water volume requirements.** The amount of deionized water used can be significant. It is normally necessary to buy equipment to produce deionized water in the plant.

**High energy demand.** The E-coat process requires a significant amount of electrical energy.

**Large-volume suitability.** Because of the elaborate equipment that is required, E-coat is generally suited only for finishing a large volume of parts.

**Coating thickness limitation.** An E-coat film thickness is limited to about 0.5–1.5 mils (12–37 microns). After this thickness is applied, the coating acts as an insulator to prevent further deposition. It should be noted that maximum thicknesses vary widely among E-coat paints.

# Autodeposition

Organic paint films can be deposited onto iron, steel, and zinc parts by an oxidation-reduction precipitation process known as *autodeposition*, or chemiphoretic or autophoretic coating. The process, which uses no external source of electricity as with E-coat, is available primarily in black; however, several colors such as brown, orange, and light blue have been reported. The paint film has a dull or low-gloss appearance and is primarily protective and not particularly decorative. The largest application areas for autodeposition coatings have been nonappearance and under-hood parts for cars and trucks. Excellent anticorrosion properties and the black color make it highly appropriate for this application. It is also used on drawer slides for office furniture, replacing zinc-plating.

An important advantage of autodeposition is its 100% coverage of all surfaces of a part that are wetted by the coating bath. Faraday cage areas, which hinder E-coat deposition, are nonexistent with autodeposition. Where it wets, it coats.

Figure 10-12 shows the main process stages in an autodeposition system. The process begins with a heated aqueous alkaline spray cleaning for about 1 min at 160° F (71° C). Then a dip cleaner follows for 1–3 min at 185° F (85° C). These cleaners are the only stages in the entire process that are heated; the other stages operate at normal room temperatures.

Thorough cleaning is extremely important because the process tends to be especially intolerant of contaminants. After multiple rinses, ending with a deionized water rinse, the parts go into the autodeposition bath for about 2 min. Two autodeposition paint resins are available: vinyl and acrylic. The acrylic is less common since a chromic sealer rinse must be used on coated parts before oven curing. The acrylic resin has a much higher thermal stability than the vinyl so it can

PAINT APPLICATION II 153

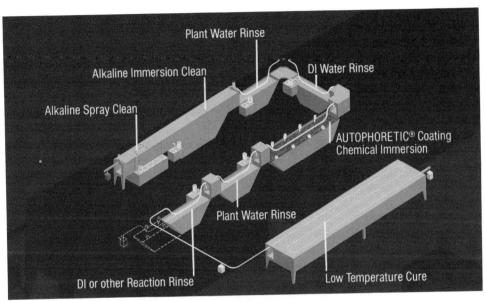

**Figure 10-12. The Main Process Stages in an Autodeposition System**

be used as a primer for powder coat paints, whereas the vinyl cannot. The bath, typically held at 68° F–72° F (20° C–22° C) is a waterborne material containing about 10% of a resin emulsion, hydrofluoric acid, and hydrogen peroxide. The coating deposition reaction is nonexothermic (does not give off heat), and the rate of deposition slows as increased coating covers the metal surface.

Coating deposition begins immediately upon immersion of the parts into the bath, which is maintained at a pH of 2.5–3.5. After about 2 min, 0.75–1.0 mil (19–25 microns) of coating is deposited. Hydrofluoric acid etches and removes metal ions from the part by chemically attacking its surface. Hydrogen peroxide converts the solubilized iron ions from the +2 to the +3 (ferrous ion to ferric ion) oxidation state. The $Fe^{+3}$ metal ions combine chemically with the emulsion polymer to form an insoluble resin that precipitates onto the surface of the metal. Even after parts are removed from the bath, unreacted resin is soon insolubilized by the continuing chemical action. Thus, no unreacted paint needs to be rinsed from the parts. The inorganic chemicals remaining on the parts are removed from the paint film by a 30-s water immersion rinse. Figure 10-13 shows automotive brake parts emerging from one of the baths in an autodeposition process. When zinc is coated, a recent possibility with autodeposition paint, close control of the metal ion concentration becomes necessary.

After the coating is deposited, several options exist. One option is the use of a dilute chromic acid rinse, followed by an oven bake at about 250° F (121° C) for 15 min. Another option is to cure the coating in water at 180° F (82° C) for about 8 min. The water cure tends to yield a coating with less corrosion resistance than a coating cured in the hotter bake oven temperatures. The hot water cure method may include the use of chromic acid in the water for greater corrosion resistance if desired, but it is not required.

154 INDUSTRIAL PAINTING

**Figure 10-13. Automotive Brake Parts Emerging from Autodeposition Bath**

An autodeposition coating cannot be recoated by a second autodeposition process because no exposed metal is left for the acid to attack. A major advantage of autodeposition is that no organic solvents are needed and thus no VOC is emitted. Autodeposition is used both by various OEM product manufacturers and also by a few custom-coating shops. The current processes work only on zinc and ferrous metals.

# Chapter 11

# Paint Application III—Spray Guns

## Components of Air-Atomizing Spray Guns

The methods of paint application described in the two previous chapters rely on various forms of spreading or splashing paint over the parts and immersing parts in various types of paint baths. As you have learned, each method has its own unique advantages and disadvantages, and any one of these techniques can become the perfect choice for a particular end use.

A different and also very common group of methods for applying paint involves atomizing the paint into tiny mist-like particles and directing them to travel onto the surface to be coated. If a sufficient number of the particles is applied, they will create a continuous coating film. Paint can be atomized to various degrees and by several different techniques. The most common of these is with an air-atomizing spray gun, often referred to as *conventional air spray*.

The essential components of an air-atomizing spray gun are:

- Gun body
- Fluid inlet
- Fluid nozzle or fluid tip fluid
- Needle and seat assembly
- Air inlet
- Air nozzle or air cap
- Air valve
- Fan or spray pattern control
- Trigger

**Gun body.** The gun body consists of the handle for the operator to grip and the barrel. The main thing that the almost countless models of air-atomizing spray guns have in common is a handle designed for operator hand comfort. The handle and barrel house are attached to the gun's various components.

**Fluid inlet.** The fluid inlet is an opening, usually below the tip of the barrel, that allows the paint to flow into the gun. The opening is threaded to allow attaching either a siphon cup or a paint hose attachment. In some instances, a gravity cup mounts above the barrel to feed paint into the top of the barrel just ahead of the gun tip.

**Fluid nozzle or fluid tip.** The fluid nozzle is a small device with a precision opening to permit the paint to flow out of the gun at a determined rate. One end of the nozzle is externally threaded and screws into the internally threaded barrel tip. An assortment of nozzles is available with different diameter fluid orifices to accommodate paints of different viscosities and delivery rates. The fluid needle seat is at the back of this part.

**Fluid needle and seat assembly.** This assembly inside the gun serves as a valve to control the flow of paint through the fluid nozzle. The needle is attached to the external trigger. By using this trigger, the sprayer can start and stop the paint flow and also to a degree control the quantity of paint released from the gun. By turning a knurled knob at the back of the gun, the sprayer can alter the extent to which the trigger moves back with a full trigger pull.

**Air inlet.** This is a threaded opening at the bottom of the gun handle to allow the attachment of a hose connected to a source of compressed air.

**Air cap or air nozzle.** The air nozzle is a small device that allows compressed air to be directed at the paint exiting the gun to produce optimum atomization. The nozzle, also called the air cap, is internally threaded to attach to the externally threaded gun barrel tip. Air nozzles are available with many different configurations of openings to allow various atomization patterns. Air nozzles typically have two opposed "wings" or "horns" with precision openings that can be utilized to vary the spray pattern into variably elongated configurations. Figure 11-1 shows the details of an air nozzle on a common type of spray gun.

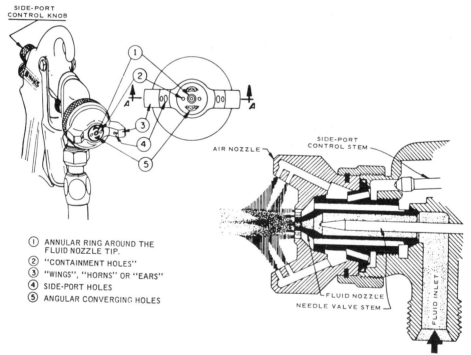

Figure 11-1. Common Type of Spray Gun

**Air valve.** A simple on-off air valve gives the operator a means of controlling the flow of compressed air used for atomization and spray pattern control through the gun.

**Fan or spray pattern control.** This control permits the operator to regulate air flow through the air nozzle horn openings to vary the spray pattern. Figure 11-2 shows how a spray pattern can

# PAINT APPLICATION III   157

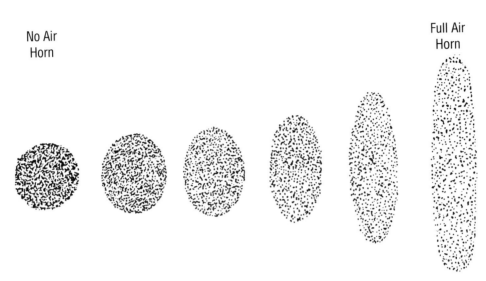

**Figure 11-2. Altering Spray Pattern by Varying Air Flow**

be altered by turning a knurled knob at the back of the gun, which varies the air flow to the gun's air cap "horn" or "wing" openings.

**Trigger.** The trigger is connected to both the compressed air and fluid flow controls in the most commonly used type of spray guns called "nonbleeder guns." When triggered, the air valve opens first, and then the fluid valve; when the trigger is released, the paint valve closes just before the air valve closes. This assures that all paint exiting the gun will be atomized by the air flow. Partial triggering activates just the air valve and allows the sprayer to blow dust off parts before painting if desired.

In "bleeder guns," the trigger activates only the fluid controls; compressed air flows continuously, even when the trigger is released. Bleeder spray guns are far less prevalent than nonbleeder guns, except for heated air high volume low pressure (HVLP) spray guns. These are discussed later in this chapter.

## Compressed Air Supply

To prevent paint contamination, the air supplied to a spray gun must meet three requirements.

1. The air must be dirt free. The air must be filtered to remove dust, lint, and other foreign contaminants. The air is generally filtered at the inlet port of the air-compressing device and again after oil and moisture removal.
2. The air must be oil free. Air compressors usually send out oil vapor with the compressed air because oil is used to lubricate compressor rotors. Such oil vapor is generally removed with an oil-absorbing or oil mist coalescing filter.
3. The air must be moisture free. The solubility of water vapor in air is such that warm air is able to hold far more moisture than cool air. When air is cooled, moisture tends to be wrung

out of the air in the form of condensation or water. Since air increases sharply in temperature when it is compressed, it normally holds a considerable amount of water vapor. Enroute to the spray gun, the compressed air temperature drops, which produces moisture condensation that must be removed from the air lines. This can be accomplished with refrigerated air-cooling systems and water cold traps, or by chemicals that hold water by absorption and adsorption. Figure 11-3 shows the capability of deliquescent, refrigerated, and desiccant air dryers to remove moisture from compressed air streams.

| Drying Medium | Dew Point | Moisture Downstream |
|---|---|---|
| Deliquescent | 40° F to 70° F | 30 to 80 gal/day |
| Refrigerant | 35° F | 20 gal/day |
| Desiccant | -40° F to -100° F | 0.04 gal/day |
| Compressor inlet air 90 F, 3000 SCFM generated at 100 psig. | | |

| Drying Medium | Dew Point | Moisture Downstream |
|---|---|---|
| Deliquescent | 4.5° C to 21° C | 115 to 300 L/day |
| Refrigerant | 1.7° C | 75 L/day |
| Desiccant | -40° C to -73° C | 1.5 L/day |
| Compressor inlet air 32 C, 85 standard cubic m/min generated at 690 kPa. | | |

Figure 11-3. Dryer Performance Chart

Air can be supplied to an air-atomizing spray gun by either an air compressor or an air turbine. Air compressors are of various types and may include diaphragm, rotary, and reciprocating construction. They generally operate in a range of 75–160 psi. The compressed air from one air compressor can be further compressed in a two-stage system. In contrast, an air turbine is a fan-like device that operates at pressures usually much lower than those generated by air compressors. Turbines are used to supply relatively large volumes of air to paint application equipment at pressures rarely exceeding 50 psi.

Compressed air generated by the compressor is generally passed through an air regulator to maintain a steady pressure for the gun(s). For example, the air line pressure in a manufacturing plant may be 100 psi, but perhaps only 50 psi is desired at the spray gun. The air regulator would drop the pressure to 50 psi and consistently maintain it at this level, even though the plant air pressure might fluctuate, so long as the inlet pressure never dropped below 50 psi.

## Paint Supply

Paint to be fed to an air-atomized spray gun can be supplied by gravity, siphon, or pressure systems. Gravity and siphon cups are acceptable for applying small amounts of paint at a time; for continuous painting, a large container avoids the need to stop frequently for refilling the paint supply. In gravity feed, as the name suggests, the paint supply is above the gun and flows down by gravity. The paint container must be covered to keep out dirt and dust, and it needs an air inlet vent to allow paint to flow out of it.

Paint in a *siphon cup* is pushed upward into the gun by atmospheric pressure. This is because the atmospheric pressure is reduced at the exit of the fluid tip by the flow of compressed air out of the air cap. Thus, paint flows only when compressed air flows through the gun. Siphon cups

**Figure 11-4. Siphon-Feed Spray Gun**

need to be vented. They come in various sizes ranging up to about a liter (1.06 qt) or slightly larger. Weight becomes a negative factor for the sprayer's arm and hand if too much paint is in the siphon cup. Figure 11-4 is a drawing of a spray gun using a siphon cup paint supply.

Because of the limitations of gravity feed and siphon feed, the paint supply system in predominant use is pressure-feed. *Pressure-feed* is of two types: (1) compressed air is applied to paint in a pressurized container, forcing paint out through a hose to the gun; and (2) paint is pumped to the gun, either in a dead-end system or in a recirculating system, by various types of pumps.

The pressurized container may be a cup that is mounted at the bottom of the gun, but it is usually a *pressure pot* (tank) connected by a 5–10 ft (152–305 cm) flexible hose to the spray gun. Pressure pot sizes range from about 2 qts (1.9 L) to 50 gal (190 L). Figure 11-5 shows a drawing of a pressure tank with a double regulator so that the atomizing air and the air inside the paint tank can be controlled independently. Figure 11-6 illustrates the cross-sectional difference in an air nozzle for a gun using a siphon cup and one using pressure feed.

Paint can also be pumped to the spray gun out of any unpressurized container such as a pail, drum, or tote tank. Paint pumps that supply air-atomizing spray guns can be grouped into four basic categories:

- reciprocating piston (single- or double-acting)
- diaphragm
- rotary (cam or gear)
- centrifugal

160  INDUSTRIAL PAINTING

Figure 11-5. Pressure-Feed Paint Tank

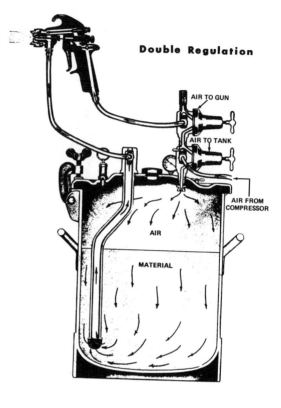

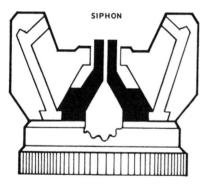

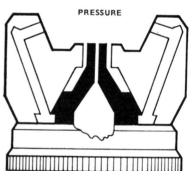

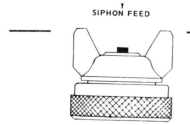

Figure 11-6. Cross-Sections of Siphon Feed vs. Pressure Feed

The pumps may deliver paint to just a single gun or to a system from which paint can be piped to one or more guns. The paint delivery can be dead-ended or recirculating with either style. Paint recirculation tends to be used with paints that have pigments that settle out rapidly. This method is also used to prevent localized overheating with heated paint systems.

The paint delivered to a gun is generally filtered to remove contaminants that might cause rejects if allowed to be sprayed onto a product. The micron rating of the filter must be high enough to permit passage of pigments and metallic particles.

## Gun Operation

The rate of flow of paint through a spray gun is a function of the fluid pressure (driving force), paint viscosity, size of the fluid nozzle orifice, and the degree to which the gun's fluid needle is retracted from the seat in the fluid tip. The fluid pressure is set to deliver the required amount of paint through an appropriately sized fluid nozzle orifice. High production requirements need high rates of paint flow and large orifice paint nozzles in order to apply sufficient coating on parts moving past the spray gun at conveyor speeds. For lower paint-flow requirements, lower paint pressures and smaller fluid tips would be used.

The degree of atomization produced by an air-atomized gun depends on how efficiently the atomizing air breaks up the paint particles. It isn't only the amount of air pressure alone that breaks up the paint particles, but instead is a culmination of air pressure, air volume, and the precise merging configuration of the air and fluid streams. Increasing air pressure and increasing air volume both increase the degree to which a paint is atomized.

### Types of Guns

Air-atomizing spray guns can be categorized into two general types according to the volume and pressure of the atomizing air:

- Low-volume high-pressure (conventional air spray)
- High-volume low-pressure (HVLP)

#### Low-Volume High-Pressure (Conventional Air Spray) Guns

This category of air-atomized spraying has been called "conventional" to distinguish it from the modified gun versions that have appeared over the last 10 to 20 years. Figure 11-7 is an exploded drawing of a conventional air spray gun. Figure 11-8 shows a conventional gun being used. This style spray gun (now on rare occasions called an LVHP gun) was developed early in the 1900s. It has remained essentially the same over the years except for refinement in gun construction materials and in the design of air and fluid nozzles.

These spray guns use compressed air supplied from an air compressor. The air pressure may range from roughly 25–30 psi (172–207 kilopascals or kPa) to about 90–100 psi (620–690 kPa), and the air volume from about 5 cubic feet per minute (cfm) to 20 cfm (142 to 566 dl/min). The air volume tends to be low when using low air pressures and rises somewhat proportionately as the pressure is increased. The air pressure selected is tied in closely with the air nozzle that is used. The size of the openings in the air nozzle orifices must not deplete the air compressor capacity. The relatively high air pressure typically used with these guns gives exceptionally fine atomization up to a maximum paint flow of 1 qt/min (0.95 L/min). It allows high rates of

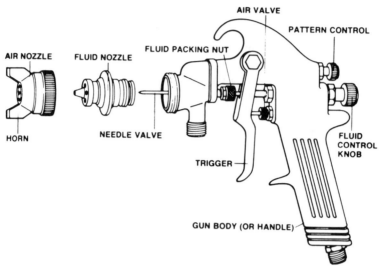

Figure 11-7. Conventional Air Spray Gun

paint flow to meet high production requirements, but it can also be used for situations in which only a few drops per minute are needed. Both flow rates can be accomplished with outstanding finish quality.

Paint flows and air pressures must be balanced and proportional. Should insufficient air pressure be used, the finish will be rough and tend toward runs and sags. If the atomizing air pressure is too high, droplets as small as five microns in diameter can be formed and this may create a paint fog that will decrease overall application efficiency. The greater surface area of tiny droplets will also increase solvent evaporation, producing "dry" spray, and giving the parts being painted a dusty appearance.

## *High-Volume Low-Pressure (HVLP) Air Spray Guns*

The HVLP category of air spray gun uses appreciably lower atomizing air pressures than conventional air spray guns, most often 10.0 psi (69 kPa) or less. The figure "10 psi" has become important as a maximum pressure because air quality regulations in many states stipulate that to be considered as part of certain compliance efforts, the air-atomized guns may not exceed this pressure. Actual values for industrial painting are typically in the 2–5 psi (14–34 kPa) range. Excellent atomization can be obtained at these pressures. Air volumes range from about 15–65 cfm (425–1,845 dl/min) or even higher.

HVLP guns are classified according to whether the air is supplied by an air compressor or by a turbine. The type of gun that uses an ordinary air compressor typically works in conjunction with an air-regulating device located in the air line or inside the spray gun itself to ensure that no more than 10.0 psi (69 kPa) of air pressure reaches the gun tip. In this case, a nonbleeder style HVLP gun is used. The tendency is to use specially chambered guns that will not exceed 10 psi (69 kPa) air pressure at the gun tip rather than regulators on the compressed air lines.

PAINT APPLICATION III 163

Figure 11-8. Conventional Air Spray Gun in Use

The other type of HVLP gun uses warm air supplied by a turbine. An air turbine can typically put out about 200 cfm (5.7 cubic m/min) of air at 10.0 psi (69 kPa). This means that up to eight HVLP guns can operate off a single turbine. An advantage of a turbine is that its air output can be heated to as high as 180° F (82° C), which helps provide easier atomization by heating the paint in the end of the gun to lower its viscosity. Bleeder style spray guns are used with heated turbine air so that the tip of the gun remains warm. An air heater would need to be used with the air compressor type of HVLP gun to provide the same heated air—an option which was offered by an equipment supplier but it did not find acceptance in the market. When hot atomizing air is used, it can make the handle of a manual gun uncomfortably hot. Introducing the hot air into the gun at the tip (as seen in Figures 11-9 and 11-10) circumvents this problem.

Figure 11-9 compares the air source for a conventional gun with an HVLP gun using a turbine. As mentioned, air to the HVLP gun can also be supplied by an air compressor.

164  INDUSTRIAL PAINTING

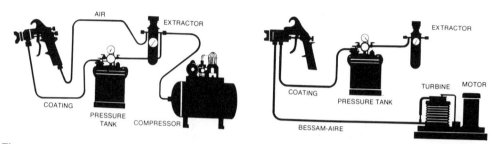

Figure 11-9. Air Source for Conventional Gun vs. Gun Using Turbine. Note that the supply tank in the turbine system still requires pressurized air from a compressor (not shown).

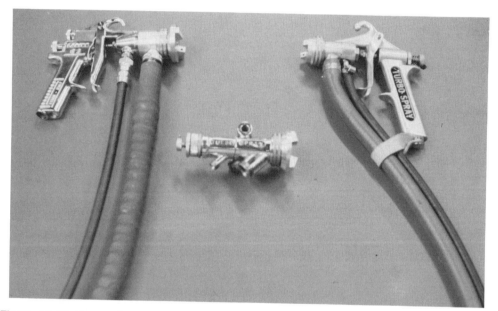

Figure 11-10. Comparison of Air Hoses Used in Conventional vs. Turbine Type Gun

Both the air compressor and turbine types of HVLP guns are characterized by an air nozzle with a relatively large-diameter opening for atomizing air. At 10.0 psi (69 kPa) the air compressor typically supplies 15-30 cfm (425-850 dl/min). Turbine-type HVLP guns are recognizable by the large-diameter air hose connecting to the gun. This is evident in the guns shown in Figure 11-10. Compressor-type HVLP guns are usually not easily distinguishable from conventional guns by their outer appearance.

# PAINT APPLICATION III

The low-atomizing air pressure of an HVLP gun tends to minimize the amount of "bounce-off" paint fog and reduces the amount of atomized paint that is blown past a part to be painted as overspray. The improved transfer efficiency helps hold down operating costs by reducing paint waste. As the solids percent of paint is increased, the need to minimize overspray increases accordingly to hold down costs. However, such reduced-atomizing air pressure also tends to decrease the fineness of atomization, which reduces the finish smoothness capability. The low atomizing air pressure of HVLP guns also tends to require reduced paint flow to the gun, which limits production speeds.

Figure 11-11 summarizes the differences between high- and low-pressure air-atomized spray guns.

| | Advantage | Disadvantage |
|---|---|---|
| **High Pressure Air Spray** | • Excellent atomization permits fine finishes. | • Extensive overspray wastes paint. |
| | • High production rates can be accommodated. | • The overspray increases booth cleanup costs. |
| | | • The overspray increases exhaust filter replacement costs. |
| | | • The overspray increases waterwash reservoir treatment costs. |
| **Low Pressure Air Spray** | • Minimal overspray reduces paint waste and increases transfer efficiency. | • Atomization may not be sufficient for fine finishes. |
| | • Minimal overspray lowers booth cleanup costs. | • High production rates may not be possible. |
| | • Minimal overspray cuts exhaust filter replacement costs. | |
| | • Minimal overspray decreases waterwash reservoir treatment costs. | |

Figure 11-11. Comparison of High Pressure and Low Pressure Advantages and Disadvantages

### Low-Volume Low-Pressure (LVLP) and High-Efficiency Low-Pressure (HELP) Guns

If conventional high-pressure guns are modified to operate below 10 cfm (285 dl/min) with an internal mix air cap, LVLP and HELP guns may be suitable for some lower viscosity paints applied at rates of less than 10 oz/min (300 cc/min). The coarser atomization of such guns compared to the conventional and HVLP guns is acceptable for some applications.

## Spraying Techniques for Air-Atomized Guns

Proper spraying techniques with air-atomized guns are extremely important because of the cost of the paint being applied and the expense of having to rework reject parts. Good spraying techniques will minimize overspray, ensure uniform film builds, and result in the correct paint application for optimum appearance.

Before a person begins to spray paint onto a product, a number of points should be reviewed. The spray gun should be clean and in perfect working order. It should yield a spray pattern with clearly defined boundaries.

A proper spray pattern is achieved with the least amount of air pressure for correct atomization and the minimum amount of fluid pressure to provide enough paint to meet production requirements.

Good spraying technique requires adhering to the following basic principles:

- The gun should be put in motion before the trigger is squeezed.
- The gun should be kept a uniform distance from the surface being coated.
- The gun should be moved across the surface to be coated at a uniform speed.
- The gun should be triggered at the beginning and end of each stroke.
- Each part sprayed should be started at the same vertical or horizontal location for that particular product.
- Each previous stroke should be overlapped by the same amount (50%).
- The same number of strokes should be used on identical product surfaces.
- The final stroke on identical products should be ended at the same location.
- For an optimum coating, the gun distance from the product surface being coated should be 6–8 in (15–20 cm). Moving the gun closer will increase the wetness and film build. Backing the gun away will decrease wetness and minimize film build.

Some general principles should be followed when manually spraying products with different shapes. For example, the ends of vertical flat panels should be sprayed first, followed by back-and-forth horizontal spraying, beginning at the top. Large panels should be sprayed this way in strokes up to about 5 ft (1.5 m) long. When edges are being sprayed, the gun should be aimed so that as much overspray as possible lands on uncoated surfaces. Exterior edges should be sprayed first. Spray should not be directed straight into internal corners; each side of the corner should be sprayed instead.

Gun variables should be monitored closely while spraying. These variables are the paint flow rate, fluid pressure, paint viscosity, air pressure, fan pattern, and distance of the gun from the work.

The evaporation rate of the solvent from the atomized paint particles moving from the gun to the product being painted needs to be considered. Excessively slow solvent evaporation will yield an applied coating that might be excessively wet and cause paint to run or sag on vertical surfaces. Excessively fast solvent evaporation can produce a coating that is too dry.

## Air Spray Characteristics

Air spray can be suitable for a single-gun system applying only 2–3 oz (55–85 cc) of paint per hour and can be equally suitable if scaled up for use by a system involving a dozen painters who are each spraying a quart or liter of paint per minute working out of a common paint supply line. Air spray is readily adaptable to any size coating operation and rate of coating application.

Quick-disconnect fittings provide fast color change capability. Automatic color changers are also available, principally for use with multiple- or automatic-gun systems.

The readily interchangeable fluid nozzles and air caps permit the application of paints having a fairly wide range of viscosities. Heaters can be added to reduce the viscosity of highly viscous coatings. It is easy to regulate the air and fluid pressures at numerous points in the system.

The spray guns can be mounted onto various devices for automatic application. These can include reciprocators that move the guns back and forth or up and down as parts are conveyed in front of them and robots that move the guns through programmed paths. Also, various rotational or indexing systems can be incorporated to move the part to improve application efficiency.

# Chapter 12

# Paint Application IV—Airless Spray and Air-Assisted Airless-Spray

## Components of Airless Spray Guns

In addition to using compressed air as the paint-atomizing force as in air spraying, another much used method of atomizing paint is to sharply increase the paint's fluid pressure in an airless spray gun. This fluid pressure may range into the thousands of psi and so requires the fluid control to be a heavy-duty type such as an on/off ball-and-seat style control valve. A redesign of the fluid tip nozzle to a much finer orifice is also necessary. With these changes, the paint is then atomized without introducing a pressurized air flow. This type of spray gun is termed an *airless spray gun*.

The overall appearance and parts of the construction of the airless spray gun (Figure 12-1) are much the same as that of the air-atomized spray gun. The main differences are those for fluid control and the complete elimination of the air inlet, air nozzle, air valve, and fan control, all of which are unnecessary because of the absence of any air supply to the gun. Like the air-atomized spray gun, the airless paint gun also has the following:

- Handle and barrel
- Fluid control (on or off only) assembly
- Fluid inlet
- Trigger
- Fluid nozzle

**Handle and barrel.** These parts perform the same function as in the air-atomized gun: The handle allows the operator to grip the gun, and the barrel provides a support for the fluid nozzle and trigger.

**Fluid inlet.** On some guns the fluid inlet is located under the end of the barrel, and on other guns it is located at the bottom of the handle.

**Fluid nozzle.** The fluid nozzle on an airless spray gun differs substantially from the fluid nozzle on an air-atomized spray gun. The airless fluid nozzle orifice is elliptical in shape but is rated in equivalent circular diameter, typically from 0.007–0.072 in (0.18–1.83 mm). The orifice is beveled or fanned out at various angles, typically in increments from 10–80 degrees. This determines the fan spray pattern width.

170  INDUSTRIAL PAINTING

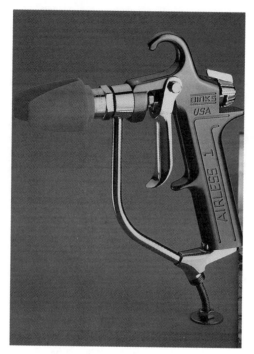

Figure 12-1. Typical Airless Spray Gun

**Fluid control (on/off) assembly.** This part starts and stops the flow of paint through the fluid nozzle. A rather powerful spring is used with a ball-and-seat valve rather than a needle-and-seat style valve. This type valve is necessary because the fluid pressure is so high—often several hundred times higher than air spray pressures. Some airless guns use a tungsten carbide ball and seat, but steel and aluminum are more common.

**Trigger.** This part gives the operator a convenient means of operating the fluid-control assembly. While triggering will control the amount of paint flow in an air spray gun, the airless gun has only an on/off capability. A trigger lock is a normal part of the gun construction that should be used to prevent the accidental release of paint from the airless spray gun.

A duckbill device (see Figure 12-2), shaped like its name suggests, or a similar device is mounted at the end of the barrel as an additional safety device. It is designed to prevent the operator from accidentally touching the high-pressure paint emerging from the fluid nozzle. The paint stream exits the nozzle with great force and can penetrate the skin and cause serious injury so this protection is vital. Manual airless guns should never be used without these in place.

Paint is supplied to an airless spray gun typically by an air-driven multiplier reciprocating piston pump. The pressure exerted by the pump is directly proportional to the ratio of the area of the pump piston and the area of the air piston. For example, if the pump piston is 20 times as large as the air piston, and if 100 psi (690 kPa) is applied to the air piston, then 100 × 20 or 2,000 psi (13,800 kPa) will be applied to the pump piston or to the paint. Frictional losses through the paint lines lower the actual pressure at the gun tip, however.

# PAINT APPLICATION IV 171

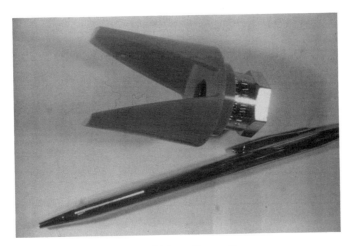

**Figure 12-2. "Duckbill" Safety Guard for Airless Spray Gun**

Airless spray paint delivery systems are of two types: dead end from pump to gun(s) and circulating. In the dead end type, paint is pumped directly from a container to the gun without any return line. In a circulating system, paint is circulated from the paint through a paint line continually. Some systems circulate paint all the way to the gun; others circulate it only through a main line, and the guns are connected to the circulating line by a dead end paint hose. A fluid filter may be located just ahead of the gun or elsewhere in the paint line to minimize plugging of the fine tips used in airless spraying.

## Gun Operation

Paint is pumped to airless guns typically at about 900–1,200 psi (6,210–8,280 kPa), although this may also range anywhere from 500–6,500 psi (3,450–44,850 kPa). When the paint exits the fluid nozzle at these high pressures, it expands slightly and encounters air resistance. Thus, it is atomized into tiny droplets without the impingement breakup from atomizing air pressure. The high velocity of the exiting paint propels the droplets toward the work being painted.

The width of the spray fan from the fluid nozzle is determined by the nozzle's fan angle. With the gun tip 12 in (30.5 cm) from the part being sprayed, the spray width (at the part being painted) may vary from about 5–17 in (13–43 cm), depending on the fan angle of the fluid nozzle being used.

Airless fluid delivery rates are high and not amenable to low volume flow. It ranges from about 25–75 oz/min (0.71–2.14 L/min) as compared to one drop to nearly one qt/min (0.95 L/min) for air spray. It is the paint pressure plus the size of the tip orifice that determines the quantity of fluid sprayed. The tip orifice size and shape affects the fan angle. While spray techniques are the major control, all of these factors help determine the thickness of the coating. Two nozzle tips having different spray angles—but the same orifice size—will deposit the same amount of paint, but over different sized areas.

## 172  INDUSTRIAL PAINTING

## Spraying Techniques for Airless Guns

Recommended spraying techniques for airless spray guns are nearly the same as for air-atomized spray guns. The basic difference relates to the high paint flow of airless guns, to the larger droplet size in airless atomization, and to the absence of compressed air pressure propelling the droplets toward their target.

The high paint flow requires a consistent spray procedure to prevent uneven paint film build and the associated problems of runs and sagging. It is extremely important in spraying to point the gun directly perpendicular to a surface and to move the gun laterally or vertically in strokes that as closely as possible are parallel to the surface. Overlapping of the spray pattern from one spray stroke to the next must be consistent to prevent fluctuations in film thickness.

The absence of pressurized air flow in the vicinity of the target allows airless guns to spray more readily into corners and hard-to-reach areas. When air-atomized spray is directed into these restricted areas, the air flow builds a cushion of air turbulence that noticeably repels the move-

**Figure 12-3. Air Spraying and Airless Spraying Demonstrated on Enclosed, Box-like Products**

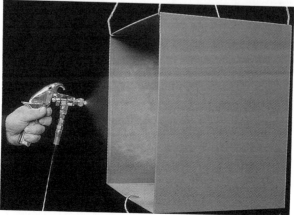

ment of atomized paint particles toward the target. Figure 12-3 compares the efficiency of an airless spray gun with an air-atomized spray gun in painting a product interior, such as a box-like part. The paint sprayed from an airless gun nicely penetrates the cavity, while the air-atomizing spray gun has considerable atomized paint blown back out of the cavity.

Figure 12-4 shows the identical phenomenon happening to a lesser degree when comparing airless and air-atomized spray on a flat surface. The high air pressure associated with an air-atomized spray (conventional) gun creates air turbulence and atomized paint *bounceback* (Figure 12-4A). The absence of air pressure along with the larger droplet size in airless spray eliminate to a substantial degree much of the turbulence and bounceback (Figure 12-4B and C).

A.

B.

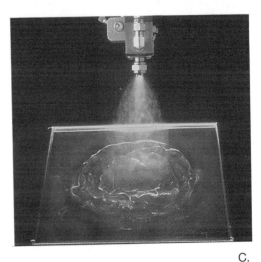

C.

**Figure 12-4. Spraying Velocity Differences in Droplet Size and Spray Bounceback Between Air Spray (A) and Airless Spray at 1500 psi (B) and 600 psi (C)**

## Advantages of Airless Spray Guns

The advantages of airless spray guns include:

- High rates of paint flow
- Relatively high transfer efficiency capability
- Gun-handling versatility
- Ability to apply highly viscous fluids

**High rates of paint flow.** The high fluid pressure and the availability of nozzle tips with various-size openings permit high rates of paint flow. This allows airless guns to be used advantageously on very high-speed production lines and where surface areas to be coated are large.

**Relatively high transfer efficiency capability.** The availability of nozzles with various fan openings helps optimize application efficiency. The absence of blowing compressed air associated with the gun operation simplifies application. Most important in the high transfer efficiency achieved by airless spray, however, is the relatively coarse paint droplet atomization, resulting in the near absence of fine spray droplet "fog" from atomizing air usage and droplet "bounce-off" from the target.

**Gun-handling versatility.** The absence of an air hose improves gun-handling versatility. One fewer hose connected to the gun lightens the gun and facilitates improved gun maneuverability, both of which reduce sprayer fatigue.

**Ability to apply highly viscous fluids.** Fluids that are simply too viscous to be applied with air spray guns are often readily sprayable with airless spray guns. Obviously, there are limits beyond which even airless cannot spray if the material is too thick to flow well in the lines. But even thick, syrupy materials, such as urethane roof sealing mastics, are frequently put on as thick as 50–65 mils (1.25–1.63 mm) by airless spray.

## Disadvantages of Airless Spray Guns

The disadvantages of airless spray guns include:

- Relatively poor atomization
- Nozzle wear
- Limited fan pattern control
- Coatings limitation
- Tendency for tip plugging
- Skin injection danger

**Relatively poor atomization.** The atomization with an airless spray gun is distinctly less fine compared to that obtained with an air-atomized spray gun. The appearance quality of the final film is noticeably less smooth, which restricts the use of airless spray application to surfaces or parts that do not require fine finishes.

**Nozzle wear.** The hundreds of pounds or kilopascals of fluid pressure used with an airless spray gun delivers a high rate of paint flow through the very tiny nozzle opening, which tends to cause wear that enlarges the orifice, increase the flow rates, and change spray pattern characteristics. This is especially true at very high pressures and particularly with paints that contain large concentrations of pigments or paints with highly abrasive-type pigments.

**Limited fan pattern control.** The only real control of the fan pattern is in the design tip put on the airless spray gun. To change fan patterns with airless spray guns requires the comparatively slow process of changing out one fluid nozzle for another.

**Coatings limitation.** To maintain the fluid pressure at the needed high pressure for atomization, the fluid tip orifice must be much finer than with air spray guns. The generally small fluid nozzle openings limit the materials that can be sprayed to those in which all components are fluid or else quite finely ground. This rules out fiber-filled coatings, coarsely pigmented paints, and similar finishing materials.

**Tendency for tip plugging.** The small orifice of the nozzle is easily plugged during spraying.

**Skin injection danger.** Probably the biggest disadvantage of airless spray is the considerable danger of injecting paint through the gun operator's skin. The high fluid pressure creates a paint stream that can possibly penetrate the skin if the gun is triggered directly against or close to the skin. Sprayers have accidentally injected paint into their fingers, hands, arms, and other parts of their bodies. As a consequence, some have required amputation of fingers, hands, and arms. Some have even died after their bodies went into shock from the physical insult of paint injection. When paint injection occurs, only a tiny opening may be noticeable in the skin. This can be deceptive, for the injury can be severe, despite the minor-appearing wound.

Total flush out and surgical removal of such injected paint is extremely difficult. The body reacts to the injection by the formation of considerable amounts of fluid that may cause further tissue damage if the pressure is not relieved. Figure 12-5 shows a tiny airless injection injury at the end of a forefinger. Note the small entrance wound and that the entire finger is swollen and discolored. To remove as much of the material as medically indicated, the finger will need to be surgically opened throughout its length, possibly causing further nerve and tissue damage.

Physicians knowledgeable in airless spray paint injection recommend immediate surgical examination. Keeping the affected area immobile after an injection will minimize the spread of the injected paint deeper into the body. Physicians report that paint injected into a finger can work its way past the wrist surprisingly rapidly if the accident victim continues to manipulate the affected hand and fingers. This obviously complicates the surgery and extends the patient's recuperation period. Unfortunately, many victims try to move their injured hands and clinch their fists repeatedly in attempts to assess the extent of damage and physical motion impairment.

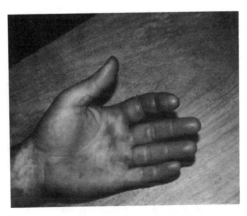

Figure 12-5. Injection Injury Caused by an Airless Spray Gun

As aids to preventing injection accidents, devices such as the duckbill have been helpful. The extremely high pressure at the gun tip decreases rapidly with distance from the fluid nozzle. A pressure of 3,000 psi (21,000 kPa) at the tip will decrease to 25–200 psi (173–1380 kPa) several inches or centimeters away. Safety tip guards should always be used. The safety on the gun trigger should always be activated when the spray operator moves the gun to a new location and at all similarly appropriate times.

Some manual airless guns are designed to be inoperable without the duckbill. However, sometimes operators will cut off the safety device because the tips of the duckbill have a tendency to collect paint that then drips onto newly painted parts. This reckless behavior is rarely tolerated in larger plants, but this author has observed it—with horror—in more than a dozen locations.

Using the duckbill is a small inconvenience for the greatly added safety that it brings. Even with the duckbill in place, operators of airless spray guns still need to use extreme caution. Sprayers should be reminded frequently that fluids under great pressure can be dangerous, as evidenced by the use of high-pressure water jets to cut thick steel plates. Nor is injection danger limited to the gun. Airless hoses that burst when being moved during normal painting operations and forced paint into the victims' hands have caused permanent physical damage. It is redundant but certainly not at all superfluous to state again with emphasis: **You must be extremely careful with airless spray equipment!**

## Air-Assisted Airless Spray Guns

The air-assisted airless spray gun looks almost exactly like an air-atomizing spray gun. The handle, barrel, trigger, and tip look the same. An air hose attaches to the handle, and a paint hose connects to the bottom of the end of the barrel.

But beyond the similarity in looks, the air-assisted airless spray gun is much different in operation than the air-atomizing spray gun. The difference lies in the amount of fluid and air pressures that are used. An air-assisted airless gun uses from about 150–800 psi (1,035–5,520 kPa) of fluid pressure and only 5–30 psi (35–210 kPa) of air pressure. The fluid pressure of an air-assisted airless gun is far greater than an air-atomized gun but considerably less than in an airless gun. The air pressure is far less than a high-pressure low-volume (conventional) air-atomized gun, and, of course, higher than in the airless gun, which uses no atomizing air pressure at all. When the assisting air is 10 psi (69 kPa) or lower, the guns fall into a special compliance category in some states, just as the HVLP-type spray guns. One manufacturer typifies its gun as a high-efficiency low-pressure (HELP) air-assisted gun.

The major difference in gun construction between an air-assisted airless gun and an air-atomized gun is in the atomizing tip. The air-atomizing gun incorporates a juxtaposed fluid nozzle and an air cap. The round fluid orifice in the center of the tip is surrounded by a concentric ring of atomizing air. Contrast that with the air-assisted fluid tip that delivers a flat fan spray of partially atomized paint. Jets of atomizing air, exiting from ports in small projections on each side of the tip (similar to the wings of air-atomizing guns), impact at a 90° angle into the spray. The air jets further break up the large coarse droplets from reduced fluid pressure and complete the atomization, "assisting" the airless spray in this process.

The tips are available with various size fluid orifices and fan angles. These may range from about 0.009–0.036 in (0.23–0.91 mm) and 15°–80° angles. With only fluid exiting the tip, an air assisted airless spray gun has a spray pattern with heavy fluid "tails" on each side of the pattern. The paint-heavy tail portions of the spray pattern are eliminated by gradually increasing the assisting air atomizing pressure.

The gun tip also has two additional air vent holes, called "shaping air holes," located outboard of the slotted fluid opening. The width of the overall pattern can be refined by varying the air flow from these shaping air ports. The choice of tips and shaping air pressure variation enables the gun to achieve a spray width from about 4–19 in (10–48 cm) at a distance of 12 in (30.5 cm) from the target.

The paint flow rate can be varied from about 5–20 oz/min (0.14–0.56 L/min) by changing the fluid pressure. The selection of tip and fluid pressure would be determined by production requirements. Paint is typically delivered to the gun from an air-driven reciprocating pump with about an 8:1 ratio of air piston to fluid piston size. Triggering is the same as with other guns. Fluid on/off control typically is with a ball-and-seat valve inside near the end of the gun.

## Air-Assisted Spray Gun Advantages

The advantages of the air-assisted spray gun include:

- Low equipment maintenance
- Good atomization
- Varied fluid delivery
- Low bounceback
- Reduced paint injection danger
- High paint transfer efficiency

**Low equipment maintenance.** The reduced fluid pressures in comparison with airless spray guns cut down on pump and fluid nozzle wear, a real but somewhat minor benefit.

**Good atomization.** The atomization quality of an air-assisted airless gun is rated as superior to an airless gun but not nearly as good as with an air-atomized gun. In actual practice, however, air-assisted airless spraying is mostly used just like airless spraying would be—not to improve atomization and thus obtain a better paint appearance, but for rapid film build and reduced paint waste.

**Varied fluid delivery.** The paint flow rates can be varied considerably from about 5–50 oz/min (0.14–1.4 L/min). This greater fluid control (airless delivers at least five times this minimum amount of paint) results in the ability to paint with more film uniformity, in reduced chances for runs and sags on small and complex-shaped parts, and with less mess and booth cleanup required.

**Low bounceback.** The extremely low assisting air pressure allows air-assisted airless guns to spray into corners and hard-to-reach areas.

**Reduced paint injection danger.** The danger of accidental paint injection into the skin with air-assisted spray is not eliminated, but it is considerably less than with the very high fluid pressures used for regular airless spraying. However, since the fluid pressure is still relatively high, it still requires great caution on the part of the operator in order to avoid any chance of an accident.

**High paint transfer efficiency.** With a low-end delivery rate of just 5 oz/min (0.14 L/min) for air-assisted spray compared to 25 oz (0.71 L) minimum for airless spray, the added flow control allows the air-assisted spray to accomplish a transfer efficiency that is even higher than airless spray (which is already significantly higher than for air spray).

# Chapter 13

# Paint Application V— Electrostatic Painting

## The Basics of Electrostatics

In spray application, the propelling force that pushes the atomized paint to the part to be coated includes various combinations of fluid pressure and air pressure. In the electrostatic application of coatings, the small coating particles are provided with an additional driving force—an electrostatic attraction that is made possible by electrically charging the coating droplets.

All types of coatings application can be designed to include electrostatically charging the particles of atomized paint. The spray coatings that were applied by the previously discussed spray methods—air atomizing with high or low air pressures, airless, and air-assisted airless spray—can be applied with or without electrostatic charging. In addition, paint can be applied by electrostatically charged disc or bell rotary applicators discussed in the next chapter.

The principles of electrostatic charging apply equally to liquid or to powder coatings; however, this chapter will deal only with the electrostatic charging of liquid coatings. Electrostatic powder charging was covered in Chapter 6 on powder coatings.

Sooner or later everyone has a personal encounter with static electricity and electrostatic charging. Who hasn't experienced a small, yet startling, electrical shock in low humidity after walking on a certain type of carpet and touching either another person or an object? Who hasn't observed that after combing his or her hair, the comb can attract small pieces of paper? Who hasn't noticed flashes of lightning? These are all personal experiences and observations of electrostatics. Scientists have discovered the following basic electrostatic principles:

- All matter is electrostatically chargeable to various degrees.
- Matter is composed of positive protons, neutrons, and negative electrons.
- Matter with an excess of electrons is therefore negatively charged.
- Matter with a deficiency of electrons is positively charged.
- Objects with like (both positive or both negative) charges repel.
- Objects with opposite charges attract.
- A grounded (neutral) object and a charged object attract.

*Static electricity* is the name given to the transmission of surplus electrons from one object to another, frequently as a spark that jumps the air gap between the objects. Static charges can accumulate on all conductive materials of any size; the larger and more conductive the object, the greater the number of electrons that can be retained on it. The surplus electrons will seek any opportunity to hop onto grounded machinery, cabinets, people, liquids, and even air itself. Liquids can easily build up surplus electrons when flowing through ungrounded pipes and by being poured or mixed in ungrounded containers. Static electricity can then jump between any nearby grounded object and the liquid, creating a spark which can easily ignite flammable solvent vapors. Safety grounding of containers holding flammable liquids is an important rule always, but especially so around electrostatic equipment of any sort.

Keeping in mind the basic electrostatic principles, it becomes easy to understand what happens in the three examples of static electricity mentioned previously. In the first example, the friction of a person's shoes moving across a carpet transfers an excess of electrons either to the person or to the carpet. The carpet and the person become oppositely charged. If the charged person fairly quickly (before the charge drains away) touches another person or a metal object, each of which has a neutral charge, the person's excess charge becomes neutralized, or equal to the charge on the other person or object. Electrons flow from one object to another to neutralize the charges, sometimes with a visible spark. This is less likely in humid weather because electrical charges are drained away rapidly by the water in moist air.

The forces of electrostatic attraction and repulsion are noticed more strongly, of course, on objects that are low in weight. In the second example, running a comb through hair results in its becoming charged. When the charged comb is brought near to the neutrally charged pieces of paper, the difference in the charge on the comb and paper creates an attraction, and the paper is drawn to the comb. After making contact, the charges tend to become neutralized, and the paper will fall from the comb. In the third example, clouds moving across the face of the earth similarly can become so highly charged that finally the excess electrons jump between the clouds and the ground in a dramatic lightning strike.

## Electrostatics in Painting

In electrostatic painting, each of the droplets in the atomized paint cloud are given an excess of electrons, thus, every droplet becomes negatively charged. The part to be painted is grounded through the hook and conveyor, and so is electrically neutral, resulting in electrostatic attraction forces between the neutral (grounded) part and the negative paint droplets.

With an electrostatic spray gun, the droplets pick up the charge from an electrically charged electrode at the tip of the gun. The charged droplets are given their initial momentum from the fluid pressure/air pressure combination. As the charged droplets approach the electrically neutral part, the charge tends to attract the droplets toward the part. This attraction toward the part reduces the number of droplets that would otherwise travel past the part, increasing transfer efficiency. The attraction is so strong that many charged paint droplets hurtling past the part will actually curve, turn around, and be drawn back to the part (wraparound), as shown in Figure 13-1. This tendency allows electrostatic painting to fully coat the edges of a flat part and even paint a portion of the part facing away from the spray gun.

PAINT APPLICATION V    181

**Figure 13-1. "Wraparound" Effect in Which Electrostatically Charged Paint Droplets are Attracted to the Front and Back of a Grounded Target**

Charging the spray gun tip is achieved by an electrical power supply that continuously provides a source of electrons. The electrons are pushed by the power supply to a needle-like electrode at the gun tip, as shown in Figure 13-2. With the power supply energized, the excess of electrons leaks off the tip of the electrode to the air in the immediate vicinity, creating a charged cloud of air molecules. This is called the *ionized air cloud.* When paint begins to flow and atomized paint droplets are ejected from the gun tip, they must pass through the charged air cloud. In doing so, the previously neutral paint droplets pick up an excess of electrons.

The electrical pathway is typically completed as follows: The part to be painted is suspended from a metal hook or hanger attached to an overhead metal conveyor. The conveyor is electrically grounded since it is connected to the building's steel supports that connect to the foundation in the earth under the building. The negative side of the power supply is connected to the gun electrode, and the positive side of the power supply is connected to ground via the building's steel beams and foundation. Thus, the electrons drawn from the ground by the power supply travel to the gun electrode, to the ionized air cloud, to the paint droplets, to the part being painted, to the conveyor, to the building's steel beams and foundation, and back into the ground from where they started their journey. An electrical power supply can be viewed as a generator or pump of excess electrons. All electrons that flow through the power supply eventually make the external electrical circuit complete and return back to electrical ground, that is, the earth itself.

Except for the metal handle, electrostatic guns are constructed in large measure of nonmetallic materials. This is especially necessary where the needle electrode protrudes from the gun tip,

# 182 INDUSTRIAL PAINTING

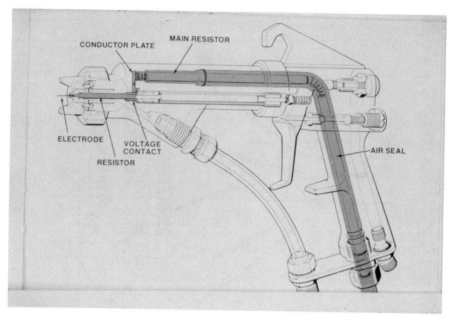

**Figure 13-2. Electrostatic Gun**

which is physically separated from the handle by a long insulated plastic barrel. Nonconductive plastic is necessary so that these parts do not hold electrostatic charges and do not carry charges from the ionized air cloud around the needle electrode at the gun tip back to the grounded gun handle. This avoids both arcing and shorting to ground.

The gun's electrode is in the shape of a needle to help the excess of electrons from the power supply drain to the air molecules that comprise the charged cloud. It is the nature of a sharp point or edge to easily allow electrons to be attracted to it or drain away from it.

The quantity of flow of electrons from a gun's electrode during electrostatic painting is intentionally extremely small, frequently in the order of 5–100 millionths of an ampere (5–100 microamperes). An *ampere* is a unit of current flow and is a measure of the strength of electron flow. An energized 100-watt light bulb has about 1 amp flowing through it. A lightning bolt may have a flow of millions of amps. A part being electrocoated (Chapter 10) may draw 100–300 amps.

The force in an electrical power supply that provides the push for the electron flow is called *voltage*. A residential power outlet has about 110 V. A bolt of lightning may have millions of volts. A typical electrostatic power supply for painting operates in the range of 30,000–120,000 V (30-120 kilovolts or kV).

## Grounding and Safety Precautions

In electrostatic terminology, a building's steel frame is said to be grounded. This is because the frame is either embedded into the ground or is physically connected to something that is embedded. The term *grounded* means that the object is electrically neutral with respect to the ground (earth).

One of the important requirements for success in electrostatic painting is to be sure that the positive side of the power supply is grounded and that the part to be painted is grounded. The positive electrode of the power supply is hardly ever a problem because it is connected to ground by a tight mechanical connection such as a bolt or a soldered welded connection. The part to be painted may not be properly grounded, however, due to a poor electrical connection between the part and the hanger, between the hanger and the conveyor, or between the conveyor wheels and the conveyor I-beam. For efficient electrostatic painting, all of these connections must be electrically sound. If they are not, electrical current flow will be restricted, and poor electrostatic attraction of paint droplets will result.

For top efficiency, the part to be coated should be the closest grounded object to the charged paint spray droplets coming from the spray gun. The charged paint particles are attracted to the nearest electrically grounded item; the larger the item, the greater the attraction.

Paint buildup on hooks and hangers can act as an insulator and block the flow of electrostatic charges from completing the circuit to the conveyor and on to ground. The greater the paint buildup on the hook, the more severe the problem. If the grounding loss is only slight, wrap-around may be only partially reduced. When paint buildup on hangers is heavy, the paint droplets that hit the part cannot lose any of their negative charges. From the repulsion forces between like charges, the incoming negatively charged particles are actually pushed away by the negative charges already on the part. The result is thin paint. These parts may fail to pass inspection because they have film builds below specification.

Ungrounded objects in the vicinity of the charged gun electrode can accumulate electrons and pick up a considerable electrical charge. The charge buildup can then arc over (spark) suddenly if a grounded object is brought near. The intense heat of the arc may be sufficient to ignite solvent vapors coming from the freshly painted parts, and fire can then spread to other items in the spray booth. For fire safety reasons, grounding of all persons and equipment—of everything in the booth—is critically important!

Should complete loss of grounding be experienced, the danger of arcing becomes great, and fires can result. Numerous instances can be cited where fires have occurred this way. To ensure good electrical connection between parts and hangers, some plants use square rather than round rods for hangers. The rod is formed into hangers edge-up so that a sharp edge meets the hung part to be painted. The sharp edge tends to cut through any paint that may accumulate on the rods. Hangers and hooks should be regularly stripped or otherwise cleaned of paint buildup to maintain good grounding contact to both the parts and the conveyor.

People in the vicinity of the charged electrode (or any part of the spray gun) should also be grounded. Spray paint operators should wear conductive footgear or at least leather-soled shoes to drain charges to ground. In some plants, operators are required to wear conductive safety shoes or to attach a conductive, carbon-impregnated rubber strap around one ankle above the sock. The strap trails on a metal grate to make certain the sprayer is continuously grounded. Sprayers need to have a gloveless (bare) hand on the grounded spray gun handle, a special conductive electrostatic sprayer's glove, or a glove with its palm removed. The metallic handle of the gun is also grounded through the power cable and will ground the sprayer if the skin is in direct contact with the gun. Metal rings, watches, and other jewelry worn next to the operator's skin will not accumulate electrons on a grounded sprayer. These precautions will ensure that the operator does not accumulate electrostatic charges.

To complete the electrical circuit satisfactorily, the part to be coated must be an electrical conductor. The part need not be a highly efficient electrical conductor since the amount of charge is small, but at least some conductivity is essential. This is required to maintain current flow and to prevent a charge buildup on the part, which would repel additional charged paint droplets.

The conductivity of the paint can affect the path of the electron flow from the electrical power supply, to the gun's electrode, to the part, and back to ground. Solventborne paints tend to be rather poor conductors; but waterbornes and some metallic paints are excellent conductors. This increased paint conductivity, however, introduces a new and shorter pathway by which the electrons can reach ground. The conductive paint can carry the full supply of electrons reaching the electrode back to the fluid in the gun tip and then further back through the paint line to the grounded paint tank. Thus, the spray droplets never have an ionized air cloud to pass through; all the charges follow the short circuit to ground. If this occurs, electrostatic painting is not possible.

To prevent highly conductive paint from shorting (grounding out) the system through the paint hose and the paint pot, it is necessary to isolate the hoses and paint supply from ground using nonconducting plastic supports. Once the isolated paint supply gains enough electrons, the additional electrostatic charges flow off the gun tip as with poorly conducting paints. This isolation of the paint supply from ground is called an *isolated system* and is one of the best ways to enable electrostatic application of conductive paints. (Isolated systems should never be used with solventborne paints because the fire danger from current arcing to ground would be extreme with such materials.)

Notice that an isolated system still permits electrons to travel back to the pressure pot, turning it into a very dangerous source of a potential high voltage electrical shock to persons in the vicinity. As a safety precaution with isolated electrostatic application, the paint line and pressure pot are confined in a caged area. The poorly conductive solventborne paints do not transfer the electrostatic charges back to the pressure pot.

To ensure that solventborne paint remains a poor conductor, paint formulators recommend using nonpolar solvents. Polar solvents tend to be conductive because of the nature of their covalent and ionic atomic bonds. Nonpolar solvents tend to be poor conductors. Examples of nonpolar solvents are mineral spirits, VM&P naphtha, xylol (xylene), toluol (toluene), and N-butyl acetate. Examples of high-polarity solvents are acetone, methyl ethyl ketone, isopropyl alcohol, methyl cellosolve, diacetone alcohol, ethyl alcohol, methyl alcohol, and methyl acetate.

The electrical hazard associated with electrostatic conductive paint spray tends to cause some plants to limit it to automatic application where operators are not involved. A manual electrostatic gun used to spray conductive paint must, of course, not permit current to reach the person spraying paint. While the isolated fluid becomes charged and carries the full supplied voltage, the gun handle held by the sprayer is grounded and insulated from the charged paint. Although with the proper equipment it is possible to safely spray conductive coatings manually, for sprayer protection the manual systems operate at lower voltages and amperage draws than the automatic systems.

The electrical shock and fire dangers inherent with high voltages used in electrostatic paint spraying have led to the development of devices to lower the voltage at the gun's electrode when a person approaches or as the electrode approaches an electrical ground. This prevents electrical shock and eliminates the chance of a spark starting a solvent fire.

A number of patented protective systems are used by various equipment manufacturers in their power supplies as safety devices to restrict the total amount of current that can flow. Some of these current-limiting devices proportion voltage and current. If the current begins to rise, the voltage drops correspondingly. This is known as a *resistive system*. Figure 13-3 depicts how the voltage at the tip of a current-protected electrostatic gun declines to zero as electrical current flow increases. This occurs because the increasing current flow causes the device to decrease the voltage at the gun tip. A so-called "stiff" system maintains the voltage until a preset current flow limit is reached. At this point the circuit is tripped off. Generally, stiff systems are not approved for use with manual spray guns but are appropriate for some automatic electrostatic spray operations. Other sophisticated protective devices measure how fast the change in current draw occurs and can shut off all current almost instantly, even though current draw has barely begun to rise.

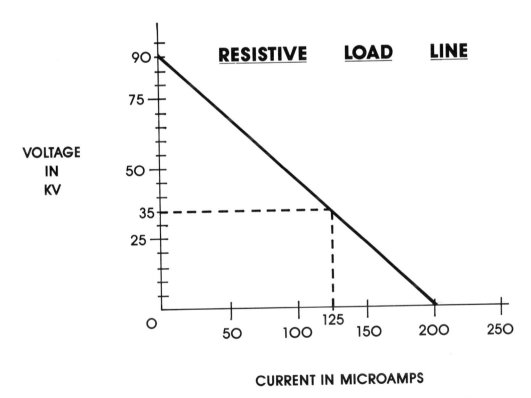

Figure 13-3. A Safety Feature in the Power Supply Reduces Voltages As Current Draw Increases

# 186  INDUSTRIAL PAINTING

## The Effects of Humidity

Recall that two of the examples of static electricity—the charging of the comb (after combing one's hair) and a person's body (after walking across a carpet)—tend to be more pronounced in low humidity. This is because the highly polar water molecules in humid air readily accept electrons. Thus, humid air will tend to speed the leaking away of charges from the comb and from a person's body. In the case of lightning, the air in the vicinity of a thunderstorm tends to be humid, increasing the tendency of a cloud's charge to discharge to earth or to another cloud in the form of lightning.

Humid air in the vicinity of a gun's electrode tends to increase electron flow away from the electrode. This is why electrostatic painting is sensitive to humidity fluctuations. Amperage draw from the electrostatic gun is often noticeably higher in humid weather.

## Advantages of Electrostatic Spray

The advantages of electrostatic spray include:

- Higher transfer efficiency
- Good edge coverage
- Paint wraparound
- Uniform film thickness

**Higher transfer efficiency.** It is a rule of thumb in paint spraying that electrostatics will increase transfer efficiency by 10%–20%. Much of this is due to the wraparound effect.

**Good edge coverage.** Excellent edge coverage is achieved with electrostatic painting because electrons (and droplets with electrons) are attracted more strongly to edges than to flat surfaces, as shown in Figure 13-4. Sometimes this can be a mixed blessing. Occasionally, so much paint is drawn to edges that they will undergo solvent popping in the oven due to excess paint buildup. At times edge buildup can be so severe that paint runs are caused. As would be

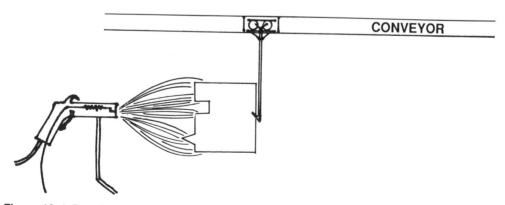

Figure 13-4. Drawing of Electrostatic Force Lines Showing Attraction to Product Edges

expected, this is most likely to cause problems on parts with vertical edges. Turning off the electrostatic voltage prior to the spray reaching the edges and altering the spraying technique can avoid such runs.

**Paint wraparound.** The attractive force of electrostatics will to some extent draw paint droplets around to the back side of the parts, a phenomenon called *wraparound* or simply *wrap*. This reverse side coating will extend around on parts about 1–3 in (2.5–7.6 cm) away from the edge, with diminishing thickness. Because of the wraparound effect, complete coverage of small "C"-shaped parts is often possible with just a single gun pass.

**Uniform film thickness.** A uniform coat is somewhat self-produced because the incoming paint droplets experience the best grounding where the paint is the thinnest. As the paint wet film thickness builds, it forms an electrically insulating layer. Therefore, newly arriving paint droplets are attracted most strongly to areas on the part where grounding is best, that is, the spots where the paint is the thinnest, as shown in Figure 13-5. In this way an excellent overall film thickness uniformity results.

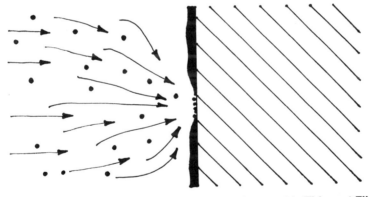

Figure 13-5. Grounding and Electrostatic Attraction Are Greatest in Thinnest Film Areas

## Disadvantages of Electrostatic Spray

Although the advantages of electrostatic spray are impressive, the disadvantages also tend to be pronounced, and relate to:

- Safety/fire hazards
- Gun construction
- Equipment cleanliness
- Faraday cage effect
- Conductivity
- Metallics
- Solvent selection

**Safety/fire hazards.** These are lumped together as a disadvantage because they tend to go together. A high voltage is hazardous and can cause a spark that could ignite a fire in a solvent-laden atmosphere.

The phenomenon of *capacitance* enables any ungrounded object that can conduct electricity to store extra electrons (at least temporarily). The better the conductor and the larger the mass, the greater the quantity of electrons that the object can store. Thus, everything near the tip of the gun must be grounded to prevent accumulating electrostatic charges. This includes machinery, personnel, and anything that can build up enough charge so that arcing of electrons to ground can occur. The connection to ground provides a path for the electrons to flow from the objects. If ungrounded, charge accumulation can continue until suddenly the electrons jump to a grounded object in the form of a spark. The energy of the spark, as the electrons drain off in a miniature form of lightning, can be more than enough to ignite solvent fumes. For this reason, spray gun operators should be cautious about ungrounded metal items on their person. Coins and keys in pants pockets, metal pens in shirt pockets, metal buttons, large decorative metallic belt buckles, and similar items can cause this danger, although in these cases, the charge usually discharges through the fabric of clothing to the wearer's body in a very noticeable shock.

Ignition of paint caused by improper operations during electrostatic application are not at all uncommon. A number of people, almost always painters, have died from the resultant fires. When the proper safety procedures are followed, fires will not occur. But the danger is always there and must be recognized. Unfortunately, doing things properly hundreds of times does not afford protection the one instance when a mistake is made.

**Gun construction.** Electrostatic guns tend to be bulky and delicate. Their bulkiness reduces maneuverability; their relative susceptability to physical damage compared to nonelectrostatic guns requires more careful handling. The bulky feature is inherent with the long barrel.

Bulkiness is increased by the necessity of some electrostatic guns to have an attached high-voltage power cable. The higher the voltage carried in the cable, the bulkier is the cable. Some guns require the full high voltage to be brought to the gun. Others utilize a low-voltage cable and rely on electronic transformers within the gun to step up the voltage. Other electrostatic guns do not require an electrical cable at all because they have an air turbine inside the gun which generates the high voltage, as illustrated by Figure 13-6.

The needle electrode protruding from the gun tip can be damaged easily. Therefore, the gun requires careful handling. A bent electrode will reduce the efficiency of the electrostatic charging. In addition, the relatively soft plastic parts, such as the fluid tip and air cap, tend to wear and require frequent replacement.

**Equipment cleanliness.** Extra cleanliness is essential for efficient electrostatic painting operation. This is important for hooks, hangers, and for the outer surfaces of the paint applications devices. Dirt or oversprayed paint can form a conductive track on the surface of the plastic gun tip and barrel back to the grounded metal handle, thereby shorting out the system. Sometimes people think they are painting electrostatically; but because of accumulated overspray on the application device, the system actually is not applying paint with any appreciable electrical potential. Sometimes spray operators, not recognizing the problem, will simply turn off the power supply and use the electrostatic gun as a nonelectrostatic gun because no electrostatic

**Figure 13-6. Cartridge Electrostatic Spray Gun**

charging appears to be taking place. When this occurs, the reason for the lack of charging taking place is obvious almost every time: dirty guns (charge leaking to ground) and/or dirty paint hooks and hangers (charge not getting to ground) are the ordinary suspects.

**Faraday cage effect.** Electrostatic application is not able to coat recessed areas as well as nonelectrostatic application. This is due to a phenomenon termed the *Faraday cage effect*, as shown in Figure 13-7. Charged paint particles always seek the nearest grounded surface. Thus, when coating a recessed area, the charged droplets tend to be attracted toward the sides of the recess, and especially toward sharp edges that may be present, instead of traveling to the bottom of the recess. Nonelectrostatic manual touch-up may become necessary to get coating into these problem areas. A common procedure is conveyorized painting first with automatic electrostatic application, followed by manual nonelectrostatic touch-up. Faraday caging becomes severe at high voltages. Some degree of Faraday caging is inherent in electrostatic application and cannot be avoided. For this reason not all parts are suited for electrostatic application. This is particularly true for parts with complex geometries, sharp edges, points, deep ridges, or cavities.

**Conductivity.** In electrostatic paint application, the parts being coated must be somewhat conductive to complete the electrical circuit. The part need not be conductive throughout its construction; only the outer surface needs to be conductive. The surfaces of nonconductive materials, such as most plastics and wood, can be made conductive by the application (by dipping, usually, or spraying) of alcohol that contains ionic organic salts or with water that contains potassium and calcium chlorides. Surface conductivity is achieved by the ionic material remaining on the parts after the alcohol or water evaporates. Precoat applications for conductivity are neither costly nor time-consuming. However, it is nevertheless an extra step in the painting process and an operation to be avoided if possible.

# 190 INDUSTRIAL PAINTING

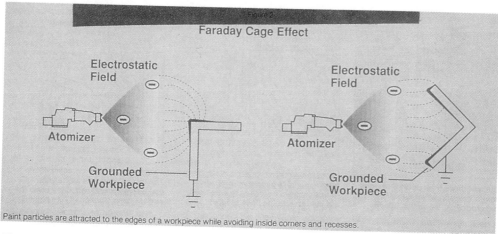

Figure 13-7. Faraday Cage Effect

**Metallics.** Metallic paint contains millions of tiny metal particles (usually aluminum flake or aluminum powder) that are mixed uniformly throughout. When metallic paint is atomized and applied without electrostatic charging, the particles tend to lie flat or "flop" on the surface being coated. This flat orientation provides maximum metallic surface area for reflecting light, giving a bright, sparkled effect.

When metallic paint is charged electrostatically, most of the charge is held by the metallic particles, which are the most conductive portions of the atomized droplets. This extra charge tends to make the metallic particles stand on edge (instead of lying flat as in nonelectrostatic application). Standing on edge helps the particles release their negative charge. The edge orientation reduces the surface area of the particles that can reflect light, causing a noticeable darkening compared to surfaces painted nonelectrostatically, as shown in Figure 13-8.

Metallic coatings are used extensively on automotive finishes, as high as 90% for some models. Automotive metallic topcoats are usually applied on the assembly line with nonelectrostatic spray for the final color coat because it gives a brighter, more appealing appearance to the finish. To a lesser extent it is done nonelectrostatically also because almost all automotive repair shops tend to use nonelectrostatic spray. If a car were given a metallic finish on an assembly line with electrostatic guns and then later be "spot" repaired in a body shop with nonelectrostatic guns, even with the same identical paint, the metallic sparkle of the repaired spot would be very noticeably different in appearance from the rest of the car.

The electrostatic darkening effect of metal flake and the increasing economic and environmental desirability of electrostatic application have led some paint manufacturers to use mica flakes as a metal flake replacement. Mica, a complex sodium aluminum silicate identical in chemical composition to asbestos but with none of the health hazards, is a nonconductor and does not sink into the wet film or orient itself on edge during electrostatic application as do metal flakes. This prevents mica flakes from exhibiting unwanted darkening when applied electrostatically.

Pearlescent effects can be achieved with mica. Pearlescence refers to the multiple-color rainbowlike appearance noticed in a thin film of oil on water, an effect much used in recent years.

## METALLIC FLAKE ORIENTATION
### (EXAGGERATED FOR EFFECT)

**WITHOUT ELECTROSTATIC**

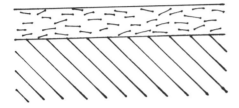

FLAKE IS DISTRIBUTED UNIFORMLY AND PARALLEL TO THE SURFACE

**WITH ELECTROSTATIC**

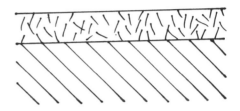

FLAKE GOES DEEPER AND ALIGNS PERPENDICULAR THE SURFACE

Figure 13-8. Comparison of Flake Orientation in Nonelectrostatic and Electrostatic Application of Metallic Paint

**Solvent selection.** The solvent selection is considerably more critical with electrostatic painting. The paint must be polar enough to accept electrons well. If the paint is too low in conductivity, the droplets are unable to gather electrons as they travel through the ionized air cloud, and the electrostatic effect is minimal. With "dead" paints, adding a minor amount of polar solvent can correct this shortcoming. However, this is rarely required. Nearly all coatings have no need for additional polarity. In fact, many high-solids paints are inherently almost too polar due to the high reactivity of the resin molecules. For this reason the preferred solvents for electrostatically applied paints are the nonpolar or slightly polar solvents. Paints that are polar accumulate so much charge while passing through the ionized air cloud that they cause excessive Faraday cage difficulty.

# Chapter 14

# Paint Application VI— Rotary Atomizers

## Introduction to Rotary Atomization

Rotary atomization involves the breaking up of paint into atomized droplets using a spinning device. Instead of air or fluid pressure to atomize the paint, as with spray guns, rotary atomizers called *discs* and *bells* use what may be referred to as centrifugal force, which tends to project an object outward from the center of rotation.

Some spray guns use electrostatic charging while many others do not. Rotary atomizers, however, always use electrostatic charging except in rare instances. With spray guns the electrostatic charging is not believed to have much effect on the actual atomizing; with rotary atomizers, electrostatic charging can play a key role in the atomizing. The older rotaries showed a sharp reduction in droplet size when no electrostatic charge was used. Figure 14-1 shows a low-speed disc painting room air-conditioner housings. With the high-speed rotation of the more modern versions, however, almost all the atomization is the result of rotation. Droplet diameter size decreases only minimally when the electrostatic charge is turned off.

The rate of rotation of low-speed discs is typically 900–8,000 rpm. Figure 14-2 shows low-speed bells, which rotate from 1,000–10,000 rpm, and feature shrouds to catch solvent flushing during cleaning and color changes. The rotation of high-speed discs and bells can be varied from about 10,000–60,000 rpm. Figure 14-3A through C shows three different models of high-speed bells. Summing up the difference between the low- and high-speed rotary atomizers, the low-speed units tend to be rather large and rotate fairly slowly; the high-speed rotaries tend to be small and spin very fast.

In 1956 Ransburg first offered a rotary disc called a number 2 process disc. General Electric was the first company to use it in manufacturing shortly thereafter, painting refrigerator shells with this then-new device. These old, slower rotaries are no longer used much because they only give fair atomization even with low-solids paints. They are just not able to handle the higher viscosities of today's high-solids coatings that can be applied readily with the newer high-speed discs.

194 INDUSTRIAL PAINTING

Figure 14-1. Low-Speed Disc

Figure 14-2. Low-Speed Bells in a Paint Line

# PAINT APPLICATION VI    195

Figure 14-3. Three Different Models of High-Speed Bells

All modern rotary atomizers are alike in that they feed paint to the center and hurl paint droplets off the perimeter of a spinning device. However, they differ in two principal ways:

- Shape of the atomizing device
- Mounting configuration

## Shape of the Device

Rotary atomizers are usually made from a quality aluminum alloy and come in two basic shapes: discs and bells (cups). The discs, as their name suggests, are thin, relatively flat, and round. "Dish-shaped" is a good description. Discs come in diameter ranges from about 4–8 in (10–20 cm).

The bells are not, however, shaped like a bell. Their shape more closely resembles a cup, truncated cone, or shallow sauce dish. Figure 14-4 shows a bell rotary atomizer in use; Figure 14-5 illustrates how it works. The bells range in diameter from about 1–5 in (2.5–13 cm). Newer bells are constructed of plastics coated with a thin conductive coating.

## Mounting Configuration

Discs are mounted in a horizontal plane to a vertical shaft and rotational drive that is generally attached to a *reciprocator*, a device that moves the rotating disc slowly up and down as shown in Figure 14-6. Parts to be painted by a rotary disc are hung from an overhead conveyor

196  INDUSTRIAL PAINTING

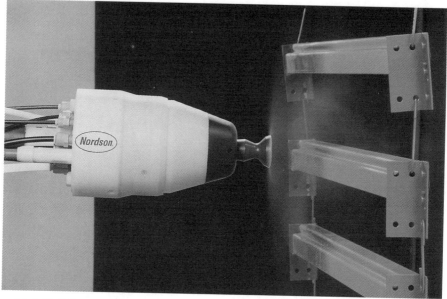

**Figure 14-4. Bell Rotary Atomizer**

that is looped around the disc, the parts being about 16–20 in (41–51 cm) from the disc perimeter. The vertical reciprocating stroke distance is determined by the length of the parts to be painted. Extremely long vertically hung parts, such as extrusions, may incorporate floor- and ceiling-mounted discs, each with a painting stroke approximately half the length of the parts being painted. Very small parts may require no reciprocation.

Bells can be mounted in a vertical (see Figure 14-7) or horizontal plane, or at any angle in between. They can be mounted in fixed positions, onto vertically or horizontally operating reciprocators, or onto robots. If a reciprocator is used with bells, the stroke is determined by the size of the part being painted.

Whereas a disc requires parts to be conveyed around it, a bell can function somewhat like a spray gun and be aimed at the part being painted. This allows bells to be used with conveyors much like spray guns. The typical distance from a metal bell to the part being painted is about the same as with discs, but plastic bells can be set much closer without danger of electrostatic arcing.

## Rotational Speed and the Degree of Atomization

As a general rule of thumb, the faster the rotational speed, the greater the centrifugal force and the finer the atomization. However, two exceptions stand out: rotational speeds below 2,500 rpm and between about 5,000 and 15,000 rpm. Figure 14-8 graphs atomized particle size as a function of disc or bell rotational speed. (Figure 14-9 compares atomized particle delivery velocities and particle sizes for air spray guns and bells.)

Atomizational efficiency occurring with low speed rotaries is due both to centrifugal force and the "pushing apart" of paint particles due to electrostatic charge repulsion. The low speed

PAINT APPLICATION VI    197

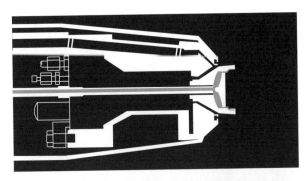

1. Coating material is fed through a tube in the center of the turbine shaft and flows into the rear of the cup.

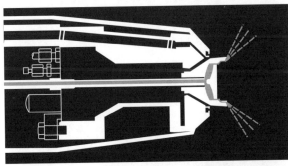

2. Spinning at speeds up to 40,000 RPM, the coating material exits through small holes in the cup and is centrifugally atomized into very fine particles.

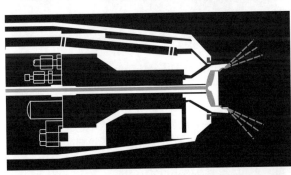

3. Shaping air is directed from behind the cup to provide excellent pattern control. It is also used to increase penetration into corners and deep recesses.

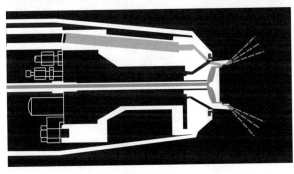

4. At the cup, an electrostatic charge is supplied to the coating material, creating a strong attraction between the paint particles and the grounded part being coated. This provides maximum transfer efficiency, uniform coverage and excellent wrap onto part surfaces not directly in the spray path.

Figure 14-5. Bell Rotary Atomizer Operation

198  INDUSTRIAL PAINTING

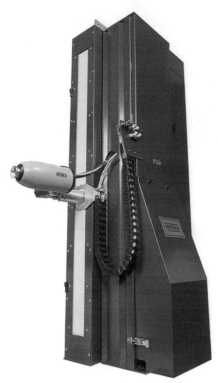

**Figure 14-6. A Reciprocator Moves the Rotating Atomizer Up and Down While Painting**

centrifugal force isn't great enough to hurl the paint off the perimeter in finely atomized droplets, but only as coarse particles of paint. It is the electrostatic charge that atomizes the paint to a finer droplet size. This is termed *electrostatic atomization*. Without the electrostatic charging, atomization is not effective enough to use low-speed rotation to apply coatings.

Atomization quality on low-solids paints with low-speed electrostatic discs and bells is intermediate between the extremely fine atomization of air spray guns and that of airless spray guns. The low-speed rotary atomization was used very successfully on low-solids coatings for some years before EPA regulations brought about the development of high-solids coatings. Now they are disappearing, having been supplanted by the high-speed versions.

Atomization is fair at speeds of 2,500–6,000 rpm (low-speed discs) and excellent at about 15,000 and higher (high-speed discs). Strangely, atomization occurring between 5,000–15,000 rpm tends to be extremely poor. This is because the centrifugal force hurling paint off the disc forms stringlike paint filaments instead of discrete droplets. The filaments tend to intertangle and be flung from the rotating head as large sized "blobs" rather than droplets. As spin speed increases to above 15,000 rpm, the long filament formation ceases. Atomization is accomplished almost totally by centrifugal force at these speeds; droplet size decreases steadily as

Figure 14-7. Vertically Mounted Bells Apply a High-solids Singlecoat Epoxy to a Metal Tool Chest

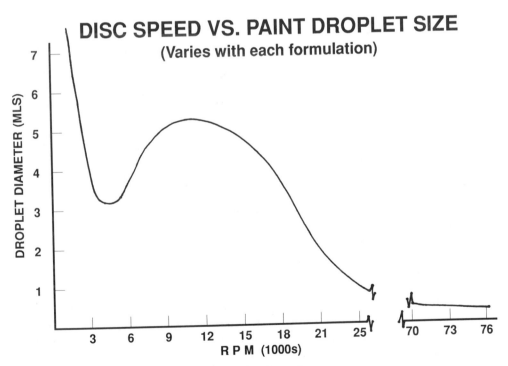

Figure 14-8. Droplet Size As a Function of Disc Speed

| Method | Typical Particle Delivery Velocity | Average Particle Size | Particle Size Range |
|---|---|---|---|
| Airspray | 10 Meters/Second | 3.0 Mils (Diam.) | 0.5-5.0 Mils (Diam.) |
| Bell | 0.7 Meters/Second | 0.8 Mils (Diam.) | 0.6-1.0 Mils (Diam.) |

Figure 14-9. Comparison of Particle Delivery Velocity, Average Particle Size, and Size Range for Air Spray Guns and Bells

rpm's increase. The function of the electrostatic charge at these high rotational speeds is no longer helping atomization, but only placing a charge on the atomized droplets so they will be attracted to the grounded parts being painted.

High-solids and waterborne paints are difficult to atomize at slow rotational speeds because of their high surface tension. High-speed discs and bells are almost always operated at above 20,000 rpm to ensure a fine atomization. Speeds up to 60,000 rpm are common. The particular operating speed is selected for the degree of atomization desired, the product being coated, the characteristics of the paint being applied, and the rate at which paint is fed to the rotor. As expected, the higher the viscosity and the flow rate, the faster the speed necessary for atomization.

## Paint Application

The atomized paint leaving a rotating disc is hurled outward in a 360° pattern. The droplets are guided to the parts to be painted only by electrostatic attraction. No directional air is forthcoming from the disc vicinity to give impetus to the atomized particles. With electrostatic attraction as the major propelling force, the air in disc booths must be relatively calm. Strong air currents would tend to carry the atomized paint droplets away from the parts to be painted.

With the bells, directional air concentric with the bell is used to reduce the circular size of the atomized paint particle cloud, helping the electrostatic attraction guide the particles to the parts to be painted. The air, varying in pressure from about 10–40 psi (69–276 kPa), is directed through a circular groove or a series of small, closely spaced holes around the bell perimeter. The air forces the atomized particle cloud into a smaller "doughnut" of droplets whenever a reduced size pattern of paint delivery is needed. With small parts the shaping air helps considerably to reduce the atomized pattern size and thereby minimize paint waste. The shaping air does not assist in atomization or delivery of paint, but only controls the size of the delivery pattern.

The discs and bells are insulated from ground to accept the electrostatic charging voltage (about 100,000 V) and transfer these charges to the paint as it travels across the rotating head. As with electrostatic spray guns, the parts to be painted should be electrically grounded to establish maximum electrostatic attraction. The parts to be coated, being the nearest electrical ground, attract the electrostatically charged particles. The electrical circuit is completed as with electrostatic spray: The paint particles give up their charge to the grounded parts on contact, and the negative charges flow to ground through the grounded hooks and conveyor.

## Configuration of the Disc System

Various arrangements are possible with disc systems to bring about painting of the total part. Unless the part is narrower than about 1 in (2.5 cm), only the side of a part facing the disc will be completely painted as it is conveyed around the disc. The back will get some paint on it from the electrostatic wraparound. Both sides of flat parts can be completely painted by incorporating two booths with reciprocating discs and by having the conveyor line form a loop around each in a "figure S" configuration. Alternatively, parts can be painted, then rotated 180° and passed through a second disc booth. A circular part, such as a water heater jacket or propane tank, can be painted all around by automatically rotating the part slowly on its hanger as it is conveyed around a single reciprocating disc.

Flat parts to be coated by being conveyed around a disc must not be excessively wide or else the leading and trailing ends of the part will receive less coating than the center. This is because the part must be conveyed around a circular path. The maximum width of a part that can be painted by being conveyed around a disc is about 4 or 5 ft (1.2 or 1.5 m).

Disc application booths are shaped like a letter "C," that is, circular in shape with a notch out. Up to about one third of the circular booth is open for the conveyor entry and exit. The conveyor to and from the entry and exit portions are linear, but the conveyor curves in, then circles in the opposite direction around the disc. This configuration of the conveyor yields a shape resembling the capitalized Greek letter omega, and so it is often called an *omega loop*.

The air exhaust from disc booths must be at a low intensity to prevent generating strong air flows within the booth which could interfere with the attraction of the charged atomized paint particles to the parts being conveyed around the disc. The booths are always the dry-filter type; the lower portion of the circular wall of the booth consists of paint filters. Air is drawn in through the open ceiling and booth opening, and it is exhausted through the filters.

## Disc and Bell Rotation

Low-speed discs and bells are spun by electric motors. The high-speed discs and bells, however, are rotated by an air-driven turbine. They operate on the same principle as a high-speed air-driven dental drill. Much higher speeds can be achieved with turbines than with an electric motor drive.

The high-speed rotary units require disc and bell turbine drive systems of exceptionally fine quality. Bearings that support the rotation are of two types: mechanical and air. The mechanical bearings must be precision-machined and lubricated to withstand the high rotational speeds. Some manufacturers still sell mechanical bearing disc and bell rotary applicators, but the air bearing types are preferred and will provide superior bearing life if treated properly.

*Air bearings* substitute a steady stream of air for the mechanical bearings, and the air prevents the rotating element from touching the housing. Some manufacturers of air bearings insist that air be turned on to support the shaft **at all times**, even when the device is not in use. This prevents the weight of the shaft from causing an indentation deformation that would adversely affect the unit for obvious reasons.

A frequent cause of high-speed rotating system failure is poor air quality and dirty bearings. Insufficient filtration of the air and failure to replace filter elements on schedule will permit oil, moisture, or particulates to gel or harden the lubricants in mechanical bearings. These contaminants can build up in air bearings as well and eliminate the thin cushion of air between the sleeve and the rotating shaft.

The dynamic balance of high-speed bells with mechanical bearings is of crucial importance. Any abnormality can disturb the balance at high speeds and strain the bearings, shortening their life. These defects include nicks and dents in the disc or bell heads caused by improper handling and uneven paint buildup. An imbalance with air bearings generally has no adverse effect until it causes the rotating elements to contact the housing. Then bearing wear becomes rapid.

To preserve rotational balance, exercise care in handling and cleaning discs and bells. In some plants normal booth maintenance includes spraying most of the booth equipment with cleaning solvent. Other plants use a cloth or brush and a pail of solvent to remove overspray from rotary units. Either of these practices can cause problems by forcing paint buildup into the bearings. Solvent getting into mechanical bearings will wash away the lubrication and also can carry paint solids into the bearings. Nearly all rotaries have an air barrier to protect the bearings from paint or solvent incursion. This air should always be turned on during cleaning operations, but even this may not provide complete protection from pressurized solvent spray cleaning.

Automatic cleaning machines for bells have been designed by several automobile assembly plants to avoid rotary atomizer head damage. Nicks are easily put into the delicate bell or disc edge if it should be struck by a part on the conveyor or by a metal tool when the rotary atomizer head is being replaced or removed for cleaning.

## Rotary System Operation

Paint is delivered to the spinning surface of rotary atomizers via delivery holes around the periphery of a smaller concentric inner cup. Center delivery feed and uniform distribution are needed to maintain precise head balance. Solvent is fed through this line to purge paint for color changes and to clean the head as necessary.

On older model bells, retractable flush shrouds were used to collect color-change purges and solvent rinses. Collection shrouds for side-mounted bells were fitted with gravity drain lines; overhead-mounted bells used siphon tubes. Capture of paint and solvent during these operations reduced the plant's overall VOC output and in many instances lowered maintenance costs significantly. Internal dump valves are now used for line purges and flush outs with solvent or air/solvent mixtures. Both alternating air/solvent pulses and integrally "foamed" air/solvent mixture systems are available for flushing. Mixing or pulsing air with solvent not only uses less solvent, but also scours the fluid lines more effectively.

The rotational speed of a high-speed bell slows down when paint flows into it. This slowing is the result of the increased weight of rotating material. At a typical fluid flow of 10 oz/min (0.28 L/min), the reduction in spin speed is roughly 10%. For a fairly high fluid flow of 20 oz/min (0.57 L/min), the speed dropoff is about 20%. If one is near the bell, one can even hear the abrupt lowering of the pitch emitted by the spinning head when this occurs. Depending on the paint being used and the particular application, this slowing could be a significant problem. Therefore, the bell spin rate is preset high enough so that the spin-speed reduction when paint starts flowing does not become a negative factor in atomization. Spin-speed governing controls have been used by some rotary manufacturers to maintain rotational speed irrespective of fluid loading, but such devices have not been universally used. A third method uses a rather simple dual-air pressure system with a "high/low" setup. In this type two different air pressures are applied to the turbine. A high pressure is used when paint is being applied; a low pressure is used during nonpainting or bell "idle-speed" times. A dual-pressure system works well with stable fluid-flow rates. Electronic speed control is preferred for painting operations that require frequent on/off triggering, those demanding extremely fine finish quality, or whenever fluid flow rates are varied.

Early discs and bells had no convenient, fast method to monitor rotational speeds. Systems that permit digital rpm readouts of spin speed are now almost standard. The spin rate of the bell can be continuously monitored in several ways. Two systems involve electronic speed regulation: one with fiber-optic pickup and the other with magnetic impulse pickup. The latter type offers the advantages of extra durability and low cable replacement cost.

Serrated (mechanically grooved) edges on some bells eliminate air entrapment "microbubbles" in the wet paint film. Entrapment of air is most likely to occur with high–surface-tension coatings, including both waterbornes or high-solids. Serrated edges enhance atomization so that bells can be operated at slightly reduced rotational spin rates. Each serration line forms a paint filament to enhance atomization.

The fineness of atomization from a high-speed rotary depends on the paint viscosity, paint flow rate, and rotational speed, but it is virtually unaffected by the electrostatic charging voltage. Bell spin speed can affect not only atomization and pattern size, but also the distribution of mica and metallic flakes in the film. Rotational speed may affect the paint color until a certain maximum speed is reached, above which virtually no color difference is noticed.

Different coating application methods used on separate parts of assemblies can cause appearance mismatches. Paint films applied by spray guns can take on a slightly different shade of color from those applied by high-speed rotary atomizers. This is due to the differences in droplet particle size and the velocities at which droplets are delivered to the part. This possibility should be checked to avoid color matching difficulties when interplant painting is introduced.

## Hand-held Bell

The bell rotary atomizers described so far in this chapter have wide use in applying finely atomized paint on industrial painting lines. These bell systems are mounted as permanent capital equipment; thus, the parts to be painted are brought to them.

A portable bell rotary applicator is used, however, for low overspray finishing of selected items. One widely used application is in repainting metal office furniture, as shown in Figure 14-10. It produces so little overspray that items to be painted need not be transferred to another location; they can be refinished right in their normal location. Another use is to repaint installed metal fences or grillwork with minimal mess and paint waste.

The portable unit has a rather long configuration but is surprisingly well balanced and easily maneuvered. The operator holds the unit roughly 6 in (15 cm) away from the part being painted and moves the unit slowly back and forth or up and down in even strokes, or in circular strokes of 4–6 in (10–15 cm) in diameter.

The bell electrostatically charges the atomized paint particles as they travel across it. An electrical grounding wire is affixed with a metallic clip onto office furniture being painted,

## 204  INDUSTRIAL PAINTING

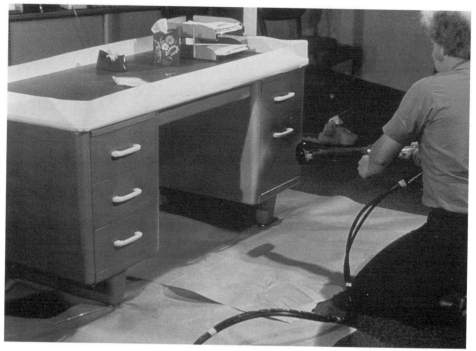

Figure 14-10. Portable Bell Rotary Atomizer

which enables the part to attract charged paint droplets. Protective canvas or plastic sheets can be positioned around other objects in the vicinity and on the floor to catch any overspray. The painting is usually done after business hours. Air dry paint allows the parts to be ready for use when business begins the next day.

The bell can only apply up to 5–6 oz (142–170 cc/min) of paint per minute. This rate of flow, while severely limited, is still satisfactory for applications such as in-office furniture repainting but is just not adequate for industrial production line painting.

## High-speed Rotary Atomizer Advantages

High-speed rotary disc and bell atomizers have several important advantages:

- Fine atomization
- High-solids, low-solids, metallics, two-component, and waterborne versatility
- Viscosity flexibility on both high and low VOC coatings
- High transfer efficiency

### Fine Atomization

The high rotational speeds and electrostatic charging voltage team up to do an exceptional job of atomizing paint into extremely fine particles to ensure a high-quality finish.

| APPLICATION TYPE | TYPICAL TRANSFER EFFICIENCY RANGES |
|---|---|
| | 10    30    50    70    90 |
| AIR SPRAY High Pressure | 25 ⊢―――⊣ 50 |
| AIR SPRAY Low Pressure | 40 ⊢―――⊣ 70 |
| AIRLESS | 35 ⊢―――⊣ 65 |
| ELECTROSTATIC AIR SPRAY | 35 ⊢―――⊣ 60 |
| ELECTROSTATIC AIRLESS | 40 ⊢―――⊣ 70 |
| AIR-ASSISTED AIRLESS | 40 ⊢―――⊣ 70 |
| ELECTROSTATIC AIR-ASSISTED AIRLESS | 45 ⊢―――⊣ 75 |
| ROTARY (All are Electrostatic) | 60 ⊢―――⊣ 90 |

Figure 14-11. Typical Transfer Efficiency Ranges for Various Application Methods

## High-solids, Waterborne Versatility

High-speed rotary atomizers can apply high-solids and waterborne coatings as easily as low-solids, metallic, and two-component coatings. Virtually all coating can be applied with rotary atomization.

## Viscosity Flexibility

The speed of rotation can be adjusted to compensate for coatings with varying viscosities, which eliminates the need to add solvent to adjust viscosity in many cases.

## Transfer Efficiency

Fine atomization, electrostatic charging, and minimal air turbulence in the vicinity of the rotary atomizer and part being painted ensure a high transfer efficiency. Figure 14-11 compares typical transfer efficiencies for the various types of spray guns with the rotary atomizers.

# High-Speed Rotary Atomizer Disadvantage

High-speed rotary atomizers have only one noted disadvantage. Although their capability to paint broad open surfaces is outstanding, the absence of air or fluid pressure to push the paint droplets to the target limits the ability to paint into Faraday cage areas. Manual touch-up in these hard-to-paint areas is usually required. However, the use of nonmetallic plastic or ceramic bells coated with an electrically conductive material eliminates the possibility of a spark jumping from the bell to a grounded part. This permits the operator to safely position the bell much closer to the part, which overcomes much of the Faraday cage problem.

# Chapter 15

# Coating Types and Curing Methods

## The Curing of Coatings

The curing of a coating is the process whereby the freshly applied liquid or powder coating is transformed to a finished solid paint film. Coatings are generally of little value as they are applied; they must be converted to a more durable finished paint film to perform well, either for appearance or function, or for both. Liquid coatings are converted to their cured state either by solvent evaporation only or by solvent evaporation plus resin cross-linking.

Powder coatings are cured by heating to melt the individual powder particles and thus form a continuous film on coated parts. In roughly 98% of powder coating formulations, the heat also initiates chemical cross-linking of the resin(s), although a few powder coatings simply melt without any cross-linking taking place. On cooling, the molten film becomes a solid once again, but now it is a continuous paint film instead of separate particles.

## Coatings That Cure by Solvent Evaporation Only

A liquid coating that cures by solvent evaporation only contains one or more resins with long-chain polar molecules dissolved in a solvent. As the solvent evaporates, the molecules are drawn together by polar attractive forces at numerous sites along their chain lengths; no chemical reaction takes place, however. Lacquer-type powder coatings likewise do not undergo any cross-linking during curing; they only melt and resolidify upon cooling.

Coatings that contain these long-chain molecules and cure by solvent evaporation only are designated as *lacquers*. Lacquers can be made from any soluble linear polymer, such as chlorinated rubber, cellulose, and acrylic resins. They characteristically have a high molecular weight, which gives good paint film properties. Their main weakness is a lack of solvent resistance to the type of solvent originally used to formulate the liquid coating. This so-called weakness can also be an asset, however. In the refinishing of a lacquer film, the paint can be made to reflow with solvent to increase the degree of surface smoothness.

Although liquid lacquers can often air dry, they can also be processed (cured) in an oven. The only function of the oven is to speed the solvent evaporation. Because of the high–molecular-weight resins of liquid lacquers, their solvent content is necessarily high. A lacquer may be as much as 90% solvent and only 10% solids. Because of EPA VOC emission limits, industrial use of liquid lacquers has declined considerably, and lacquer powders make up only about 1% of total powder usage.

If solvent is added to the dry (cured) paint film, the resin molecules will again go into solution. The solvent molecules will interfere with the polar attractive forces, which allows the resin to become fluid (dissolve). The addition of heat to this type of resin can also reduce the polar attractive forces, which softens the resin. *Thermoplastic* is used to describe such materials.

## Coatings That Cure by Cross Linking

All coatings that cure by cross-linking are categorized as *enamels*. This applies to both powder and liquid coatings. A liquid coating that cures by cross-linking normally contains one or more partially polymerized resins. Although there are 100% solids liquid enamels, most enamels contain resins that are dissolved in a solvent. As the solvent evaporates or as heat is applied, a chemical reaction occurs either among the resin molecules or with molecules of another cross-linking resin so that chemical bonds are formed. The newly formed chemical bonds cannot be dissolved by adding solvent. This does not mean that enamel paints cannot be softened or crazed by any solvent, but enamels will tend to be less soluble than lacquer paints in all solvents.

Coatings that cure by cross-linking can be divided into at least six categories:

- Oxidizing
- Moisture cure
- Heat cross-linking
- Reactive catalytic cross-linking
- Catalyst vapor
- Radiation cure

### Oxidizing Coatings

These liquid coatings use resins based on drying oils. The resins may be alkyds, phenolics, oil-modified urethanes, epoxy esters, or various oleoresinous systems. Some have drying oils added expressly to facilitate air-dry curing. After the coatings are applied, the solvent begins to evaporate, which exposes more and more of the unsaturated resin to the air. This air contact triggers a cross-linking reaction between the resin and atmospheric oxygen. These resins and their modified chemistries contain molecules with carbon-to-carbon double bonds that react with oxygen to form peroxy bonds, and subsequently to ether cross-linkages which comprise the oxidative cross-linkages. This cross linking occurs slowly at room temperatures but progresses rapidly at bake oven temperatures.

### Moisture-Cure Coatings

This category of liquid coatings functions in a manner similar to the oxidizing systems. After the paint is applied, moisture in the air will begin to react with the coating resin, resulting in resin cross-linking. An example of this type of coating is a moisture-cured urethane, in which isocyanate groups react with atmospheric moisture to form urea linkages. As with oxidizing systems, elevated temperatures can speed the curing process.

### Heat Cross-linking Coatings

Some liquid coatings will either cure extremely slowly or not cure at all by air drying, and therefore require heat for curing. Their essential composition consists of an uncross-linked primary resin, such as acrylic, plus a secondary cross-linking resin, such as melamine. The two resins are blended with solvent, other additives, and possibly pigments as well.

Once the container is opened and the coating is applied, the solvents will begin to evaporate, but little or no curing will occur at room temperature. The applied coating will merely remain tacky (sticky) and uncured unless heat is applied. The paint temperature must reach or exceed a level that triggers cross-linking of the resins. In the example of the acrylic resin and the melamine cross-linker, the resultant cured paint film is called a heat-cured melamine-acrylic.

Heat cross-linking coatings can have various types of primary resins and a variety of cross-linking resins. Whatever the resin types, however, the basic principle is the same: a certain minimum temperature must be reached to trigger the reactions that cause cross-linking and paint curing. For many high-solids coatings, the cross-link temperature is well above room temperature. Such paints can remain tacky almost indefinitely without curing unless heat is applied. Heat can be introduced several ways including hot air ovens, induction heating, and infrared or microwave radiation.

Nearly all enamel-type powder coatings, which includes nearly 99% of them, fall into the heat cross-linking category.

## Reactive Catalytic Coatings

The coatings in this category cross-link cure. Each of these coatings consist of two separate parts: the major coating component and a lower-volume catalyst component. The major coating component contains mostly resin, various additives, solvent, and perhaps pigment. The catalyst component contains essentially only solvents plus the catalyst, or it may contain solvents, a catalyst, and a cross-linking resin. As long as the two paint parts are kept separate, each remains essentially unchanged in a fluid state. Once the two components come together, cross-linking reactions begin between the resin and cross-linking agent. The catalyst increases the rate of cross-linking, and film curing is the ultimate result. The applied coating will begin to cure just before or during application without applied heat. However, elevated temperatures will speed the rate of cross-linking and solvent evaporation.

An example of a catalytic coating is a two-component urethane consisting of a polyester resin as one part and an isocyanate cross-linker plus amine catalyst as the second part. When the two are intimately mixed, a cross-linking occurs that forms a urethane linkage, thus the resultant coating is called a urethane. The coating can be mixed in a single container and then applied, or the two components can be brought together in the atomizing portion of a spray gun and mixed (see Figure 15-1).

When the two components are mixed together, the resultant mixture must be used within a limited amount of time, called a *pot life*. If the mixture is allowed to remain in the container, total cross-linking will occur over a predetermined period of time, and the mixture will become hard. Formulating chemists are able to control the rate of cross-linking and the resultant pot life to various degrees. The pot life can be controlled to be as brief as a few minutes or as long as 16 h. If the pot life is exceeded, excessive cross-linking lowers the film quality even though the paint is able to be applied normally and appear correct.

A word of caution must be stated here. The terms *catalyst* and *cross-linker* are often replaced with the terms *accelerator* and *catalyst* by paint manufacturers, so the true catalyst and cross-linker are not always obvious. However, most manufacturers use terms consistently within their literature, so you should be able to follow their instructions and achieve correct results.

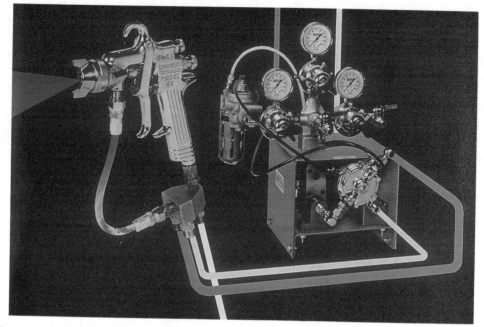

**Figure 15-1. Spray Gun Mixing Metered Ratios of Each of Two Components**

## Catalyst Vapor Coatings

Known as *vapor curing*, this process uses a room-temperature vapor catalyst to speed the curing of certain two-component liquid formulations. After the two components are applied, the coated parts are conveyed into a chamber filled with the vapor of a catalyst such as diethanolamine. The catalyst concentration is held at 1,200–1,500 ppm. Air seals at the entrance and exit of the curing chamber prevent loss of the catalyst. The vapor accelerates the ambient-temperature rate of the reaction of the isocyanate and hydroxyl groups, which forms urethane cross-linkages. The catalyst reduces the cure time from 40–60 min, down to about 5–10 min. The vapor and liquid of diethanolamine are safe and nontoxic, but they have an unpleasant fishlike odor. Catalyst is scrubbed from the oven exhaust with water to prevent its release into the environment.

A low amount of heating is usually employed after the vapor chamber to drive out solvents. This low-heat requirement can be an advantage when painting large or thick castings and similar parts. It avoids the long, slow heat-up and cool-down times needed when oven-baking coatings on massive metallic substrates or on parts with large heat capacities. Catalyst vapor curing can also be used to advantage on heat-sensitive substrates such as paper, cloth, wood, and plastics.

A variation of the catalyst chamber process, called catalyst vapor injection curing (VIC), injects catalyst vapors into the atomizing air supply of air spray and air-assisted airless spray guns, as shown in Figure 15-2. It can also be introduced into the shaping air (air shroud) of rotary bell paint applicators. In this way a separate catalyst chamber stage can be eliminated from the coating process.

COATING TYPES AND CURING METHODS   211

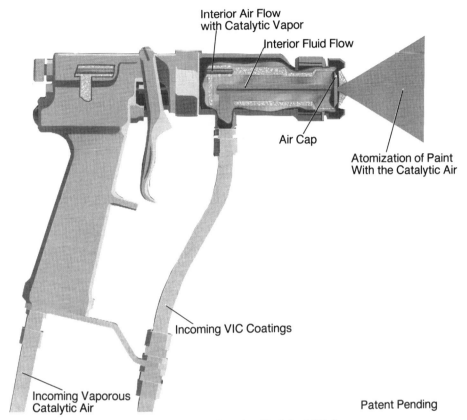

Figure 15-2. Vapor Injection Curing (VIC) Coating/Catalyst Mixing

## Radiation-cure Coatings

*Radiation-cure coatings* contain various accelerators or catalysts that are dormant until acted upon by either ultraviolet (UV) light or electron beam (EB) radiation. The UV light or electron bombardment triggers a free-radical reaction among chemical groups that results in cross-linking (curing) of the paint resins.

**UV-cure coatings.** UV-cure coatings contain chemical photoinitiators that are sensitive to UV light, which causes the chemical bond structure of the photoinitiators to form free-radical groups that trigger resin cross-linking. Curing happens in a two-step sequence: first, a photoinitiator absorbs UV rays and resin free radicals are formed; then, polymerization cross-linking reactions of the resin free radicals produce a cured paint film. Any heat from the UV lamps will accelerate the polymerization curing reactions. Some radicals often remain for a brief time after UV exposure, which gives a small degree of postcuring to the film. Both liquid and powder paints can be UV cured.

UV coatings may or may not require solvent or other fluidizing media to reduce their viscosity and promote flowout. If solvent is used, a flash-off time is allowed after application prior to

UV cure. If the fluidizing media is also a cross-linker, it is called a *reactive diluent*. For reactive diluents, no flash-off time is required because they become part of the cured film. Rapid, extensive resin cross-linking can be initiated with UV light, so that often extremely low–molecular-weight resins with very low viscosities are possible in the coating formulation. These resins flow out so well by themselves that no solvent is required, which allows a zero VOC coating for UV curing.

Cure by UV is accomplished in shielded and enclosed chambers saturated with high-intensity, electrically generated UV light. For total curing to take place, the UV light must activate all of the photoinitiator molecules, which means that the light must "see" them. This is fine for unpigmented coatings, but only about 1 mil (25 microns) dry-film thicknesses of pigmented coatings can be UV cured because the pigment molecules will block UV light rays from some of the photoinitiator.

The energy of UV light decreases with the square of the distance between the light source and the surface receiving the light. So, doubling the lamp-to-paint distance drops UV light intensity to 25%; tripling the distance cuts light intensity to just over 10%. Therefore, the UV light source must be kept as close to the painted part as possible. For this reason, UV cure is used mostly on flat surfaces, which can be kept very close to the light source. However, highly polished parabolic reflectors enable certain types of three-dimensional items to be UV-coated. These include wall emblems, golf balls, guitar bodies, and wooden stocks for firearms.

UV curing is fast, usually in 10-60 s, which permits UV ovens to be confined and compact. The quick cure minimizes substrate heating, which is a great advantage when curing films on heat-sensitive substrates such as printed circuit boards, wood, and thermoplastics. Since the UV lamps become rather hot, it is necessary that they be turned off whenever the production line stops to avoid harming the product being coated. In the past, many UV lamps could not be restarted quickly once they had been turned off, but now fast "on/off" UV lamps that cool off rapidly can be used to enable starting and stopping the coating line quickly.

**EB-cure (Electron Beam) coatings.** EB-cure coatings are formulated with resins that undergo cross-linking when radiated with beams of electrons. As with UV curing, only a limited variety of EB-cure coatings is available. The electron beams emitted by a radiation source are sufficiently strong that they can be dangerous. Therefore, the radiation source must be shielded, and the electron emission area must be designed with safety interlocks to prevent humans from accidentally getting close enough to be harmed.

EB systems require an inert gas flush, such as nitrogen or argon, to keep air out of the electron radiation zone. This is necessary because electron radiation passing through air (which you may recall contains approximately 20% oxygen and 80% nitrogen) creates ozone and mixed oxides of nitrogen. These are harsh irritant gases that present severe health hazards for humans even at low concentrations. People must also be excluded from the inert gas atmosphere because of the danger of asphyxiation from a lack of oxygen.

The method of initiating chemical cross-linking by electron beams is in many ways similar to UV curing. Not surprisingly therefore, some of the same advantages accrue with both UV and EB curing. The quick line start and stop capability, the low floor space requirements, the minimal substrate heating, the low or zero VOC emissions, and the rapid production rates possible for UV curing are also advantages of EB curing. While both are fast-curing methods, EB curing is much faster than UV curing. EB curing is often possible in less than one second, for example. The thickness of the coating is not a problem in EB curing as it is in UV curing; electron beams are far more penetrating than UV light.

The electron beams lose their energy rapidly as they travel away from the generating source. A 6- to 8-inch (15- to 20-cm) target distance is about the maximum possible. It is mainly a "line-of-sight" curing process when metallic substrates are involved; electron beams will not penetrate metal. A slight amount of reflected beam curing can occur, but that is not significant in this process.

At first it might be expected that EB curing would be ideal for curing paints on heat-sensitive substrates such as wood, paper, and plastics. This is only somewhat true. EB curing is possible for some papers and wooden materials, but it is not well suited for the general curing of coatings on plastic parts because polymeric substrates do not readily block the passage of electron beams. The penetrating electron beams can cross-link the underlying plastic substrate as well as cure the paint resin. Substrate cross-linking can dramatically reduce the plastic's impact strength by excessive embrittlement.

## Types of Ovens

The most commonly used coating curing ovens are the convection and infrared types. Both have the same goal: to heat the applied coating to its curing temperature. The heat can cause both solvent evaporation and resin cross-linking reactions to occur. Natural gas and propane gas are the fuels most frequently used for convection ovens; electricity is most frequently the energy source for infrared ovens. Of course, any of these energy sources could be employed for convection or infrared ovens. In most locations electricity is more costly for convection heating than gas, but electric infrared ovens are easier to regulate in temperature and also have a much faster start-stop response (faster heat-up and cool-down) than gas infrared ovens.

### Convection Ovens

Convection ovens heat the applied coating by first heating the oven air, which in turn transfers the heat to the coating. Convection ovens are of two types: direct-fired and indirect-fired. In both types the hot air is normally moved around within the oven by fans and directed through an arrangement of vents. Some hot air is lost through the parts entry and exit openings, and a portion of the oven atmosphere is exhausted with a fan. Fresh air makeup into the oven, normally filtered to remove contaminants, is also supplied by a fan.

Direct-fired ovens warm the oven air directly with a source of heat, generally a gas flame. This flame's products of combustion are therefore also present throughout the oven, including where the coated parts are being conveyed through the oven. The combustion products may discolor or wrinkle some types of coatings, but in most cases they cause no harm. However, for fine finishes it may be necessary to separate the oven burner combustion products from the curing chamber. Since it is more efficient to heat air directly than indirectly, direct-fired ovens tend to be preferred if their use causes no problems with curing or paint appearance.

Critical appearance requirements, such as those on automobiles, necessitate the exclusion of oven combustion gases from the areas in which coated products are being cured. Indirect-fired ovens confine fuel combustion—the source of heat—to a separate enclosure. The products of fuel combustion are exhausted outside the plant by a vent or stack. The walls of the "fire box" get hot and warm the enclosed air contained in the plenum section of the oven. A blower circulates the hot air out of the plenum and into the oven's curing section. Thermostatic controls can be used to maintain a constant temperature throughout the cure zone. A constant measured amount of fresh air makeup is introduced into the curing section, and an equivalent amount of oven air escapes through the oven openings or is blown out through the oven exhaust stack. This air exchange prevents a potentially dangerous buildup of solvents within the oven, which could otherwise cause an explosion.

Normally the maximum allowable concentration of flammable vapors inside an oven is limited by fire prevention codes to about 25% of their lower explosive limit (LEL). LEL refers to the leanest possible combustible air/solvent mixture expressed in percent vapor by volume. In actual practice, the volume of solvent vapor in most industrial paint ovens is generally far lower than 25%. Many ovens are found to be operated at only 2%–7% of their LEL. Oven air volume turnovers per minute generally range from 2 to 10, depending on solvent loading levels in the oven.

To save energy, some plants have installed monitoring systems to minimize the fresh air makeup needed in the oven. Insurance companies may permit ovens so equipped to be operated as high as 50% of the LEL value. The problem with such high LEL levels is that paint curing is not always efficient even at 20%, not to mention at 50% of the LEL. Nevertheless, hot air dilution of oven solvent vapor to lower concentrations than necessary is an outright waste of fuel. A study by a U.S. automaker revealed corporate wide energy waste in this regard. The company's paint ovens were operated at an average of merely 2% of the LEL; the highest single vapor concentration found in any of their 58 paint ovens was only 8% of the LEL. Fresh air turnover rates corporation wide were reduced sharply, producing an enormous annual dollar savings.

Many other techniques can be used with curing ovens to conserve energy. One consideration when ovens are designed is to use sliding doors, air seals (see Figure 15-3), or "bottom entry and bottom exit," which confines the oven's hot air inside where it belongs because hot air rises. Bottom entry/exit is better than air seals, but a combination of both is the most energy efficient design. Another help is to fully insulate the oven. A commonly overlooked measure is to operate the oven at no higher temperature than what is required to elevate the applied coating to its curing temperature for the specified amount of time that parts remain in the oven. Excess oven temperatures may not overbake the coating, but it may very well be wasting heat. Of course, it is likewise important that underbaking be avoided; too low an oven temperature will cause even greater problems and expense. Two or three oven zones at increasingly higher temperatures can be effective in preventing solvent boiling (popping) with both solventborne and waterborne paints if the oven is constructed to make this possible.

Indirect ovens with a full top or bottom plenum are best for efficiency and temperature uniformity. A variation on conventional indirect-fired ovens uses high-velocity recirculating hot oven air that is directed by nozzles or vents to impinge rather directly onto the applied coating. These designs are termed "high-velocity ovens." The rapid exchange of air across the part surfaces heats the coating more quickly and sweeps away solvent vapors at a faster rate. This causes fast curing and allows the conveyor speeds to be increased by as much as 50%.

COATING TYPES AND CURING METHODS 215

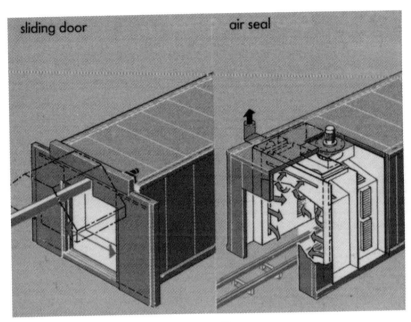

Figure 15-3. Oven Ends Designed to Conserve Energy

Gas is usually the source of heat for indirect-fired convection ovens. Oil or electricity can be used, just as with direct-fired ovens. The ideal oven type and shape varies with parts volume and the specific product or products being coated. The oven may be square or long and narrow, and it may be conveyorized to handle any number of parts. Square ovens use fewer panels and are the easiest configuration to balance. Low-volume finishers often use a batch (usually unconveyorized) oven. These may have a single door for in/out access, or they may be fitted with doors at both ends for flow-through of parts.

New ovens must be cleaned thoroughly during construction and at completion to avoid built in "dirt" problems. Paint on some new enclosed light gauge conveyor tracks can degrade to the extent that it also contributes to "dirt" defects.

## Infrared Ovens

Infrared (IR) ovens use radiant energy rays that pass freely through the oven air, but without heating the air. Some air heating occurs in infrared ovens, but this is because heat-up of the coated parts and the oven walls generates a modest amount of convection heating of the oven air. The IR energy is converted to heat when its rays strike a surface, such as the coated part and oven walls. The source of infrared rays can be electrical heating elements of various types and design configurations, but it also can be surfaces heated by combustion of fuels such as natural gas. Hot surfaces of many types will emit infrared rays. Gas-fired black-surface infrared ovens are not as common and are less easy to control but can be just as effective as electrical infrared ovens in curing paint films. IR ovens have been shown to have up to 90% efficiency if both heat and IR emissions are included in the calculations.

A properly designed infrared oven directs the radiant energy as efficiently as possible onto the coated surfaces. Most IR ovens use either tubular lamps, quartz glow bars, or filament-wound ceramic cones to emit the energy. The wavelength output of the radiant energy varies slightly with each source.

Short-wave infrared rays, having wavelengths of 1–2.3 microns, appear to the eye to have a white or yellowish color. These waves are absorbed well by black and dark colors but are reflected or only absorbed poorly by white and other light colors. Short-wave infrared rays heat up bare metals, unlike the medium- and long-wave IR.

Rays with 2.3–4.5 micron wavelengths are classed as medium-wave infrared. They have a distinctively red color. Most colors absorb medium-wave IR so it can be used to cure a variety of paints; however, dark colors are able to be cured more readily than paler shades. Medium-wave infrared is used to good effect with waterborne paints, which absorb these wavelengths strongly. Spot or panel repair paints on motor vehicles are frequently cured with portable, medium wavelength IR lamps.

"Black wall" radiation, named for its visual appearance, has wavelengths greater than 4.5 microns on up to several hundred microns and is therefore termed long-wave infrared. This is the most colorblind region of IR radiation and will be absorbed rather completely by all hues of paint.

Tuned infrared ovens utilize mainly the IR wavelengths that can be most fully absorbed by the paints being cured. This increases efficiency and thus lowers cost. The energy of the waves decrease as their wavelength lengthens, but any difference in safety among the three categories is insignificant.

Each of the three types commonly will use polished reflectors behind them to direct the reverse-side radiation toward the painted parts being cured. Studies have shown that gold-coated reflectors, although more costly initially, surpass the others in infrared reflecting efficiency and are ultimately the most economical choice. Figure 15-4 shows some painted parts being conveyed through an infrared oven equipped with gold-coated reflectors.

Infrared radiation is often the best method to cure powder coatings because no air blows through the oven to disturb the delicate powder layer before it melts and cures. Fast heat-up rates can be achieved with infrared, a highly desirable procedure for generating optimum powder coat appearance. Ideally, powder coatings should melt and flow out completely before any cross-linking takes place. If heated slowly during curing, rather than melting, powder coatings may undergo considerable cross-linking. Complete melting and full flowout may never occur. Fast initial heating of powder coatings gives a noticeably smoother and more uniform paint finish.

## Other Curing Methods

Various other curing methods have been devised and have been used. For that reason they are explained here, but none of these has anything close to widespread use. These minor paint cure processes include induction heating, microwave and radio frequency (RF) curing, and heat of condensation curing.

# COATING TYPES AND CURING METHODS

Figure 15-4. Infrared Oven Equipped With Gold-coated Reflectors

## Induction Heating

For induction heating, a source of low-frequency alternating current is placed close to the freshly coated metal. The rising and falling electromagnetic lines of force from the alternating current induce powerful circulating currents in the metal, which creates a rapid temperature rise both in the metal and coating. Carefully regulating the alternating current can control the induced current and the resulting temperature rise in the metal and coating. To be successful, the source of alternating current must be positioned very close to the coated metal.

Induction heating has been used successfully to cure coatings applied to a moving strip of metal (coil coating). An advantage in induction heating is that metal can be heated almost instantaneously, eliminating the need to have long extended convection ovens.

## Microwave and Radio Frequency (RF) Curing

These types of curing are similar in nature to induction heating. The only difference is in the frequencies used, which are in the thousands of cycles per second. Both microwave and radio frequency curing require shielding and are far too slow or too costly for nearly all applications. The further disadvantage of RF curing is that the source of RF energy is a radio transmitter, creating a powerful radio signal at the particular frequency being used. The signal must be totally squelched to satisfy Federal Communications Commission (FCC) regulations.

218  INDUSTRIAL PAINTING

## Heat of Condensation Curing

Heat of condensation curing is based on this scientific principle: When vapor is allowed to condense to the liquid state, a large quantity of heat is given off, as shown in Figure 15-5. At least two companies reported exploring the possibility of using the heat of condensation of a nonflammable vapor converting back to the liquid state in order to generate heat for curing paints. In practice, freshly painted parts would be placed into a chamber containing hot liquid vapors. These vapors would then condense on the cool parts and give off heat to cure the paint. The problem is that the condensed liquid material runs down the parts, creating unsightly grooves in the paint.

One possible solution to this problem is a heat pump oven that to my knowledge nobody has tried, but which would work. In a refrigerator the evaporator allows a liquid compressed refrigerant gas to expand. This process absorbs heat, keeping the inside of the refrigerator cold. The compressor forces the expanded gas vapor back to a liquid, giving off heat. This is dispersed through coils behind or under the refrigerator. Usually a fan blows ambient air across the condenser coils to help remove this heat. On a large scale, such heat could undoubtedly run a curing oven, much as a heat pump is used for residential heating. The economics are another story.

These are rare curing methods, but the future is something at which we can only guess. Some years before vapor curing was a small commercial reality, opinions were expressed by painting "experts" that catalyst vapors to cure paint would never be used outside a test laboratory. They were wrong. (But not by much!) Skeptics doubted E-coating would ever be economically feasible, too, so who knows what the 21st century will bring?

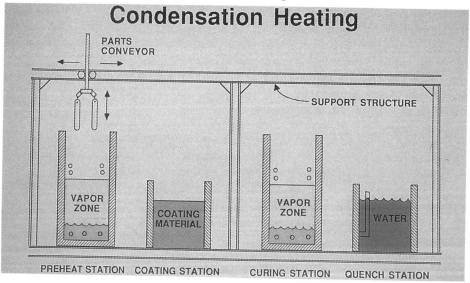

Figure 15-5. Heat of Condensation Curing

# Chapter 16

# Film Defects in Liquid Coatings

## The Root Cause of Defects

The types of defects are various and sundry. However, all paint defects have one thing in common: a cause. The cause is always due to the breakdown of good painting practice, somewhere along the line, from cleaning and pretreatment to paint application and curing.

Tracking down the cause of a paint defect ought to be pursued in a straightforward manner. A painting defect is like a warning sign along a highway—it gives cautionary information that alerts one to potential hazards and situations that may cause problems. The defect always points to a violation of good painting practice or an inadequacy of the coating or the substrate. Therefore, the secret of good "detective work" with paint defects is to build solid foundations of knowledge in good painting practices and in painting defects.

## Types of Defects

A comprehensive list of painting defects could include perhaps several hundred names. Such a lengthy list can be abbreviated considerably by limiting it to the more frequently encountered types of defects. By and large, these major defects come close to covering all types of defects experienced by most painters. The defects that will be covered are:

- Blisters
- Bubbles and craters
- Color mismatch
- Dirt
- Fisheyes
- Gloss variations
- Mottle
- Orange peel
- Runs, sags, and curtains
- Paint adhesion loss
- Soft paint films
- Solvent popping, boiling, and pinholes
- Solvent wash

### Blisters

As shown in Figure 16-1, a *blister* in a paint film is a small domelike raised area that contains, or once contained, moisture (water, water vapor, or both), solvent, oil, or grease.

**Figure 16-1. Blisters**

## *Causes on Metal*

Blisters on painted metal are traceable to contamination left on a surface prior to painting, either from water, residual soils, poor rinsing, or incomplete dryoff of rinse water. Fingerprints may contribute salts or skin oils that can lead to blisters. Any unrinsed cleaners, pretreatment chemicals, or unremoved greases and oils can permit moisture to get under a paint film in a matter of days and form a blister. The moisture can also come from poor water removal in a compressed air supply; however, when that occurs, it tends to cause a large, obvious flaw in a paint finish.

Moisture in or under a paint film expands and contracts with temperature change, thereby producing rapid blistering and lifting of the film. The moisture strains a paint film to the point that cracks may appear. A paint film is unable to resist this movement because of reduced adhesion at the part surface. Once the film cracks, moisture easily reaches the underlying surface, and corrosion can begin.

Inadequate rinsing or rinsing with water containing high amounts of dissolved solids can sometimes cause a problem. When these difficulties are experienced, it may be necessary to use ultrapure rinse water containing as little as 10 ppm total dissolved solids (TDS) that can be tolerated in rinse water. In some cases 200–300 ppm dissolved solids may be adequate to avoid rinse water deposits. Tap water rinses may be followed by a brief misting with deionized water to eliminate blisters from this source.

Blisters can be produced on painted parts by osmosis. When a semipermeable membrane separates solutions of different concentrations, the solution with the lesser concentration will seep into the solution with the higher concentration because of osmotic pressure. Such a scenario is formed with a paint film when solid residues remain on the surface of a substrate beneath the paint film. When the paint film gets wet, moisture will seep through the paint film, contacting the residues and forming pockets of concentrated solutions. Osmotic pressure will draw additional moisture from the top of the paint film through the paint film to the concentrated solution sites. When a sufficient amount of moisture has been drawn through the paint film to the pockets of solutions below, the paint film will be lifted from the substrate at the solution sites in the form of blisters. (Osmotic pressure can be amazingly high; osmotic pressure forces water to the topmost leaves of tall trees.)

# FILM DEFECTS IN LIQUID COATINGS   221

### *Causes on Wood*

Blisters on painted wood are usually caused by moisture escaping from the wood to the wood/paint film interface. When the moisture gets warm, it expands, which exerts enough pressure to raise the film into a blister.

### *Prevention*

To prevent blister formation on painted metal, be absolutely certain that the surface is free of contaminants before you apply the paint. Be sure that the gun air supply is clean and dry. The TDS in rinse waters must be excessive if a final deionized water rinse mist is not used.

To deter blister formation on wood (in addition to the cleanliness precaution for metal), make sure that moisture can find another exit route out of the wood instead of pushing up through the paint film.

## Bubbles and Craters

A paint *bubble* in a film is a small domelike raised area that contains, or once contained, solvent vapor. A bubble closely resembles a blister (which contains water and/or water vapor).

A crater is a small, concave depressed area that formerly was covered by a bubble, as shown in Figure 16-2. The breaking of the bubble contributes to the crater's rounded bottom and built-up sides.

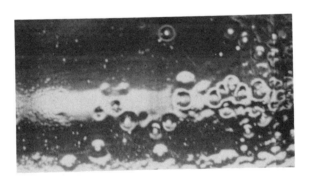

**Figure 16-2. Cratering**

### *Causes*

Bubbles tend to form during a bake cycle when the top layer of the paint film skins over before most of the solvent in the film has had a chance to escape. The rising solvent pushes up portions of the skinned-over top layer, forming bubbles. If the bubbles break, the raised skinned-over areas constrict to the sides, forming craters. These defects are rarely seen on air-dry finishes because ambient air tends to allow most of the solvent to escape before the film skins over. Bubbles (and craters) are usually traceable to the following causes:

— Inadequate flash time before the bake cycle. This leaves an excessive amount of solvent in the paint film. The bake oven heat may skin over the film before most of the solvent has escaped.

— An extra heavy wet film application. This can trap large amounts of solvent in the film. The bake oven heat may skin over the top layer of film before most of the solvent has escaped.
— A solvent blend that evaporates too slowly. This will invite skinning over of the top layer of film before most of the solvent escapes.
— Insufficient primer bake. This can leave an excessive amount of solvent in the primer, which may form bubbles in the subsequently applied topcoat during the topcoat bake cycle.

### Prevention

The best ways to prevent bubble and crater formation are to allow enough flash time before a bake cycle, apply a coating in several thin layers (rather than one thick layer), use the proper solvents, and maintain specified primer bake cycles.

## Color Mismatch

*Color deviations* from one part to another are shifts in the color of a paint film that occur in applying the same batch of paint. These color deviations are totally independent of *metamerism*, which is a color shift due to a change in the nature of the incident light. Color mismatch between separate batches of the "same" paint are the fault of the coating manufacturer.

### Causes

The causes of color deviation can include:

— Variations in the degree of film wetness. The degree of film wetness can affect the shade of a metallic paint. The wetness affects the relative distribution of color pigment particles and metallic flakes. In general, the drier a metallic paint is applied, the lighter it appears.
— Inadequate agitation. Improper agitation of the paint can leave uneven pigment distribution, which will cause color deviation. The paint may not have been thoroughly agitated in its container before being put into the paint application circulation system. The pigment portion of nearly all paints will settle during use if the paint is not agitated at least occasionally throughout use. Inadequate mixing will result in the alternate application of resin-rich paint and pigment-rich paint.
— Low film builds. Excessively thin films may fail to hide a substrate thoroughly. Some colors are far more effective in hiding than others. Light colorcoats often require a greater film thickness to hide a given substrate completely than do darker shades.
— Different application procedures. Various application procedures can produce apparent differences in color with the same paint. Paint applied with air spray may have a different shade than the same paint applied by a centrifugal rotary applicator due to different degrees of atomization and particle delivery velocities. Electrostatic versus nonelectrostatic application can produce color deviations as well.
— Different substrates. Even if the same paint is applied with the same application device at the same time, the color may look different if the substrate material varies. If an assembly of plastic and metal is painted, a variation in the shades of color may occur. This is thought to be due to the different heat-up rates (heat capacities) of metals and plastics during the

curing process. The paint on the plastic could remain wet for a longer period than the paint on the metal, or vice versa, allowing variations in the rates of pigment separation. This is less likely to be a problem with an air-dry paint than with a baked finish.
— Different surface textures. Surface texture differences can cause perceived color deviations. However, color detection instruments may not indicate a color difference. No correlation seems to exist in the perceived shift of light and dark colors on smooth and rough surfaces.
— Overbake. Baking either too long at normal temperatures or at overly high bake temperatures can darken light colors. Some paints seem to be much more sensitive to this than others. Oven loading rates can complicate the situation. Since the heat removal is less when only a few parts are in the oven, the tendency to overbake is greater when the conveyor line is fully loaded.

### *Prevention*

The prevention of color deviations requires absolute consistency in paint agitation, degree of film wetness, film thickness, and application procedures. Color shifts with widely different substrates such as metal and plastic can be prevented in some instances by baking each substrate separately with different bake cycles. Perceived color shifts with different surface textures can be prevented by eliminating the differences in texture (changing the product design). These so-called "solutions" to the problems of color mismatch are often impractical, however.

## Dirt

*Dirt* in paint or in painted surfaces is defined as any and all contaminants, including lint, dust, small clusters or improperly mixed pigment, tiny particles of overspray paint debris, and oil mist.

### *Causes*

The causes of dirt in paint or in painted surfaces can almost always be traced to inadequate facilities, to poor housekeeping, or to poor painting practices. Some of the causes of such dirt include:
— Lint. This type of dirt is one of the most frequently found paint contaminants. Lint can originate from cardboard package interleaving, cartons, boxes, masking paper, shop cloths, clothing, etc. A large plant conducted a detailed study on the nature of the contamination in its painted products and discovered that over 50% of the dirt was attributed to various types of lint fibers.
— Overspray. A common source of dirt is overspray paint that dries and accumulates on various objects in the spray booth. The overspray can disintegrate into tiny particles that can work their way into the applied paint. Improperly balanced air flow in a spray booth may contribute to extensive overspray accumulation.
— Ruptured, loose, or missing filter elements. Poor filter maintenance, whether for paint or for air makeup/air exhaust, can contribute to dirt in paint. Clogged air filters, for example, can add dirt by reducing either makeup or exhaust air flow, and can themselves be a source of dirt by having debris collect on and fall off the filters.

— Dust. Airborne dust can be generated in the painting area by many sources, which may include dirty floors, dirty conveyors, fans, forklifts, and sanding processes. Sanding rework is often performed just outside the paint booth because of the convenience of location. Sanding should be done as far from the spray booth as possible and kept in a confined area with separate and filtered ventilation to contain the sanding dust. Sealing concrete floors with epoxy or urethane can help contain the debris that otherwise will arise from abrasion of the surface.

— Inadequately cleaned paint delivery piping and tubing. Paint piping and tubing that are seldom cleaned can be a source of pigment cluster and resin globule formation.

— Reduction with improper solvents. Adding thinners of the wrong polarity can force resin out of solution. Soft lumps of the resin may readily be squeezed through filters and end up in the finish as visible clumps.

— Dirty paint from the supplier. Paint can be contaminated with dirt when delivered from the coating manufacturer. This is not very common, but it does happen. Occasionally, an uncleaned tote tank or dirty pails are inadvertently filled with paint. The steel balls used to disperse the ingredients into a paint formulation may give off small pieces of steel flake that do not get filtered out before the paint is packaged.

## Prevention

Eliminating dirt in paint or in painted surfaces should be a daily priority assignment for everyone in the painting department of a plant. Sharp vigilance must be maintained in the following areas:

- ☑ Lint. All possible sources of lint should be minimized in the paint department, which if possible should be a closed and restricted area. Only authorized persons properly attired with lint-free outerwear should be allowed to enter. Doors should be opened only for authorized personnel access.
- ☑ Overspray. Spray booth air balance should be checked regularly to confine overspray to the booth's interior. Booth doors should be opened only by authorized personnel and closed promptly, not propped open.
- ☑ Filters. Periodic cleaning or replacement of air filters in the compressed air line helps prevent particle accumulation and dirt. Filter booths that draw plant air into the spray zone must have tight-fitting intake filters with a pore size small enough to remove fine dirt. Flash zones must receive the same timely maintenance.
- ☑ Dust. Dust can be controlled by converting the painting area into a "clean room." Incoming air should be filtered; shoe baths and blow-off vestibules are helpful; sweeping should not be permitted—only wet mopping. Wiping products with a tack cloth prior to painting can be an asset. Ionized air blow-offs can be directed onto plastic parts, which tend to attract dust. It is prudent to cover pallets of parts that are awaiting painting to protect them from possible contamination by oil mist, dust, and particulates present in plant air. Sheets of inexpensive polyethylene are suitable for this purpose.
- ☑ Inadequately cleaned paint piping and tubing. Simply flushing thinner through paint piping and tubing until the solvent is clear often does not do an adequate job of removing the old paint. Blends of aggressive solvents can be purchased that are specifically de-

# FILM DEFECTS IN LIQUID COATINGS 225

signed to clean paint lines before they are used for a new color. Plants that use such solvents for the first time are always amazed at the material removed from supposedly "clean" paint lines.

- ☑ Inadequately stirred paint. Paint should be thoroughly agitated before and during use. If minor amounts of pigment cannot be redispersed, it is satisfactory to filter the paint carefully before using. If appreciable quantities of pigment have settled out of the paint, the coating must be discarded or returned to the supplier. Coatings that are well past their normal shelf life may have formed irreversible pigment seediness.
- ☑ Reduction with improper solvents. The only solvents that should ever be used for reducing paint viscosity are those recommended by the paint suppliers.
- ☑ Dirty paint from the supplier. Paint should always be carefully filtered immediately before use. The paint from the supplier may well be perfectly clean. Last-minute filtering before painting eliminates any possible accidental contamination and can prevent untold quality problems.

## Fisheyes

A *fisheye* defect in a paint film is a small depression (crater) with a mound (dome) in the center (see Figure 16-3). The resemblance to a fisheye is the roundness of the depression (the outer circle of the eye) and the central mound (pupil).

**Figure 16-3. Fisheyes**

### Causes

This defect is nearly always caused by residual oil or grease, especially silicone types. Silicone materials may be accidentally introduced onto surfaces of previously cleaned parts or may contaminate parts not completely washed free of silicones during prepaint cleaning operations. Silicone molecules are extremely polar and thus make very tenacious lubricants for metal surfaces. However, this characteristic makes them difficult to remove as well. Tiny quantities of silicone in sanding dust, on contaminated paper or cloth towels, or from mold releases and polishes sprayed at a considerable distance from parts yet to be painted can cause serious fisheye problems. Because the silicones are so polar, paint tends to "crawl" away from spots of silicone contamination into itself. This forms a small dome of silicone barely covered with paint sitting in the center of a depression.

So many different silicone-containing materials are used that it is impossible to identify a single solvent that is able to dissolve whichever silicone contamination might be present. Various silicones are found in products such as hand creams, hair sprays, polishes, mold releases, rubber seals, waxes, underarm deodorants or antiperspirants, gaskets, and lubricants. One corporation traced the source of its daily fisheye problem, which always occurred between 10 and 11 a.m. to a glass cleaner being used by a vending machine attendant.

### *Prevention*

To avoid silicone-caused fisheye problems, keep silicone products out of a plant that does painting. Employees should be instructed to avoid bringing silicone-containing products into the plant. Where the problem already exists, the contamination must be removed. Mechanical abrasion may be effective, but care must be taken that the sanding debris does not recontaminate other parts. Solvents that have had good success in removing various silicone materials include butyl acetate; 1,1,1-trichloroethane; and trichlorotrifluoroethane.

Materials sold as "fisheye eliminators" are designed to be added in small amounts to the paint and mixed in before application. Roughly 0.2–2.0 ml/gal are used to reduce the tendency for the paint to pull away from the oil remaining on surfaces to be painted. The fisheye eliminators tend to work well because they contain molecules that have a polar end and a nonpolar end. The polar portion is compatible with the highly polar silicone material, and the nonpolar part of the molecule mixes readily with the less polar paint molecules. (Soaps and detergents have similar molecules that are cosoluble to remove nonpolar oil and grease, dispersing them into water that is quite polar.)

Although fisheye eliminators are effective, routine use is generally not recommended. Fisheye eliminators can actually cause fisheyes by interfering with adhesion when a coating that has had fisheye eliminator added to it is repainted. This situation can be acute when touch-up is performed on rework parts. Fisheye eliminator can be nearly impossible to clean out of paint pots and lines. Sometimes continued use of the eliminator is required to prevent fisheyes, even though the silicone contamination on the parts has been corrected. In such situations, a possible solution is to wean the system from fisheye eliminator with gradually diminished amounts of the additive. Do not use fisheye eliminator on a constant basis as a means to circumvent inadequate cleaning of parts. Finding the source of contamination—not using fisheye eliminator at all—is the best way to prevent the fisheye defect.

## Gloss Variations

*Gloss* is a measure of the capability of a surface to reflect light. A painted surface with a high gloss reflects considerable light; one with a low gloss reflects little light. Gloss-deficient patches of paint film can sometimes occur and are known as *flat spotting* and *striking in*.

### *Causes*

The causes of gloss-deficient patches can include:

— Wet spots in a basecoat. Point-to-point gloss variations may be caused by wet spots in a basecoat prior to topcoating. The wet spots arise from improper application techniques.

# FILM DEFECTS IN LIQUID COATINGS

Extra wetness at any area can allow minor pigment floating that can appear as decreased gloss or increased gloss.

— Insufficient oven makeup air. Insufficient fresh makeup air in the oven may cause paint dulling. Few ovens are ever found to be operating at air turnover rates that are too low. However, occasionally solvents need to be swept out rapidly to prevent their dulling the paint. When direct-fired ovens are used for curing, the solvents in combination with oven combustion products can contribute to poor gloss formation. In fact, if the oven is fouled in any way, this can definitely produce a haze on the parts.

— Excessive humidity in the flash zone. Excessive humidity in the flash zone, particularly with paints that contain fast evaporating solvents, can add small amounts of fine water droplets to the paint film. The *blushing* is caused whenever so much evaporative cooling takes place that the surface temperature of the paint film drops below the dew point. Minor blushing may be perceived as low gloss because of its dulling effect on the paint film, as shown in Figure 16-4. Rarely do plants experience true blushing nowadays because so few low-solids lacquer paints are still used industrially. Blushing results in a distinct whitish haze due to condensation of water droplets in the wet film. Minor blushing can generally be buffed out, but in more pronounced cases repainting is necessary. Usually not enough water condenses into the film to cause a whitish blush, but only enough to lower the overall gloss levels.

— Insufficient film build. Low film builds do not permit the paint to flow sufficiently for good gloss. Variations in film thickness can therefore produce nonuniform gloss.

— Excessive oven temperatures. Excessive oven temperatures can sometimes cause a dramatic reduction in overall gloss. Each paint is different in this respect. Dark shades are most likely to exhibit reduced gloss if cure temperatures are too high.

— Molded plastic density variations. Point-to-point density variations can occur in molded plastic parts due to unequal mold filling and unequal mold pressures. These density differences cause varying degrees of solvent adsorption and evaporation with subsequent differences in gloss level from one spot to another.

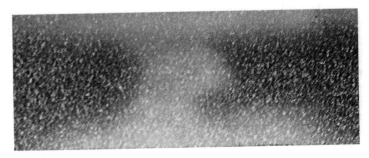

**Figure 16-4. Blushing**

### Prevention

Gloss-deficient areas can be prevented by a six-point program:

- ☑ Use proper paint application techniques.
- ☑ Maintain consistent paint film thickness.
- ☑ Avoid excessive humidity in flash zones.
- ☑ Make certain that oven makeup air is sufficient.
- ☑ Operate bake ovens at specified temperatures.
- ☑ Make sure that plastics are free of density variations by priming them first.

## Mottle

*Mottle* occurs when metallic paint is applied excessively wet and color pigments separate from the metallic flakes. The separation of pigments creates darker "rings" of color that deviate from the overall color; metallic flake forms light regions in the centers of the rings. The rings in mottle are small, typically only 0.03–0.06 inches (0.7–1.5 mm) in diameter.

### Causes

The cause of mottle is almost always traceable to applying paint too thick or extra wet (containing excessive solvent). Many colors are blends of two or more pigment materials whose densities may differ widely. When nonmetallic paints are applied overly wet (flooding), the pigments may tend to separate; one pigment may seem to float. The result is a dark visible appearance where color float occurs.

If metallic paints are applied extra wet, solvent evaporation cooling at the surface can cause minor circulation patterns known as *Benard cells*. The swirling action sweeps the finely ground color pigment to the edges of the cells and leaves the metallic flakes in the center. Because of their size and weight, the metallic flakes are not readily moved by these miniature circulation patterns in the drying paint film. This gives rise to dark rings of color pigment particles with light centers because of increased concentrations of aluminum flakes. When applied too wet, light-colored metallic paint is especially prone to this visible separation of metallic flakes and pigment particles.

A related defect in metallic or nonmetallic films is sometimes caused when extra paint, either due to surface tension or because of electrostatic attraction, accumulates on the edges of panels. This extra paint requires increased drying or cure time, occasionally allowing pigments to "float" to the surface, leaving a dark line around all four edges of panels. This defect is frequently referred to as *picture framing*. (Metallic paints sometimes undergo pigment clumping when applied electrostatically, but that is not related to mottle.) Pigment clumping is caused by charge accumulation on the metal flake pigments.

### Prevention

Pigment separation from the metallic flake can usually be prevented by not applying the paint too wet. Applying the initial coats at normal wetness and the final wet-on-wet coat somewhat dry can help prevent mottle.

Pigment clumping in metallic paint applied electrostatically can usually be prevented by using grounded paint hoses.

## Orange Peel

*Orange peel* in a paint film is characterized by repetitive bumps and valleys similar to the surface of an orange, as shown in Figure 16-5.

**Figure 16-5. Orange Peel**

### Causes

Orange peel results when the freshly applied paint film does not flow out smoothly. The causes of poor flowout are usually one of the following:

— Excessively dry spray. This occurs when excessive solvent evaporates from the atomized paint particles either en route to the target or too quickly after reaching the target so that satisfactory paint flowout is impossible. Excessive solvent evaporation en route to the target can be caused by overatomization, by applying paint in which a solvent's evaporation rate is too fast, or by too much distance between the gun and target. Overatomization creates excessive atomized particle surface area, which increases solvent evaporation. Excessive solvent evaporation after the atomized particles are deposited on the target may be caused by maintaining high booth temperatures or high part temperatures, and by an improper solvent balance. Excessive distance between the gun and target can occur from poor operator technique. For example, too wide a spray fan pattern can make the spray at the edges travel an excessive distance to the target.
— Poorly atomized spray. This can be caused by insufficient atomizing air pressure, overly high paint viscosity, and fluid pressures set too high in conventional and HVLP air spray. Fluid pressures set too low with airless and air-assisted airless application also cause orange peel. Low paint viscosity may be from heaters not working correctly or from improper mixing of paint.
— Overly thin coating. Overly thin coatings can be caused by improper application parameters, and usually have insufficient flow.
— Rough substrates. Paint films cannot level rough or uneven substrates. A metal surface or primer with a rough surface will automatically produce roughness in the topcoat.

## 230 INDUSTRIAL PAINTING

### *Prevention*

Orange peel can be prevented by taking measures to ensure that the paint film flows out satisfactorily. Steps should be taken to:

- ☑ Prevent dry spray. Atomizing air pressure should not be excessive. Solvents should not evaporate too fast. The distance between gun and target should be monitored. Proper spray procedures should be used.
- ☑ Atomize paint properly. Atomizing pressure and paint viscosity should be monitored closely. To rule out cold paint (as a cause of high viscosity) and poor atomization, bring paints inside at least 24–36 h before they are to be applied.
- ☑ Monitor paint film thickness. A sufficient film build is necessary to ensure good paint flowout. Film thickness should be monitored frequently.
- ☑ Check surface smoothness. The smoothness of the substrate should be verified before you begin to paint. Primer smoothness can be achieved by proper sanding.

## Runs, Sags, and Curtains

Paint applied on vertical surfaces may flow downward in various amounts before the curing process hardens the film and stops the flow. All such downward flows are termed *runs* or *sags* (see Figure 16-6). The term *curtain* is used because the lower portion of extended runs and sags may resemble scalloped lower edges of some styles of drapes and curtains.

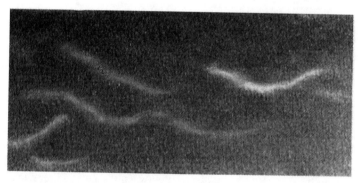

**Figure 16-6. Runs and Sags**

### *Causes*

The causes of runs, sags, and curtains are almost always due to applying a coating too thick or too wet.

Two causes of applying a coating too thick can include:

— Dirty guns. A dirty gun, especially clogged air passages in the tip, can distort the normal spray pattern and make it apply heavily in some areas. Immersing a spray gun completely into a container of solvent for cleaning can cause this situation. If the entire gun is put into solvent, the dirty solvent will be forced into the gun air passages. Paint-laden solvent can dry in the air passages, leaving behind paint solids that will clog them or reduce their operating effectiveness.

- Poor operator technique. Poor operator technique in painting complex parts may contribute to excessive application in certain areas of a part.

The causes of applying a coating too wet can be due to the following:

- Gun too close to workpiece. If the spray gun is too close to the part being painted, an insufficient "flight time" for the atomized paint will leave excess solvent in the paint when arriving at the part surface.
- Excessive solvent. Adding excessive solvent to a coating will allow too much solvent to remain in the newly deposited paint.
- Insufficient air pressure. An insufficient air pressure will atomize poorly, leaving large airborne paint particles, from which solvent evaporates poorly.
- Inadequate flash time. An inadequate flash time between coats of paint applied wet-on-wet can leave a coating excessively wet. The problem of inadequate flash time often is encountered near the end of each month when line speed tends to be increased to get products out on time.
- Cold parts. Applying paint to cold preprimed parts may slow solvent evaporation enough to cause a wetness problem. This problem can result when parts are stored in an unheated warehouse and not taken in to warm long enough before being topcoated.
- Fluid tip (nozzle) is too large. Always use a tip that gives the correct fluid amount for the areas of your parts that are the most difficult to paint properly.

## Prevention

Steps to take to prevent runs, sags, and curtains can include:

- ☑ Cleaning dirty guns. A spray gun should be cleaned properly. It should never be totally immersed in solvent; only the spray tip should be cleaned in this fashion. Leave the air on slightly when cleaning to prevent paint and solvent from entering the air passages. The gun should be checked regularly to prevent defective spray patterns caused by plugged air passages.
- ☑ Training operators. Operators should be trained to develop efficient spray procedures. They should periodically check the film thickness at different areas of a part to make sure they are not applying paint too thick in spots.
- ☑ Operating guns properly. By moving the gun farther from the workpiece, the droplet travel time increases, allowing additional time for solvent evaporation. This procedure reduces the possibility of wet paint and the resultant runs and sags.
- ☑ Using correct amounts of solvent. If excess solvent has been added, a correction should be made. In some cases, particularly in cold weather, a fast evaporating solvent blend should be used. This will increase solvent loss during application so the applied paint is drier.
- ☑ Correcting air pressure. A slight increase in air pressure will improve atomization and increase the surface area of the droplets. This will improve solvent evaporation and contribute toward a slightly drier paint.
- ☑ Providing adequate flash time. When applying coatings wet-on-wet, you must allow sufficient flash time after each application. Otherwise, solvent can be trapped under the second coating, which produces excessive wetness.

☑ Warming parts. To prevent cold parts from slowing solvent evaporation and increasing wetness, always bring the parts to room temperature before you paint them. Parts that go through a washer will be warmed in the process, but occasionally parts that have already been primed are repainted or topcoated. If these parts are stored in an unheated warehouse, they should be brought to room temperature before being hung on the line to be topcoated. A good practice is to require that parts be delivered to the paint area at least 24 hours before being painted.

## Paint Adhesion Loss

*Adhesion loss* is the premature separation of a paint film from a substrate.

### Causes

The causes of paint adhesion loss can be one of the following:

— Contaminants under the applied paint. Almost 95% of the loss of large or small areas of paint due to poor adhesion is caused by contamination under the paint. This may be oil, grease, sanding residues, water, or other contaminants.
— Excessive bake time or oven temperatures on primers (and auxiliary coats). This may render the primer (or other auxiliary coats) so thoroughly cross-linked that solvents in the next paint layer cannot microetch the surface to attain good adhesion.
— Paints that differ widely in polarity. When an applied paint film differs substantially in polarity from the next applied paint, adhesion of the second coating may fail. Paints with widely differing polarities tend to repel each other.
— Condensed moisture. High humidity in the paint shop may form condensation on parts to be painted, especially on cold parts. Applying paint over the moisture can cause fine blisters and even large-scale adhesion loss.

### Prevention

To prevent adhesion loss, the following steps should be taken:

☑ Clean parts thoroughly before painting. All washer stages should be monitored regularly to ensure that bath temperature, concentration, pH, and other factors are within specifications. Periodic tests should be conducted on cleaned parts to ensure that all soils have been removed.

☑ Maintain correct bake parameters. Maintaining correct bake cycles and oven temperatures can prevent over-baking a coating, enabling the next applied paint to microetch the coating for a proper adhesion bite. If a previous coating is over baked, a light scuff sanding may be beneficial to provide an anchor pattern for the next layer of paint. Sanding will cut through the hard outer glaze on the existing paint layer and allow solvents in the topcoat to "bite" into the previous paint layer.

☑ Maintain similar polarities for applied pants. Sealer coats of intermediate polarity must be used between paints that have widely differing polarities. In this way the adhesion of primer to sealer and sealer to topcoat will be good.

☑ Prevent condensed moisture. Making sure that cold parts are warmed before they are painted can help prevent moisture from forming on parts when humidity is high, as is often the case around the pretreatment and paint area.

## Soft Paint Films

*Soft paint films* are deposited coatings cured to a hardness below a designated specification. Soft paint films are easily marred (readily penetrated with a fingernail) and will have inadequate solvent resistance. When parts are stacked, parts with soft paint may "block" or stick together; they may also stick to or rub off onto paper and cardboard packaging materials. Another term for this blocking or sticking together is *printing*. Soft paint can happen when ambient humidity is high, especially with air-dry paints.

### Causes

The causes of soft paint films may be one of the following:

— Low oven temperature. A faulty oven temperature indicator system may result in oven temperatures below those required. Inadequate cure may be taking place without any indication that the oven temperature is incorrect.
— Low oven air makeup. Insufficient fresh air makeup in the oven may retard the extent to which oxygen molecules react with paints that cure by oxidative cross-linking. Low air turnover may keep paint from developing full hardness by not allowing the removal of sufficient quantities of solvent during the bake cycle.
— Contaminant softening agents. Solventborne paint will usually dissolve minor amounts of substrate wax, oil, and grease, and readily absorb moderate amounts of oil from compressed air. However, if significant amounts are incorporated into the paint film, the oil and grease can act as flexibilizers or softening agents.
— Excessive paint storage time. Paint that has exceeded its shelf life may not cure properly. This is especially true with some high-solids coatings, which have a shelf life considerably less than low-solids paints. High-solids paints contain small, reactive molecules, so shelf life inevitably is short. When a paint exceeds its shelf life, cross-linking in the coating progresses so far that not enough cross-linkability remains to harden the paint fully when it is cured on the parts.
— Excessive retarder solvent. Excess retarder solvent can prevent a paint from achieving normal hardness levels.
— Excessive film builds. Excessive film builds may not permit full curing during the normal cure cycle.
— Insufficient cure time. Too brief a cure time at a given temperature may leave the paint soft. This sometimes occurs when plants increase the line speed to meet production schedules without raising the oven temperature.

## Prevention

Soft paint may be prevented by taking the following steps:

- ☑ Monitor oven temperature. Oven temperature gauges and control systems should be checked periodically to ensure that the indicated oven temperature is true.
- ☑ Maintain oven air makeup. Oven air makeup should be maintained at specified levels to be certain that sufficient oxygen is being supplied for oxidative cross-linking. Oven air turnover must be high enough to remove evaporated solvent.
- ☑ Avoid softening agents. Wax, oil, and grease should be kept off parts to be painted. Compressed air should be checked for oil content.
- ☑ Check paint storage time. Paint should not be stored beyond that period recommended by the coating manufacturer.
- ☑ Use proper amounts of retarder solvent. When retarder solvents are added to improve flowout or gloss, extreme care must be exercised to avoid soft paint. Retarders should be added in small increments.
- ☑ Apply proper film builds. Operator care must be taken to avoid excessively thick paint films, which will not cure satisfactorily during normal oven cycles. If thick coatings are needed, the oven cycle should be increased.
- ☑ Cure for the correct time. If the conveyor line speed must be increased, thus decreasing time in the oven, then oven temperature must be increased. Operators should be made aware of the relation between oven time and temperature. The paint manufacturer can supply time-temperature curves to allow correct temperature compensation for cure time variations.

## Solvent Pops, Boils, and Pinholes

*Solvent pops, boils,* and *pinholes* are tiny craters on the surface of paint films, as shown in Figure 16-7. They are small versions of bubbles and craters.

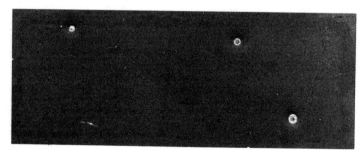

**Figure 16-7. Solvent Pinholes**

### Causes

These defects are basically caused by overly rapid solvent loss from the wet paint. Solvent may escape in small "bursts" from a paint film that has partially dried. The paint may be unable to flow back together to hide the escape ports, leaving tiny craters. The causes are exactly the same as for bubbles and craters, only to a lesser degree. Too little flash time and overly thick wet films are major causes.

Additional causes for solvent pops, boils, and pinholes include:

— Pigment clusters. Insufficiently agitated paint may form pigment clusters that can trap solvent and delay evaporation.
— Surface roughness. Some types of surface roughness can act as nucleation sites for vapor bubble formation.
— High solvent evaporation rates. High solvent evaporation rates can exaggerate solvent pops, boils, and pinholes.
— Low solvent evaporation rates. This can slow solvent evaporation during flash periods, leaving an excess amount of solvent to evaporate in the oven.
— Drafts. Strong drafts can sometimes speed solvent evaporation to the point where solvent pops, boils, and pinholes can occur.
— High oven entry temperatures. Sudden exposure to a high heat at the oven entry can heat the paint film too rapidly.
— Bell-trapped air. This is identical to solvent popping and is caused by tiny nicks in the circumferential edge of turbine-driven bells. Atomization at the nicks can be inefficient and form large atomized particles that contain more solvent than the remainder of the efficiently atomized particles.

### *Prevention*

Solvent pops, boils, and pinholes can be prevented by taking the following precautions:

- ☑ Avoiding pigment clusters. Efficient agitation will prevent the forming of pigment clusters. Filtering the paint will remove the non-redispersible pigment clusters.
- ☑ Checking surface roughness. If compatible with the product design, the problem-causing surface roughness should be eliminated by a smoothing step in the manufacturing process.
- ☑ Controlling solvent evaporation rates. Solvent evaporation rates can be controlled by using only solvents recommended by the paint manufacturer. Extra flash time is beneficial for slow evaporating solvents.
- ☑ Avoiding drafts. Direct impingement of blower-driven air onto freshly painted parts should be avoided.
- ☑ Using correct oven entry temperatures. The problem can sometimes be corrected by zoning the oven. This permits somewhat low temperatures at the oven entry to warm the part gradually and avoid overly rapid solvent evaporation. Subsequent zones are set increasingly warm to cure the paint fully.
- ☑ Preventing bell-trapped air. This can be eliminated by replacing bell heads that have any nicks along their circumferential edges. Careful handling will help avoid costly repair or replacement of bell heads. New rotary heads now cost $450-$1100.

## Solvent Wash

*Solvent wash* is the term used to describe paint voids or areas with thin paint due to solvent condensation.

### Causes

Solvent wash is caused by excessive solvent evaporation from a painted part in the entry area of an oven and the condensation of this solvent on cool areas of the part. The condensed solvent runs down the part and can wash away either some or all of the paint where the condensation occurs. Solvent wash is most likely to happen on parts with paint applied extremely wet (solvent rich). Flow coating and dip coating are especially prone to solvent wash because of the high solvent content of wet paint films applied by dipping and flow coating. Solvent wash rarely occurs with spray or rotary application of coatings.

### Prevention

Whenever paint must be applied especially wet, for example, in flow coating and dip coating, sufficient flash time must be allowed for extensive solvent evaporation before the painted parts enter into an oven.

# Chapter 17

# Paint-related Testing

## Categories of Paint Tests

Various paint-related quality tests have been devised over the years. Most are aimed at verifying the integrity of a particular property of the paint or finish. The tests can be grouped into three basic categories.

- Tests of paint "in the bulk"
- Tests on parts and paint application equipment
- Tests of the applied paint film

The test analyst should know the correct evaluation system to rate and compare each test. Test descriptions from the American Society for Testing and Materials (ASTM) generally include explanations of the preferred rating methods for most tests. Pictorial standards can also be particularly helpful.

## Tests of Paint "in the Bulk"

Paint in the bulk is often tested to determine:

- Weight per gallon
- Viscosity
- Weight percent solids/weight percent volatile organic compound
- Volume percent solids
- Electrical conductivity

**Weight per gallon.** Special metal containers holding a precise volume are used for this measurement. One such cup holds exactly 83.05 cc at 77° F (25° C). The weight of paint that fills the container, expressed in grams, is equal to the pounds per 10 U.S. gallons of the paint. The weight of paint in the container expressed in grams can be multiplied by 11.76 to equal the grams per liter.

**Viscosity.** In technical terms, *viscosity* is the ratio of the shearing stress and the rate of shear of a Newtonian liquid. Shear (in terms of liquids) is an applied force that disturbs a liquid at rest. In a Newtonian liquid, the rate of shear is proportional to the shearing stress. In a non-Newtonian liquid, the rate of shear is not proportional to the shearing stress.

To briefly explain viscosity in terms of shear and Newtonian/non-Newtonian liquids is complex and difficult. For the purposes of this book, viscosity can be defined in simple nonscientific terms as the "flowability" of a liquid. A paint with a high viscosity has a poor flowability; it is thick and syrupy. A paint with a low viscosity has a high flowability; it is thin and watery. However, more precise measures are needed than descriptions with terms such as "watery" or "syrupy." An instrumental method is required.

Various instruments, such as the Brookfield viscometer, are available for measuring true paint viscosity, but they normally are used only by paint formulators in a lab. A simple, fast device with timing system has been devised for measuring paint flowability (viscosity). The device is referred to as a viscosity cup. Names of some commonly used viscosity cups are Shell, Norcross, Zahn (see Figure 17-1), Ford, Sears, and Fisher. A viscosity cup is a small metal cup with a precision hole at the bottom. The test consists of filling the cup and timing with a stopwatch the number of seconds required for the paint to drain in an unbroken flow through the precise hole in the cup. As viscosity rises, the time increases for draining out through the orifice. Figure 17-2 provides a means of converting viscosity "seconds" from various cups to centipoise viscosity units.

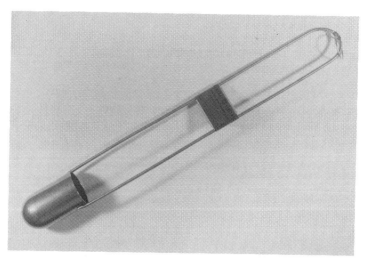

**Figure 17-1. Zahn Viscosity Cup**

Paint is normally purchased at usage viscosity, or at a higher-than-use viscosity. Solvent can be added to reduce the paint to the correct application viscosity. Usually the final adjustment to paint viscosity is done just prior to application of the paint. Because temperature has a large effect on viscosity, plant paint personnel must make allowances for this variable to avoid viscosity changes as the paint becomes warmer or cooler.

**Weight percent solids/weight percent volatile organic compounds.** For an organic solventborne paint, the sum of the weight percent solids and the weight percent VOC will be 100%, so finding one of these values readily provides the other. The usual test involves weighing a wet sample of paint, then evaporating the volatiles, and finally reweighing the remaining

| Centipoise | Fisher #1 (seconds) | Fisher #1 (seconds) | Ford Cup #3 (seconds) | Ford Cup #4 (seconds) | Zahn #1 (seconds) | Zahn #2 (seconds) | Zahn #3 (seconds) | Zahn #4 (seconds) | Zahn #5 (seconds) |
|---|---|---|---|---|---|---|---|---|---|
| 10 | 20 | | 5 | 30 | 16 | | | | |
| 15 | 25 | | | | 8 | 34 | 17 | | |
| 20 | 30 | | 15 | 12 | 10 | 37 | 18 | | |
| 25 | 35 | | 17 | 15 | 12 | 41 | 19 | | |
| 30 | 39 | | 18 | 19 | 14 | 44 | 20 | | |
| 40 | 50 | | 21 | 25 | 18 | 52 | 22 | | |
| 50 | | | 24 | 29 | 22 | 60 | 24 | | |
| 60 | | | 29 | 33 | 25 | 68 | 27 | | |
| 70 | | | 33 | 36 | 28 | | 30 | | |
| 80 | | | 39 | 41 | 31 | | 34 | | |
| 90 | | | 44 | 45 | 32 | | 37 | 10 | |
| 100 | | | 50 | 50 | 34 | | 41 | 12 | 10 |
| 120 | | | 62 | 58 | 41 | | 49 | 14 | 11 |
| 140 | | | | 66 | 45 | | 58 | 16 | 13 |
| 160 | | | | | 50 | | 66 | 18 | 14 |
| 180 | | | | | 54 | | 74 | 20 | 16 |
| 200 | | | | | 58 | | 82 | 23 | 17 | 10 |
| 220 | | | | | 62 | | | 25 | 18 | 11 |
| 240 | | | | | 65 | | | 27 | 20 | 12 |
| 260 | | | | | 68 | | | 30 | 21 | 13 |
| 280 | | | | | 70 | | | 32 | 22 | 14 |
| 300 | | | | | 74 | | | 34 | 24 | 15 |
| 322 | | | | | | | | 36 | 25 | 16 |
| 346 | | | | | | | | 39 | 26 | 17 |
| 370 | | | | | | | | 41 | 28 | 18 |
| 394 | | | | | | | | 43 | 29 | 19 |
| 418 | | | | | | | | 46 | 30 | 20 |
| 442 | | | | | | | | 48 | 32 | 21 |
| 466 | | | | | | | | 50 | 33 | 22 |
| 490 | (For poise value, divide centipoise value by 100) | | | | | | | 52 | 34 | 23 |
| 514 | | | | | | | | 54 | 36 | 24 |
| 538 | | | | | | | | 57 | 37 | 25 |
| 585 | | | | | | | | 63 | 40 | 27 |
| 657 | | | | | | | | 68 | 44 | 30 |
| 777 | | | | | | | | | 51 | 35 |
| 896 | | | | | | | | | 58 | 40 |
| 1016 | | | | | | | | | 64 | 45 |

**Figure 17-2. Viscosity Conversion**

solids. The test can be done simply: Fill a 5- or 10-ml syringe with paint and weigh carefully. Into a small preweighed beaker or disposable aluminum cup, deliver roughly 4 mls of paint. Then again carefully weigh the syringe to determine the weight of wet paint in the beaker or cup. Bake the sample for 1 h at 235° F (113° C). When cool, reweigh the beaker plus the sample to find the dry-solids weight. The percent solids and percent VOC by weight can be found as follows:

1. Percent solids equals weight of dry paint multiplied by 100 divided by the weight of wet paint.

$$\% \text{ weight solids} = \frac{\text{wt dry paint} \times 100}{\text{wt wet paint}}$$

2. Percent VOC equals 100 minus the percent solids

$$\%\text{VOC} = 100 - \%\text{solids}$$

The procedure for VOC in waterborne paints requires that the percentage of water be found by using gas-liquid chromatography, which is not readily available in most plants, or by the *Karl Fisher method*. The latter is a complex titration procedure and one virtually never attempted by painting personnel. The complete method is not detailed here for that reason. Although the Karl Fisher technique is still used, the gas chromatographic determination is faster, easier, and more accurate.

**Volume percent solids.** Determining volume percent solids requires weighing an unpainted disposable small part both in air and in water and reweighing the part in air and in water after it has been coated and cured. It is not commonly part of plant paint testing. The details of the test can be found in ASTM Test Method D-2697-73.

**Electrical conductivity.** The electrical conductivity of paint can be found by measuring its resistance with an ohmmeter adapted with wide-surface terminals to ensure test consistency. Conductivity, normally expressed in units of micromhos, is the mathematical reciprocal of resistance in megaohms, so its value is calculated that way. Notice that the higher the resistance value, the lower the value of the conductivity, and vice versa.

Some coatings need to be highly conductive, such as coatings used to provide electromagnetic interference shielding or radio frequency interference shielding. In contrast, coatings to be applied electrostatically need to have a low conductivity (high resistance) to avoid shorting out the application equipment. They cannot have zero conductivity, however, because a small amount of conductivity is needed to hold the electrostatic charges. Otherwise, these paints will not provide the benefits of electrostatic application such as "wraparound" and increased transfer efficiency. Values of 0.5–5 megaohms resistance are normal; higher resistance paint from 2–5 megaohms can be used for automatic application, which can use high voltages. Manual application is most often done with paint having about 0.5–2 megaohms resistance. Too little resistivity (too much conductivity) causes severe Faraday cage problems.

## Tests on Parts and Paint Application Equipment

Various tests can be conducted on parts to be electrostatically painted and even on the paint application equipment itself, depending on the type of equipment being used. Three tests that can be conducted are:

- Checking for an electrical ground
- Measuring electrostatic voltage
- Determining transfer efficiency (TE)

**Checking for electrical ground on parts to be painted.** Good electrical ground connections in electrostatic painting ensure maximum electrostatic attraction of paint to the product and will prevent spark-producing charges from building up on the product. The electrical path to ground goes successively from the product to a support such as hook or hanger, to the conveyor trolley, to the conveyor channel, to the grounded building metal framework, and finally into the ground under the plant. A poor connection at any of these junctions will result in a poorly grounded part and inefficient electrostatic paint application.

A quick test for an electrical ground can be made with an ordinary ohmmeter or a decreased ground detector. In both cases, one terminal is attached to the product to be painted, and the other is attached to a good absolute ground, such as the conveyor I-beam. If an ohmmeter reads 1 ohm or less, the part is grounded properly. The decreased ground detector can be made to give either an audible or visual alarm (or both) if grounding is poor.

The two most common causes of poor grounding are accumulated paint on hooks and hangers and dirty conveyor wheels and channels.

**Measuring electrostatic voltage.** Most electrical control panels will indicate the voltage of the electrostatic power supply. It is a good idea to check the voltage at the application device occasionally to ensure that sufficient voltage is reaching the applicator for good electrostatic charging. Voltage testers are available for this purpose. Instructions should be followed carefully and all safety precautions observed.

**Determining transfer efficiency (TE).** *Transfer efficiency* is a measure of the amount of paint applied to a part in comparison with the total amount of paint used. TE can be determined if the percent solids (by weight) of the paint is known by following these steps:

1. Weigh the paint container before and after use in painting a set number of parts (Value A).
2. Weigh the parts before painting, then weigh again after painting and curing is complete. The difference is the dry paint weight (Value B).
3. Multiply Value B by 100 and divide by the percent solids (Value C).
4. Multiply Value C by 100 and divide by Value A. This is the percent TE.

TE also can be determined by measuring the dry-film thickness (DFT) (in mils), measuring the surface area painted (in sq ft), measuring or acquiring the percent volume paint solids from the coating supplier, measuring the volume of coating used, and then putting the figures into the following formula and calculating the result:

$$\%TE = \frac{\_\_\_\_ \text{ sq. ft. painted} \times \_\_\_\_ \text{ mils DFT}}{\_\_\_\_ \text{ \% vol. solids} \times \_\_\_\_ \text{ gal. used} \times 0.1604}$$

(The 0.1604 is simply a unit conversion factor.)

## Tests of Applied Paint Film

Numerous tests can be performed on applied paint films. Among the most common are:

- Wet-film thickness
- Dry-film thickness
- Adhesion/flexibility (bend, impact)
- Adhesion (tape)
- Hardness
- Chemical resistance
- Stain resistance
- Extent of cure
- Water immersion
- Humidity resistance
- Accelerated weathering
- Gloss
- Distinction of image
- Color match

It is important to remember that a number of coatings, especially air-dry alkyds, will continue to cure for as long as a week or two after the initial apparent cure. For this reason a conditioning period after curing must be allowed before accurate testing can be done on the dry film. This is especially true for tests such as hardness and corrosion resistance. A conditioning time of 24–72 h before testing is a common practice for durability and adhesion tests. In most cases the test results will improve after the conditioning period. If preliminary tests show satisfactory performance, parts that have gone through the requisite conditioning period need not be retested. However, tests of film thickness, gloss, color match, and any tests not related to the extent of cure can be done immediately.

### *Wet-film Thickness*

The most commonly used wet-film thickness gauge resembles a comb with two outer teeth of equal length and about 10 teeth in between of varying lengths. When the gauge is placed into the wet film and then removed, the marking on the last tooth to be wetted by the paint gives a readout of the wet-film thickness. Because the gauge leaves a visible imprint in the wet paint, it is a destructive test. It may be possible to do the test in a noncritical spot, or a separate test panel can be used for this test if the blemish would require major rework problems.

The thickness of a wet paint film is seldom of strong interest. Sometimes knowing the wet-film thickness is helpful, such as when painters are being trained, to satisfy specifications, or when setting up automatic application equipment. Even then it is measured in order to get an assured minimum dry-film thickness on the parts.

### *Dry-film Thickness*

The dry-film thickness of coatings on magnetically responsive substrates are often measured with magnetic "pull-off" gauges or "roll-off" gauges. These substrates are primarily ferrous (containing iron) metals. (Film thicknesses on nickel-containing alloys can also be measured by magnetic pull-off gauges, but painted nickel materials are only infrequently used industrially.) These devices have a magnet suspended by a spring inside a nonmagnetic body, usually made of plastic or aluminum. When the gauge is touched to the painted surface, the magnet and the metal are mutually attracted, but the force is weakened because it must act through the paint film. The spring will be stretched whenever the magnet is pulled away perpendicularly to the surface. The magnetic attraction between magnet and substrate is reduced almost linearly as the coating thickness increases. The thicker the paint, the weaker the magnetic attraction. Thus, the spring is stretched less when the gauge is pulled away from a thick film than when it is pulled from a thin paint layer. The amount of spring extension, indicated by a scale marked on the gauge, is read by the person doing the test.

The two types of magnetic pull-off gauges are "pencil" and "banana" (roll-off). A pencil gauge, as shown in Figure 17-3, has a magnetic head attached to a spring and a pointer. Gradual manual pull-off of the gauge away from the surface gives a reading of paint film thickness with an average accuracy of ±15%. One of the easiest-to-use pencil gauges looks and operates like a hypodermic syringe. This particular model retains the readout value and thus is easy to use. With many pencil gauges, however, the person using the gauge must have a quick eye, for as soon as the gauge is pulled away from the surface, the reading disappears.

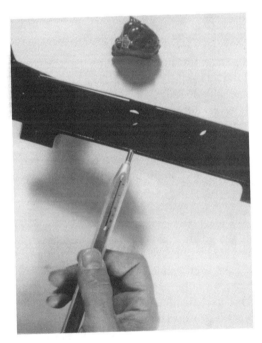

**Figure 17-3. Magnetic Pencil Gauge**

If these instruments are not operated vertically, the manufacturers state that accuracy is reduced. Accuracy can be improved by plotting a correction curve for each angle away from true vertical. Since the device is not that precise, few people bother with this type of correction factor. For usage where most readings taken are not done vertically, a Tinsley type 7000 gauge may be preferable. It contains a meterlike scale readout with a balanced pointer that is unaffected by its angular orientation. The Tinsley type 7000 gauge provides accuracies of ±10%.

A banana or roll-off style gauge, as shown in Figure 17-4, has improved accuracy. Rotating the wheel slowly pulls a magnetic head away from the surface being measured. When the magnet breaks contact, both an audible "click" and an indicator inform the tester to cease rotating the

**Figure 17-4. Magnetic Banana Gauge**

pull-off wheel. At this point the calibrations marked on the wheel allow the film thickness to be read directly, normally in units of mils or microns. The cost of the roll-off devices is approximately double that of the pencil gauges. All gauges of this type provide an easily read readout value that lasts, unlike those pencil gauges that lose the thickness reading when the magnet head releases from the surface.

One preferred roll-off model uses a battery to rotate the wheel that pulls the magnet from the surface being tested. The roll-off gauge starts and stops rotating automatically, and therefore is faster to use. In addition, errors that can occur if the tester does not stop rotating the pull-off wheel quite quickly enough are avoided. This added feature doesn't increase the gauge cost to a noticeable degree.

Both pull-off and roll-off magnetic devices are portable, easy to use, and relatively low in cost. They do not damage the coating and do not require batteries unless the model just explained is desired. However, they will not work on plastics or aluminum; they are useful primarily on iron and steel that have been painted with nonconductive coatings.

The magnet tip should be visually inspected periodically for wear and for attachment of small steel particles and other contaminants that could disturb the readings. Exposure of a magnetic gauge to a strong electrical or magnetic field can alter its strength and change the calibration. A number of manufacturers will recalibrate the device for a service fee. This should be done if it does not give correct readings when checked against a known film-thickness standard. These standards are small pieces of plastic film available from gauge suppliers for use in calibration. Roll-off gauges and pull-off gauges can both be recalibrated when that becomes necessary.

Dry-film thickness on all metals, even on nonmagnetic metals such as zinc or aluminum can be measured using "magnetometers," more commonly called "eddy current detectors." These measure dry-film thickness based on the relative strength of eddy currents in the metal induced by a varying magnetic field in a conducting coil. Differences in the current flow (because it is alternating current) within the coil are related to the film thickness separating the coil and the metal under the paint film. These compact devices are normally accurate to ±5%. Battery-operated units can be hand-held. Their accuracy, compact size, applicability to all metal substrates, and speed and ease of operation have made them popular instruments for measuring paint film thickness.

The eddy current testers must be calibrated for the particular alloy and type of coated metal substrate just prior to use. This can be done in just a few seconds using a bare sample of the substrate metal and two or three of the standard thickness plastic films always provided when the instrument is purchased. Calibration should be done at film thicknesses slightly thinner and thicker than the film thicknesses to be tested.

If extreme accuracy is needed, film thickness can be measured with beta-ray backscatter devices. They emit electrons (years ago electron beams were called beta rays and the name has stuck) toward the surface, and the relative amount of deflected rays are measured and correlated to film thickness. The beta-ray instruments provide extremely accurate values as long as an appreciable difference exists between the nature of the substrate and the paint. They work well measuring paint thicknesses on metals, wood, paper, cardboard, etc. and on many (but not all) plastics. These devices are quite costly and require careful and time-consuming calibration. Only rarely are these used for paint thicknesses. Nonpaint coatings requiring extreme thickness uniformity, such as coatings on photographic film or video and audio tapes, are often checked by beta-ray backscatter instruments. Metal film thicknesses are also measurable with these devices, and so they are used in semiconductor manufacturing.

Measuring paint film thickness on plastic parts by beta-ray backscatter instruments can often present difficulties because the paint is structurally similar to the substrate. But, of course, neither pull-off and roll-off gauges nor eddy current detectors are able to determine paint thickness on nonmetallic parts. To find paint thicknesses on plastic parts, some alternatives to the previously discussed methods used for measuring dry-film thicknesses on plastic include:

- Using an ultrasonic thickness gauge designed for non-metallic substrates. These battery-powered devices can measure DFTs to ±3% accuracy on concrete, glass, wood, and plastic.
- Measuring the plastic thickness before and after painting, using a micrometer or other means of measuring linear thickness.
- Measuring the thickness of a piece of tape before it is placed onto a plastic part, then again after the part is painted and cured. Remove the tape from the part and measure the thickness of tape plus paint.
- Measuring the thickness of a piece of steel before it is adhered onto a plastic part, then again after the part is painted and cured. Remove the steel piece from the part and measure the thickness of the paint with a magnetic gauge.
- Using a Tooke gauge. This instrument will scribe a short V-shaped groove through the paint and also into the plastic substrate. Next the magnifying glass supplied with this instrument is positioned directly over the groove. Internal scale markings projected through the viewer permit a direct thickness readout of the paint film thickness.

Since the last three of these tests are destructive, they might be performed either on scrap parts or parts in a nonappearance area.

### Adhesion/Flexibility (Bend, Impact)

Paint film adhesion/flexibility can be tested in several ways. One method puts a painted panel through various types of folds or bends. Another procedure drops weighted objects onto a painted panel from varying heights. The aim of the tests is to measure how easily paint can be made to separate from the substrate by impact or by tensile and compressive forces.

In the bend test, a panel painted on one side can be folded back 180° onto itself (with the painted surface on the inside of the fold) to test for compressive paint adhesion/flexibility film stresses. A painted panel can be folded 180° in the opposite direction with the painted surface on the outside of the bend to test for adhesion under tensile (stretch) stresses. This type of a bend is termed "0T," pronounced "zero-T." The "T" stands for thickness.

A panel bent back onto itself with nothing between the two flat parts of the fold has a 0T bend. A panel, if bent back 180° but with another section of itself (or over another piece of panel having the same thickness) between the two folded sections, is described as having undergone a "1T" bend. The "1T" stands for a bend over one thickness. Similarly, panels bent back over two sections of itself, or over two layers of the same thickness, have a 2T bend, and so on. It follows intuitively that the tighter the bend radius, the greater the severity of the compressive and tensile stresses that strain on the paint film's adhesion and flexibility.

Painted test panels may also be folded around cylindrical rods of various diameters. This will also test for adhesion/flexibility under compressive and tensile stresses. Similarly, painted panels may be tested for adhesion/flexibility by folding them around cone-shaped rods. These tests are known as "cylindrical mandrel" and "conical mandrel" bend tests, respectively.

Tests of paint film adhesion/flexibility by dropping objects are conducted by dropping steel balls or rounded-end metal impactors of specific weight, usually 3 lbs (1.361 kg). Various shape impactors are used, but all such tests are commonly termed "falling ball impact" tests because ASTM G-14 specifies a 3-lb hemispherical "tup" (impactor). The impactor may be dropped onto the painted side or the unpainted side of a test panel or onto a flat section of a production part. In some instances, tests might be specified for both sides. The test weight is dropped from various specified distances. Having the impactor strike the painted side of a panel is referred to as a "direct impact test;" striking the unpainted side would be termed a "reverse impact test." In direct impact tests, the concave-dented painted surface produced is examined for paint adhesion loss. After a reverse impact test, the convex protrusion of the painted surface is similarly examined. Overly brittle paint will crack and possibly lose adhesion in these types of tests. The degree of cracking and paint adhesion failure is identified by photographs or by a suitable rating method.

### Adhesion (Tape)

The adhesion of paint to a surface can also be measured by applying a piece of tape to the paint film and rapidly pulling the tape off the paint. A piece of 1-in (2.5-cm) wide transparent plastic tape with appropriate adhesion strength and about 3 in (7.6 cm) long is pressed onto an area free of any contamination, blemishes, or other surface imperfections. The tape is smoothed into place with a finger, then rubbed tightly to the surface with an eraser on the end of a pencil. A dark color under the tape indicates that the tape has been affixed firmly. One end of the tape is grasped and pulled steadily and quickly (but not jerked) back on itself as close to the plane of the tape as possible. ASTM D-3359 is considered to be the standard tape-adhesion test, although hundreds of variations of this test are used. The ease of the test is an attractive feature. ASTM provides a rating scale to evaluate the results. The results may vary if the tape test is performed immediately after the paint film is cured instead of waiting until after a conditioning period.

The test may be performed directly on an unaltered paint film or on a paint film that has been scored (scratched) with an "X." The paint may also be crosshatched with multiple score lines in different fashions and then tape-tested. Sometimes these test methods may be performed after the panel has been subjected to varying temperature, humidity, or salt spray exposures. Occasionally, a specification will call for a tape adhesion test to be performed on a painted panel that has been immersed for an hour in deionized water. Further details on tape-adhesion testing are provided in ASTM D-3359.

### *Hardness*

Although sophisticated paint film hardness-testing methods are available and in use by paint manufacturers, the ASTM D-3363 pencil hardness test is simplest and most convenient for the appliers of coatings. Hand-held pencils with leads of varying hardness are pushed (lead first) at a 45° angle against the paint film to determine the paint hardness. The pencils are not sharpened to a point; the full round of exposed lead is squared off. The edge of the lead is then pushed against the paint film at the specified angle. The amount of force on the pencil must be enough to cut or scratch into the film, or to break off the edge of the pencil lead. Check closely for marks in the paint film. Begin testing with the hardest lead, then decrease hardness until the hardest pencil that will not mar the paint is found. Some plants measure both "gouge hardness" and "scratch hardness," but most look only for "mar hardness" of any kind.

Paint film pencil hardness is rated on a scale of numbers and letters from soft to hard as follows: 6B, 5B, 4B, 3B, 2B, B, HB, F, H, 2H, 3H, 4H, 5H, 6H. Instead of individual pencils, one can purchase a set of assorted pencil leads that fit into a holder and are used much like an automatic pencil for this test. The pencil lead hardness test is crude, but it is quick and adequate for most purposes.

A hardness test is often used as a measure of the extent of cure of a paint film; under-cured films will be soft, and over-cured films will be excessively hard. This is allowable in the majority of cases, although the solvent rub test (see Extent of Cure section) probably is better overall for this purpose. Normally a permissible range of hardness rather than just a single hardness value is written into a coating specification. Typically, such a requirement would state "...a 2H minimum to 4H maximum paint film hardness shall be required."

### *Chemical Resistance*

In chemical-resistance tests, various chemicals are applied to paint films and allowed to remain on the surface for hours or days, as deemed suitable for the product. This is then followed by visual examination of the film for harmful effects. Popular chemicals for such tests include acids, alkali, and solvent. The film is then evaluated for the degree of harm, if any has occurred.

Softening, dissolution, loss of gloss, and degradation of the film are the four most often encountered results of chemical attack on paints.

## Stain Resistance

In stain-resistance tests, various materials are applied to coating films, followed by washing and drying, and then examining the paint for stains. Vinegar, catsup, mustard, iodine, bleach, and detergent are often used. The appliance industry is vitally interested in having a painted finish resist staining and discoloration from a wide variety of common foods and household products.

## Extent of Cure

The hardness test is one measure of how well a paint film is cured. A much preferred one is the "solvent rub" test, which measures the film's resistance to dissolution in whatever solvent is deemed to be most appropriate. In this procedure, the finish is tested for color removal after a number of manual double rubs with a cloth or paper towel saturated with the specified solvent. For example, it might be specified by a purchaser that if no color is evidenced on the cloth after 50 double finger rubs with acetone or methyl isobutyl ketone, the paint film is considered to be cured adequately. However, an ASTM test method has not been published for the solvent rub test and that is why the pencil hardness test is often used for extent of cure instead.

## Water Immersion

Moist and wet conditions can bring out poor adhesion and/or cause blister formation. The water immersion test checks on the tendency toward paint blistering. Water, even very pure water, can be a surprisingly disruptive chemical to paint films and paint adhesion. It has the ability to penetrate most paint films readily. Once it does so, it can produce undesirable effects unless the proper paint is correctly applied, at the necessary thickness, onto a correctly prepared surface. The tendency of a film to form blisters can be rapidly measured by the simple expedient of soaking the part in deionized water. This is a more strenuous test for blister formation than soaking in salt water because the osmotic pressures which cause paint films to lift (under-film water blisters) is greater with pure water than water containing dissolved salts.

## Humidity Resistance

In regular humidity testing, the parts are hung vertically in high relative humidity at around 100° F (38° C) for a specified number of hours. Steam is used to supply the moisture. With the condensing humidity test, painted panels are set at about a 30° angle from horizontal over a reservoir of warm water, as shown in Figure 17-5. The painted side of the panels faces the water; the reverse side of the panels is exposed to ambient room temperature. Water vapor condenses on the warm, moist side of the panels and drains down into the water reservoir. The heat of condensation and the continuous draining of the very pure water (actually distilled by the test process) constitute a severe test of the paint adhesion and resistance to moisture. Figure 17-6 shows paint blisters on a film test panel produced in a condensing humidity chamber. Condensing humidity can be done for as long as desired, but the maximum time is generally about 24 h. In addition to visual examination after humidity testing, the panels can be subjected to other tests, such as tape adhesion.

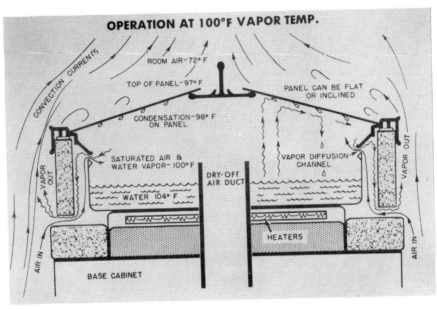

Figure 17-5. Condensing Humidity Test

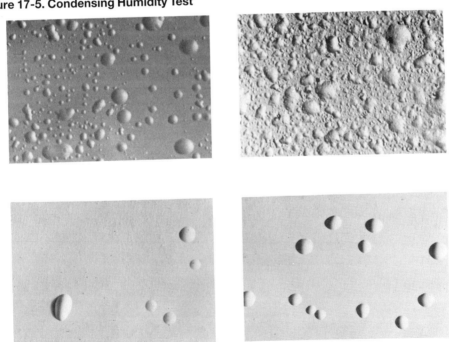

Figure 17-6. Painted Panels Exposed to a Condensing Humidity Test

### Accelerated Weathering

Various types of testing cabinets (see Figure 17-7) have been devised for accelerated weathering tests. The threads of commonality among them are heated moisture and provisions for spraying with salt solutions. Accelerated weathering tests can have two purposes:

- To compare the performance of different coatings
- To determine how a coating will perform in the field

To compare the performance of different coatings, paints are often checked in regard to their resistance to salt spray corrosion. This may be done to determine if they are better paints than a currently used coating, or to assist in the selection of the best paint from among several possible candidates. Numerous variations of this test are being utilized. Typically, test panels are exposed to dilute solutions (2%–10%, with 5% most often used) of sodium chloride in a spray mist at 100° F (38° C) for an appropriate length of time. This duration may be as little as 24 h or as long as 2,000–3,000 h. Two types of failure may be noted. *Face rust* is the appearance of rust on the surface of the part or panel where the coating seems continuous. *Creepback* is the spread of rust under a paint film and subsequent loss of the overlying paint, perpendicularly away from a line scribed through the paint into the underlying metal. Tests are usually conducted according to ASTM B-117 or B-287.

Figure 17-7. Accelerated Weathering Test Cabinet

However, nearly all salt spray tests may be misleading, and ASTM points this out very clearly. For example, salt spray tests on painted zinc and galvanized parts often appear worse than on painted steel parts, yet the painted zinc provides far superior corrosion resistance in field service. For this reason it is recommended that salt spray tests not be conducted to compare painted zinc parts against painted steel parts.

The relative performance characteristics of paints can probably be ascertained from these tests. A paint showing failure at 20 h of salt spray is probably not as durable as one showing no failure at 200 h. But there are more than a few recorded instances of one paint being better than another in salt spray tests, yet performing less well under actual field exposure conditions. This does not mean the test is not worth running, but caution is necessary here.

Accelerated testing can usually determine the best performing paint among a group of potential candidates. The most durable paint in a group can be determined with fairly high certainty by testing all the candidates under the same conditions. If the test parameters are selected carefully, the paint that performs best in testing would be expected to give the highest performance in the field under normal usage conditions. This has been found to be true at least 95% of the time. Only in exceptional instances has it been found that the results of accelerated testing are reversed in actual service. If the wrong test is selected, however, the differences among the candidates may not be discovered, and all paints may exhibit exactly the same test durability.

Unfortunately, it is not always obvious which version of salt spray testing is the correct one to use. Painted parts may need to meet a specification of 96 h in a given accelerated test condition. Others may require 500 h, or some 1,000 h, even 2,000 h under specified test conditions.

The second purpose of accelerated weathering tests is to determine in-house how a coating will perform in the field. The warm salt spray in these cabinets severely tests the integrity of a paint film, especially on metallic substrates, by speeding corrosion rates. Theoretically, the duration of a test can relate to service time in the field. Accelerated weathering tests may run from 1 h to over 2,000 h and attempt to simulate many months or years of service exposure. The question is frequently raised, "How many years of outdoor exposure is represented by each 24 hours of salt spray testing?" No reliably accurate answer to this question is available, unfortunately. The true answer will certainly be different in humid Houston than in arid Arizona.

In order to make accelerated tests as close to service life conditions as possible, evaluation groups have run tests under an extreme variety of conditions of temperature, humidity, and salt spray concentration both with and without added corrosive agents such as copper acetate and acetic acid. In some cases, ultraviolet light is added during simulated exposure tests. Whatever the test conditions, as long as they are thoughtfully selected to simulate service exposure as closely as possible and properly evaluated, they will give at least some indication of field performance.

Cyclic accelerated weathering tests have been developed that relate more closely to field conditions than continuous salt spray tests. Thus, 10 days in a cyclic test that tries to simulate a product's actual end-use conditions of varying thermal and chemical exposure conditions every 24 hours may be more realistic than 240 continuous hours at an unchanging temperature and humidity. One cycle will typically last 24 h. For example, a cycle may involve a 15-min soak in 5% neutral sodium chloride solution, 1 h at room temperature, 20 h at 140° F (60° C) with 85%

minimum relative humidity, and several hours at 30° F (-1° C). This pattern may be repeated until a specified number of cycles (days) has been completed.

In cyclic testing, the tester's imagination is the only limit on what can be devised. It is convenient if each cycle is 24 h long because it fits the work schedule of test personnel who will normally be at work for only one shift each day. Test specimens normally would be examined daily on the normal Monday through Friday workdays, skipping the weekend, unless there were special reasons for more frequent checks.

Panels may be tested unscribed or scribed with one or more lines, or the paint may be subjected to stone bombardment by a gravelometer before testing. The test results may be determined by visual examination such as measuring the amount of paint loss back from a scribe line (creepback), or tape testing may be done to determine how much adhesion will be lost after these simulated tests.

### *Gloss*

*Gloss* is a measure of the amount of light reflected from a surface. A standard polished surface is used as a calibration or reference that reflects all of the light and is given a gloss rating of 100. One that reflects no light has a gloss of zero. The gloss number is not stated as a percent, yet the number represents closely the percent of light reflected from a surface. Some samples may reflect more light than the reference sample, so gloss values greater than 100, while not common, are nevertheless possible.

Instruments that measure gloss (gloss meters) send a beam of light (incident light) at a particular angle from the vertical directed at the surface to be measured. The instrument measures the amount of light reflected at the same angle. The angle of incident and reflected light is usually specified at 30°, 45°, or 60°. The smaller the angle (to the vertical), the more the light will be reflected, and thus also the higher the gloss reading. This explains the importance of including the angle of incident light for each gloss reading, such as, "The gloss is 88 at a 60° angle." Smooth surfaces reflect more incident light than do rough surfaces, as would be expected. But the amount of reflected light is independent of the color of the surface, contrary to what one might anticipate.

To maintain accuracy, a gloss meter must be calibrated frequently against gloss standards in a range near to that being measured. Suitable standards should be included when a gloss meter is purchased. They must be stored with the instrument and used carefully to prevent scratches.

Some paint users have a need for terms not ordinarily significant for most painting operations. *Specular gloss* is the shininess or brilliance on highlighted areas of a part. *Sheen* refers to the specular gloss at very small angles of light incidence and reflection. *Brilliance* is the apparent strength of a color as perceived visually; it can be compared to loudness of sound.

### *Distinction of Image*

*Distinction of image* (DOI) is the measure of how well a surface acts as a reflecting mirror. Distinction of reflected image (DORI) and the abbreviation "DI" indicate the same phenomenon as DOI. The letters DOI should be used for distinction of image rather than the variations to avoid confusion.

A polished black glass standard has a perfect DOI; it is a perfect mirror, perfectly reflecting every object. Not all DOI measuring devices use the same comparative scale. Some DOI instruments place the highest smoothness rating at 100; at least one has a perfect rating of 20. A surface with a perfect DOI will be perfectly smooth. So, in a sense, DOI is a measure of how smooth a surface is as opposed to how rough it is. In many instances the roughness of paint surfaces is from the orange peel in the film.

Gloss and DOI are often confused. Gloss is the measure of the ability of a surface to reflect light. A surface, even though it is not perfectly smooth, may reflect essentially the total amount of the incident light impinging on it. However, a surface not perfectly smooth cannot perfectly reflect an image, that is, it scatters the incident light somewhat, even though it may reflect it all. In other words, a surface with a high DOI is always also high in gloss. But a surface with a high gloss may or may not be high in DOI. For a painted surface to be high in DOI, the substrate must be smooth, and the paint film must be free of orange peel. Moderate orange peel will not detract much from gloss, but it will reduce DOI.

Instruments that measure DOI project an image onto a painted surface and detect the distortion that occurs when the image is reflected from the surface. Some DOI devices rely on visual evaluations, and these have been shown to allow inconsistencies in readings. The evaluators rate a test panel based on the ability to see or not to see distortions in various images projected onto the surface under test. It has been shown that values thus obtained will not only fluctuate widely among evaluators, but significant variations are also perceived by the same person on the same panel if shown the panel at different times. The Landoldt Ring projector (light box) instruments suffer from this drawback.

One DOI instrument, shown in Figure 17-8A, photographs the reflected image from the paint film being tested with the test grid shown in Figure 17-8B. The operator looks at the developed picture of the reflected image. The DOI is the number of the grid where the operator decides that the parallel lines in the picture just seem not to touch. This test instrument has an additional advantage of providing a permanent photographic record of DOI.

A device for measuring orange peel, called a wave scan, is available. It is portable and battery powered. Both micro- and macro-waviness of paint films is measured very accurately with a laser beam by this cigar box-sized instrument. Orange peel can also be rated visually using a set of 10 panels that vary incrementally from mirror smooth to heavy orange peel.

### Color Match

Color does not have a real physical existence that can be handled and packaged; rather, color must be visually sensed. It is a human sensation arising from the interaction of electromagnetic vibrations on the eyes and interpreted by the brain. Almost every sighted person has seen the various colors present in "white" light separated from each other either by a prism or by raindrops that produce a rainbow. It can even be observed in a thin layer of oil floating on water. The separated colors are given the names red, orange, yellow, green, blue, indigo, and violet.

The *wavelength* of light is the tiny distance between corresponding consecutive points in the electromagnetic radiation. Wavelengths that humans can detect visually are in a small portion of the total electromagnetic spectrum. Beyond the longest visible wavelengths are the infrared waves that are associated with heat, and then the microwaves, radio, and television broadcast waves. Out past the other end of the visibility range lie the short but high-energy waves such as ultraviolet light, gamma rays, and X rays. Figure 17-9 gives a breakdown of various types of electromagnetic radiation and the associated wavelengths.

254   INDUSTRIAL PAINTING

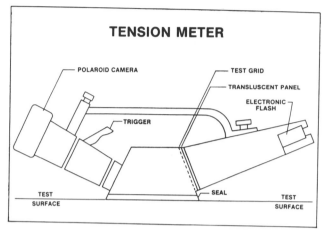

Figure 17-8A. DOI Tension Meter

Figure 17-8B. DOI Test Grid

Wavelengths in the visible spectrum lie in the tiny region between 4,000 and 7,500 angstroms (1 angstrom equals $10^{-10}$ meter). The longer the wavelength, the lower the energy associated with it. Thus, the 7,500-angstrom end of the visible spectrum, the red wavelengths, are lower in energy than the 4,000-angstrom violet wavelength end. Between the red and blue ranges lie orange, yellow, green, and blue/indigo.

The color of an opaque object depends on the wavelength(s) of light reflecting from it. A white object reflects all visible wavelengths; a black object doesn't reflect any visible wavelengths. An object with a particular color reflects the wavelength(s) required to make that color and absorbs all others. Paint pigments are able to produce distinctive color shades by absorbing some or all of the individual visible wavelengths present in the light source. Thus, a red pigment reflects red light and absorbs all the others. If several pigments of different colors are in a paint, the individual pigment colors cannot be perceived, but instead, a single color will be seen that is a blend of the combined pigment colors. Mixed white and black pigments will give a gray color; similarly, mixed blue and yellow pigments will yield a green color.

PAINT-RELATED TESTING 255

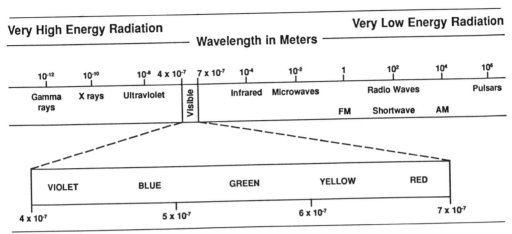

Figure 17-9. Electromagnetic Radiation Wavelength Ranges

When a broad spectrum light such as daylight hits a pigment particle, in addition to absorption of visible wavelengths in major wavelength bands, some selective absorption often occurs across the entire visible spectrum. This "muddies" the color. The most desirable pigments have less of the muddy overtones in their colors.

Matching paint colors can be a problem because the exact shade of a coating depends on many factors, such as pigment particle size and shape, spray pressure, target distance, booth humidity, and film thickness. Paint colors can often be matched fairly well visually, but they can almost always be matched more precisely using a color measuring instrument. Color matching, whether visually or instrumentally, should be done under several light sources. The light sources may include reddish (sunset type) lighting, as well as bluish (northern sky) lighting.

The reason several light sources are necessary can be explained in terms of the reflected and absorbed wavelengths. The sun radiates visible colors of all wavelengths, and an object that looks blue in sunlight absorbs all the wavelengths of visible light except the blue wavelengths. A source of light that is deficient in some of the visible wavelengths, for example, fluorescent lighting, may cause that same blue object to appear a slightly different shade of blue than in sunlight. This is because sunlight and fluorescent light contain different visible wavelengths or different proportions of these wavelengths. As a result, the visible wavelengths reflected to the observer will be somewhat different in sunlight than in fluorescent light.

Two frequently used color measurement scales are the Munsell and the CIE (L, a, b) systems. Both use a tristimulus method, meaning that three different stimuli are used to determine a color. The Munsell system uses the terms hue, chroma, and value to identify the particular shade of color (*hue*), the intensity of the hue (*chroma*), and the value (*grayness*) scale which varies gradually from pure white to pure black. The CIE scale uses a lightness, black-to-white L scale. In the "a" scale, positive values are increasingly red, and negative values become increasingly green. For the "b" scale, higher negative values are increasingly blue, and greater positive values become progressively more yellow.

## Which Tests to Run?

Tests selected for paint depend on many variables. Paint manufacturers run various tests to determine if the paint measures up to the specifications set by the end user. Finishers also conduct various tests determined by such factors as:

- Plant size
- Plant location
- Part quantity
- Type of parts
- Indoor or outdoor service location of parts
- Cost of parts being coated
- Color of parts
- Application methods
- Normal service life of parts
- Size of parts

Most finishers do not test incoming paints. A few finishers perform various tests as each batch of paint is received from the manufacturer. These incoming material inspection tests may include:

- Weight per gallon
- Percent weight solids
- Fineness of grind
- Color match to standard
- Gloss
- Hardness
- Viscosity
- Adhesion and flexibility
- Chemical and stain resistance

Tests run by finishers for production control and quality assurance are much more common, and may include:

- General overall visual appearance
- Gloss
- Color match to standard
- Tape adhesion
- Hardness
- Dry-film thickness
- Paint transfer efficiency

In a majority of finishing plants, wet-paint tests are not common practice. The exception, of course, is paint viscosity testing that is routinely done in nearly all wet-painting operations. The majority of paint tests by finishers are run on panels or parts after the coating has been applied and cured.

Before leaving this chapter, let me make an essential point on paint testing. It is highly recommended that for anyone doing even minor amounts of paint testing, the *Annual Book of ASTM Standards, Volume 06.01* (about $150) and perhaps also *Volume 06.02* (about $120) be acquired. These standards are revised periodically, and new standards are written as tests are developed. ASTM publishes all of its standards yearly, but existing copies tend to be suitable for at least 5-10 years, so it is not necessary to purchase these two volumes annually.

As long as a test is conducted identically each time, the correctness of the method is of secondary importance. It is informative, however, to read the test procedures. ASTM excels at stipulating how tests are to be executed. Until one has read the ASTM documents, it is hard to

# PAINT-RELATED TESTING 257

imagine how much detail can be specified to perform a straightforward test. Following ASTM guidelines for a test helps ensure that the same procedure is performed identically for each test. In this way, the results obtained are accurate and repeatable.

ASTM's advice and instructions are also helpful in showing how to convert observed test results into meaningful numerical ratings for comparative evaluations. Without the tester's full knowledge of precisely how to run and interpret certain paint tests, the results of a lot of hard work are actually worthless. Even worse, poorly conducted testing can convey totally misleading information.

# Chapter 18

# Stripping

## Reasons for Removing Unwanted Paint

In industrial painting, *stripping* is the term for removing unwanted cured paint film from a surface. The unwanted paint film may be on a flawed part, a conveyor hook, a paint hanger, the spray booth wall or floor, a grating, or similar equipment. Paint shop equipment is stripped regularly to maintain operating efficiency. For example, paint-free hooks and hangers are needed to ensure a positive ground for maximum electrostatic paint attraction. Clean floor gratings will optimize booth air flow and enhance painting operator safety.

Unwanted cured paint also may be on a finished product that needs to be stripped for refinishing. Painted reject parts or assemblies, and items painted in the wrong color, are usually stripped and repainted when this operation is less than the cost of a new item. Stripping is done also after paint films have undergone a life cycle or are beginning to show areas of paint loss, as on commercial airplanes (see Figure 18-1), ships, trains, and buses, not only for corrosion prevention but also to keep an attractive appearance for business reasons. Stripping and repainting is done on many military items for the added purpose of maintaining the full camouflage paint for tactical concealment.

## Types of Stripping

Stripping of paint from industrial equipment and products is done using any of these basic methods:

- Mechanical
- Blasting (abrasive grit)
- Blasting (water)
- Chemical
- Solvent
- Burn-off ovens
- Molten salt baths
- Hot fluidized sand

### Mechanical

Mechanical stripping uses physical contact devices that grind, abrade, cut, or scrape. Manual techniques are labor-intensive and potentially hazardous, but manually operated power tools can help expedite the process. Hand tools and manual power tools include scrapers, rotary grinders, rotary and orbital sanders, abrasive cloth or paper, and wire brushes. These methods are

# 260  INDUSTRIAL PAINTING

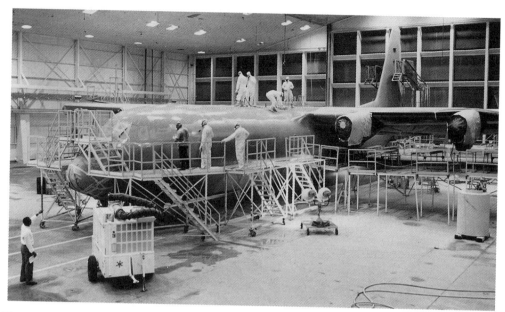

**Figure 18-1. Aircraft Can Be Stripped and Repainted**

slow, crude, and frequently leave a badly marred surface finish. If the paint is strongly bonded to the surface, manual methods are usually not completely effective in achieving total paint removal.

## Blasting by Abrasive Grit

In blasting with tiny grain or pellets, the grit is propelled at high speed against a cured paint film. Each individual grit media particle abrades a tiny fragment of the paint film. Repeated bombardment by thousands of media particles gradually removes the paint film.

Various methods are used to forcefully project the media onto the painted surface, such as a jet of pressurized air or a rapidly spinning wheel. The hurled media is almost always recovered and reused if practical. A screening filter can separate fine particles of grit and stripped paint fragments from the grit media.

Many different types of abrasive media are available, ranging from sand to particles of plastic. The hard types of media are used in stripping thick, tough-to-remove paint. Plastic media are used for "gentle" stripping and are often used in automotive and small aircraft repainting to strip only individual layers of paint, such as a topcoat from a primer.

In one type of media blasting, termed "cryogenic" stripping, the product and paint film are immersed in liquid nitrogen (about -320° F or -196° C) to embrittle the paint film, as diagrammed in Figure 18-2. The product is removed from the super cold nitrogen bath and immediately blasted with media. The cold embrittled paint film usually can be removed with blasting more readily than at room temperature. However, a large number of paint films are so durable they tend to be rather resistant to this process.

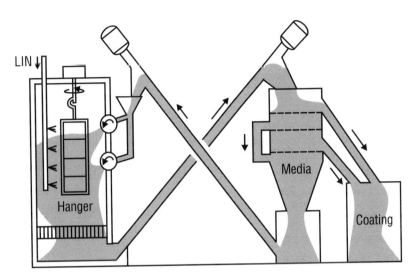

Figure 18-2. Cryogenic Stripping

## Blasting by Water

This type of stripping uses a high-pressure (3,000-10,000 psi or 20,700-69,000 kPa) stream of water to remove paint films. It is used regularly in high-volume paint booths to strip paint from floor grates. Extreme caution is necessary to prevent human injury from the high-pressure water streams, which can easily cut flesh. Wheeled machines that can be taken into paint booths for this process are designed to shroud the high-pressure water stream in order to protect the operator and workers in the vicinity. Booth floor grates also can be passed through stationary water jet paint stripping machines. They are similar to belt-type parts washers. An advantage of water blast stripping is that no chemical disposal is involved, other than proper disposition of the stripped pieces of paint that are removed.

## Chemical

Many chemicals are available for stripping paint films. The chemicals are usually categorized according to whether they are heated aqueous solutions (hot strippers) or unheated organic solvents (cold strippers). Hot aqueous strippers may be acidic, but far more often they are alkaline. Both of the hot-type strippers destroy the bonds that hold paint resins together. Organic solvent strippers employ a single solvent or a blend of solvents that will soften and/or dissolve paints.

Alkaline strippers contain strong concentrations of potassium hydroxide or sodium hydroxide (pH = 13–14) and possibly high concentrations of trisodium phosphate. In the past some strippers have contained phenols because inexpensive phenolic compounds will act as highly effective activators for reactive strippers. But the disposal of phenolic wastes is so restricted nowadays that the use of phenolic agents in strippers is rare. Yet, in parts of the world where environmental laws are not in effect, phenolic agents are still used. Proper disposal of spent stripping baths is a costly nuisance because they contain paint sludge as well as alkali. The problem and the expense are exacerbated when the paint sludges contain toxic compounds or heavy metals.

Heat is a valuable aid to achieve aqueous alkaline stripping. The boiling point of concentrated hot strippers may allow temperatures of 225° F–235° F (107° C–113° C) to be reached. Vigorous boiling action in hot tanks provides helpful bath agitation. A disadvantage of hot stripping tanks is that unless they are heated continuously, they are difficult to use on short notice. Another disadvantage is the danger to personnel from fumes and potential scalding. Condensation of water vapors that drip on people, the floor, and equipment can create a messy working environment.

Hot alkaline stripping solutions can rapidly degrade many paint resins to the point that the paint is completely removed, or nearly so, in the stripping tank. Some paints can be stripped in about 20 min; others may take an hour or more. For extremely adherent paints or durable resins, some additional physical scraping or abrading often becomes necessary on stubborn areas of the paint.

Hot caustic chemicals can be dangerous. They cannot be used on zinc or aluminum because they react vigorously, in some cases violently, with these metals. The alkali will quickly dissolve such parts. Stripping operators have been burned by hot solution when zinc parts were inadvertently lowered into caustic strip tanks. The reaction can be so rapid that some of the hot solution is blown out of the tank.

Hot alkaline stripping chemicals are consumed during usage and need periodic replenishment. In time the hot stripping action will weaken and will require the accumulated paint sludge to be removed. At this time about half of the solution should be discarded and replaced with fresh solution makeup. Water loss from evaporation and carryout on parts should be replaced regularly.

Acid strippers use nitric, sulfuric, or phosphoric acids, either alone or in combination. Acidic strippers are not used much anymore because the baths tend to be highly corrosive and give off acidic vapors that cause corrosion inside the plant. They are rarely used hot since heat would increase the emission of acid vapors that are corrosive and hazardous to workers' health.

Restrictions on land burial, liquid discharge, and air emissions have placed limitations on all forms of chemical stripping. The spent stripper solutions, whether acidic or alkaline, must be disposed of properly. The stripped-off paint forms a sludge that can be removed from the bath by filtering, then the rest of the spent bath must be neutralized. All harmful materials in the spent bath must be removed. Local restrictions may not allow dilution or draining to sewer. More and more companies have turned away from chemical stipping to thermal methods.

## Solvent

Certain solvents at ambient temperature (cold solvent strippers) have the capability of removing paint films, which eliminates the need for heated solvent. Ambient temperature solvent stripping is an excellent alternative method of stripping paint from aluminum, zinc, and other metal parts that cannot tolerate hot caustic or stripping methods using intense heat. Solvent stripping is used either as a dip, spray, or brushed gel, depending on part volume and production requirements.

Solvent strippers include highly aggressive oxygenated phenols, ketones, glycols, and esters, as well as saturated and aromatic hydrocarbons. But many solvent stripping products carry human toxicity concerns. For example, phenols are highly corrosive to skin and are toxic; numerous nonphenolic solvents are toxic and flammable; halogenated hydrocarbons have toxicity, ozone depletion, and safe-disposal concerns. Nonflammable halogenated solvents such as methylene chloride, trichloroethylene, and 1,1,1-trichloroethane were used heavily before the age of toxic and environmental chemical regulations, but few, if any, are allowed now.

Solvent strippers remove paint either through dissolving or softening, or both. Lacquer films will totally dissolve in the correct stripping solvents. Enamel coating films, however, will often swell, soften, and wrinkle when treated with solvent strippers, greatly simplifying physical removal. Some trial and error is normal to identify the best solvent blend to strip a given paint.

Vapors are a serious problem with most solvent stripping. Some vapors are hazardous to workers' health, especially if they remain in the vapors for long periods. Other solvents have odors that workers find offensive. One way of containing stripping solvent vapors is to use wax that floats on the solvent, forming a vapor seal. Another moderately effective vapor seal is water if the solvent is not water miscible and has a high enough density to enable water to float on it. Some solvents, however, cannot tolerate water. For example, water added to halogenated hydrocarbons can form hydrochloric acid, which is highly corrosive to metals such as steel.

Paint, especially soft but undissolved enamel paint, may accumulate on the bottom of a solvent-stripping tank from which it can be separated out periodically. The remaining solvent is not always reduced in stripping effectiveness. In the case of lacquer paints, the stripper solution can become increasingly viscous as more and more resin is dissolved in it. The paint solids, however, will build up on the tank bottom.

Figure 18-3 compares hot akaline strippers to cold solvent strippers.

### Waste Disposal Considerations

Probably the greatest problem with using solvent strippers is how to dispose of the solvent-rich paint sludge. Although flammable solvents sometimes can be incinerated, nonflammable solvents such as halogenated solvent strippers cannot be disposed by burning. They form the corrosive, irritant gas hydrogen chloride when exposed to flame. To compound the difficulty, stripped paint sludge that contains either flammable or nonflammable solvent residues are seldom permitted at ground disposal sites.

The solvent in paint-stripping sludges and spent cleaning solvents can be reclaimed, often for direct reuse, by distillation. Plants may do this themselves or hire a service for it. A number of solvent stills being sold for this purpose use replaceable plastic bags. The used solvents are placed in the bag and heated to distill out the volatiles. When all solvent is evaporated, the bag containing the solvent-free dry residue is removed for convenient and safe disposal. If no toxic substances are present, ground burial of bag and residues may be acceptable at waste disposal sites.

The use of halogenated hydrocarbons for solvent strippers has declined because they cannot be burned and are no longer allowed at most disposal sites. Many plants have no place to get rid of halogenated solvents at a reasonable cost. However, the increased use of distillation to recover solvents of all types has brought about a slight return to using environmentally acceptable halogenated solvent paint strippers.

## Comparing Hot and Cold Strippers

### Hot Alkaline Strippers

**Advantages**
1. Lower cost per gallon in solution.
2. Vapors pose less health hazard.
3. Requires less ventilation.
4. Can be less expensive for stripping parts in volume.
5. Needs no evaporation retardant; does not lose strength from evaporation.
6. Introducing water does not kill solution.
7. Removes rust.

**Disadvantages**
1. Active ingredient consumed with each part stripped.
2. More difficult to rinse.
3. Usually requires heat and agitation.
4. Not recommended for most non-ferrous metals.
5. Requires longer stripping time.
6. Not as effective on tough, modern finishes.
7. Must be diluted with water before use.
8. Can raise grain and discolor wood surfaces.

### Cold Solvent Strippers

**Advantages**
1. Usually do not need to be diluted.
2. Active ingredient not consumed by dead paint.
3. Faster stripping action.
4. Removes all finishes; more effective on tough finishes than hot solutions.
5. Needs no heat.
6. Strips ferrous and non-ferrous metals.
7. Needs no agitation.
8. Easier rinsing.
9. Cold strippers compatible with wood, do not discolor surface or raise grain.

**Disadvantages**
1. Higher cost per gallon.
2. Solvent evaporates rapidly, causes solution to lose strength.
3. Fumes can pose health hazard.
4. Easily killed by introducing water.
5. Vapors can be corrosive to nearby equipment.

Figure 18-3. A Comparison of Hot Alkaline and Cold Solvent Strippers

The cost of having a private disposal firm handle paint and solvent waste is frequently prohibitive. For small-quantity waste producers, however, contracting with an outside firm to perform waste management may be cost-advantageous. Companies of any size should look into the relative economics of such an approach to disposing of paint and solvent. A number of plants have been pleasantly surprised to find that it is less costly to use a disposal service company than to distill their own used solvent.

In general, however, many plants have turned completely away from any solvent stripping because of the disposal problems. Another reason for this change is that so many of the newer powder, high-solids, and waterborne coatings are resistant to chemical stripping. With such paints, both chemical and solvent stripping are slow and less than fully effective. Often, therefore, paint burn-off becomes a far more desirable method.

## Burn-off Ovens

Burn-off ovens are batch style enclosures designed to operate at about 1,000° F (538° C). The high temperature burns away the organic portion of the paint film, leaving an inorganic ash residue. The oven design may include various methods of eliminating the emission of smoke generated from paint combustion.

Some burn-off ovens include an auxiliary unit for washing ash residue from the stripped parts. A wash unit confines the ash, which prevents its spread throughout a plant. If care is not taken, ash can easily be spread to areas where it gets into paint finishes or becomes a nuisance in other ways.

With all heat-stripping processes, the loss of favorable metallurgical properties of parts and paint hangers must be considered. Some metals lose their strength (temper), if heated excessively, or will soften; as a result they may deform under load. Such metals are obviously not suited for this type of stripping process. Burn-off ovens are frequently used for stripping paint from conveyor hooks and hangers, so the problem is avoidable by proper design and material selection for the racks.

## Molten Salt

Molten salt baths can strip paint films almost instantaneously. The baths are composed of a mixture of various salts. Different salt mixtures are available with melting points ranging from 600° F–900° F (316° C–482° C). All salt baths strip by burning the paint film, usually in less than 10–25 s. As one would anticipate, the higher the bath temperature, the faster the stripping action. Large amounts of flame and smoke are immediately produced when paint-covered parts are immersed in salt baths. So much smoke forms that special emission controls are generally needed.

Hooks, hangers, and floor grates are often stripped in salt baths because they can readily withstand the high temperatures. Production parts can be stripped as well if they can withstand the heat, but that is less often the case.

The burning of the organic resin and additives in the paint leaves behind inorganic pigment residues, which settle to the bottom as sludge. Depending on how much paint is stripped, from 3–55 gal (11–208 L) of sludge may be produced a day. Each 55 gal (208 L) of paint sludge requires about 450 lbs (205 kg) of fresh salt mixture. The sludge must be separated out periodically and discarded.

Molten salt baths must be operated with extreme care and caution due to the danger from high temperatures and bath splashing. For handling items to be stripped, an overhead hoist that allows operators to stand a safe distance away is used. Conveyorized hot salt stripping is being done only at a limited number of locations. Conveyorized part entry and exit is ideal so that stripping operators need not be near the hot salt during dangerous parts of the operation.

## Hot Fluidized Sand

Beds of fluidized hot sand can effectively strip paint film. The sand is kept in a fluidized state by currents of air at 700° F–1,000° F (371° C-538° C). The combination of heat and the abrasive action of sand provides thorough and rapid stripping action.

# Selecting a Stripping Process

Various factors need to be considered before choosing any stripping process. These include:

- Cost and volume of items to be stripped
- Substrate characteristics
- Safety
- Environmental regulations

## Cost and Volume of Items to be Stripped

Purchase and installation of stripping equipment is probably not cost effective for a majority of plants. Contract stripping companies have succeeded in marketing their services to plants unable to justify in-house stripping systems, or who are unwilling to do their own paint stripping. If a contract stripper has a facility located reasonably close, it is possible that they can be used to do the paint stripping for a plant. However, the availability of a contract stripper is not of much benefit if items to be stripped need to be transported long distances; then overall costs can become prohibitive.

For many plants, stripping can turn out to be more expensive than scrapping parts. In some plants, hooks are not totally stripped, but instead, are periodically spot cleaned by grinding or sanding at contact points in order to maintain good electrostatic grounding.

## Substrate Characteristics

The nature of the substrate must be taken into consideration before selecting a stripper. Heat can cause metals to lose their temper, and various chemicals can destroy or damage materials sensitive to stripping agents.

## Safety

Personnel must be required to wear the prescribed protective equipment and clothing for all types of paint stripping. In addition, plant areas surrounding the stripping need to be checked thoroughly for damage susceptibility from the stripping. Proper ventilation must be available if stripping is to be done inside the plant, as is normally the case. Occasionally, stripping is conducted outside in geographic locations where ordinary weather conditions allow this, such as the less affluent, warm weather countries of the world. Where this is possible, it is still prudent that wind, temperature, and precipitation conditions be monitored to prevent mishap.

## Environmental Regulations

Environmental and waste disposal limitations must be considered when selecting a stripping method. It is not unusual for a plant to discover that waste chemical disposal costs are the largest single component of total stripping costs. Chemical stripping methods have been less popular since EPA restrictions on landfilling were enacted. Any stripping wastes that contain solvents or heavy metals require disposal as toxic materials, which greatly multiplies the cost.

# Chapter 19

# Special Considerations for Painting Plastics

## Understanding Plastics Chemistry

*Plastic* is the generic name for a broad category of materials produced from petroleum-based and agricultural plant-derived organic polymeric resins. Plastics may contain various additives to produce certain physical and chemical properties. Plastics may also have pigments added to them to yield particular shades of color. Plastic products have been molded into practically every conceivable size and shape with an array of textures and properties to facilitate an almost infinite assortment of end uses.

Plastic resins are characterized by complex organic chemical structures, made from the chemical bonding together of varying numbers of individual molecules. Each molecule involves mainly carbon, oxygen, and hydrogen atoms held together primarily by single and double chemical bonds. Resin molecules differ greatly in size, chemical makeup, and degree of polymerization. They may be made up of any number of smaller molecules that have linked together chemically. That number can range anywhere from 7 molecules to 700,000 molecules; there is really no theoretical limit.

More importantly, all resins are categorized into just one of two types, thermoplastic and thermosetting. In actual practice, there are materials that behave much as thermoplastic resins but which technically are thermoset resins. *Thermoplastic* resins are characterized by having various lengths of polymer chains that are held together by relatively weak, mutually attractive forces among the individual molecules. Thermoplastic resins can always be reformed into a new shape with the application of heat. *Thermosetting* resins, by contrast, consist of multiple polymer chains that are bonded together by comparatively strong cross-linking chemical bonds; thus, they become to some degree permanently rigid. Thermosetting resins cannot be reformed readily into new shapes by the application of heat. Thermosetting resins can have varying degrees and types of cross-linking, however, so some heat softening is common in paint resins before the resin is cured. "Curing" for thermosetting paint resins is synonymous with "forming cross-linking chemical bonds."

Plastic resin types include acrylic, polystyrene, polypropylene, acrylonitrile butadiene styrene (ABS), polyester, polycarbonate, nylon, polyphenylene oxide (PPO), acetal, polyvinyl, urethane, cellulose acetate butyrate, phenolic thermoplastic olefin (TPO), and ethylene propylene diene monomer thermoplastic rubber (EPDM). There are countless others as well, but these are the most important in finishing (see Figure 19-1).

Resins have acquired other generic and trade names that relate to how they are used. A few examples are listed below:

- Sheet molding compound (SMC), a thermosetting polyester resin with 25%–32% fiberglass reinforcement
- Reaction injection molded urethane (RIM)
- Reinforced reaction injection molded urethane (RRIM) with 8%–32% fiberglass reinforcement
- IMR, which is RIM or RRIM containing internal mold release
- Plexiglas, a clear glazing substitute for window glass structural foam and Styrofoam

Plastic formulations can include single resins or blends of various resins. They can contain a host of different additives, such as fillers and plasticizers. One additive (chopped fiberglass) is used in a category of plastic called fiberglass-reinforced plastics (FRP), which may consist of various resin blends and possibly other additives in addition to the glass fibers.

|      | **Chemical Name**             | **Common Name** |
|------|-------------------------------|-----------------|
| ABS  | Acylanitrite Butadiene Styrene | ABS             |
| EDPM | Ethylene-Propylene Diene Monomer | EPDM         |
| PA   | Polyamide                     | Nylon           |
| PC   | Polycarbonate                 | Lexon           |
| PPO  | Polyphenylene Oxide           | Noryl           |
| PE   | Polyethylene                  | Dylan           |
| PP   | Polypropylene                 | Profax          |
| PS   | Polystyrene                   | Lustrex         |
| PVC  | Polyvinyl Chloride            | Geon            |
| PUR  | Polyurethane                  | Bayflex         |
| EP   | Epoxy                         | Epon            |
| VP   | Vinyl Polyester               | SMC             |

Figure 19-1. Commonly Used Plastics

## Why Paint Plastic?

Pigments can be added to produce plastic formulations having just about any desired shade of color. A good example is the modern telephone. Units are available in various molded-in colors, which allows the manufacturer to stock a few basic colors of molded phones. Why, then, does anyone bother to paint plastic items?

Whether to mold-in the color or to paint a nonpigmented plastic product (or both) is a decision that needs to be made for each individual case. The governing parameters tend to be cost, appearance quality, and product performance properties. A case in point: plastics with molded-in colors are used mostly indoors away from sunlight, which tends to fade the colorants in many plastics. The reasons for painting plastics are many, and some of the major ones include the need for:

- Highlighting
- Texturing
- Protection
- Appearance uniformity
- Hiding defects
- Functionality
- Improving conductivity
- Creating a second surface

## Highlighting

Selective areas of plastic products may be painted to highlight structural details. Areas may be painted either in contrasting colors for decorative effects or with identifying names, symbols, and logos. Examples are the plastic wind deflectors atop truck cabs, which commonly contain company names, identifying patterns, or advertising. Plastic hulls of boats frequently have glamour stripes painted in horizontal arrays. Plastic containers, both reusable and single use types, have brand names painted on for identification and to increase sales appeal.

## Texturing

Coatings on plastic can give desirable texture patterns or a tactile character to an entire part, or just a portion thereof to enhance the product's look or "feel." Computer keyboards are an example of this.

## Protection

Many plastics are sensitive to solvent etching, thermal crazing, and mechanical abrasion. Protective coatings are applied to enhance the durability of such products. For example, stone chipping and solvent damage to polycarbonate auto headlight lenses are reduced by painting with polyurethane clearcoatings.

## Appearance Uniformity

Gloss and textural differences are evident throughout many molded plastic parts. In addition, most molded plastic parts tend to have glossy surfaces. Painting can produce a uniform gloss across the product that can either be higher or lower than that of the original part. Toys are often painted in high-gloss colors. Plastic furniture parts usually receive low-gloss paint finishes to eliminate the sometimes unattractive appearance of glossy plastics. Wood-graining paints, another appearance altering process, can give a matching pattern to items constructed of both wood and plastics.

## Hiding Defects

Swirls and related visual defects in reinforced plastic may need to be hidden by painting. Coatings can also give a smooth appearance by concealing marks such as injection molding marks or scratches when plastic parts are sanded to remove flash and parting lines.

## Functionality

Many clear covers, such as those over instruments in an automobile dashboard or in an airplane cockpit, are painted in selected areas only. Interior sections may receive a reflective paint to increase light output; exterior areas may be painted black to stop light from escaping in certain other locations of the lenses. The interior surfaces of optical devices are normally painted black to prevent reflected light from reducing the sensitivity of these instruments.

## Conductivity

The low electrical conductivity of most plastics may require special conductive coatings on plastic parts to meet requirements for electromagnetic interference (EMI) shielding and radio frequency interference (RFI) shielding. Electrical safety grounding requirements may also require special conductive coatings. Nickel- and copper-containing paints are often used for these purposes. Conductive paints are also used to prevent the accumulation of static charges on plastic components.

## Second Surface

Second-surface finishing is unique to transparent plastic parts. For example, motorcycle helmets of clear plastic are often painted on the inside of the shell with various decorative colors, such as metallic reds, golds, and blues. The paint is still visible through the clear plastic shell, but it is protected by the shell from scratches and abrasion. The appearance paint inside the shell is normally overcoated with an opaque paint to prevent light leaks and also to protect the decorative coating from abrasion.

# Cleaning Plastic Before Painting

Just as with any other surfaces, plastics should be thoroughly clean before painting. In practice, most plastic contamination problems develop from three principal causes:

- Fingerprints
- Dust and lint
- Mold release

## Fingerprints

Fingerprints (salt and skin oils) can usually be removed satisfactorily with a simple detergent cleaning. They can be avoided altogether by requiring handlers from molding to painting to wear cotton gloves. The gloves should be changed regularly so they do not become a source of contamination.

## Dust and Lint

Because nearly all plastics are nonconductive, they often build up static charges that hold lint and dust tenaciously. Wiping with a tack cloth in many instances will not remove all of the contaminants, which can result in visible defects under the applied paint.

One of the preferred methods of removing statically attracted lint and dirt is to use a destaticizing air blow-off (see Figure 19-2). One such system generates positively charged and

# SPECIAL CONSIDERATIONS FOR PAINTING PLASTICS 271

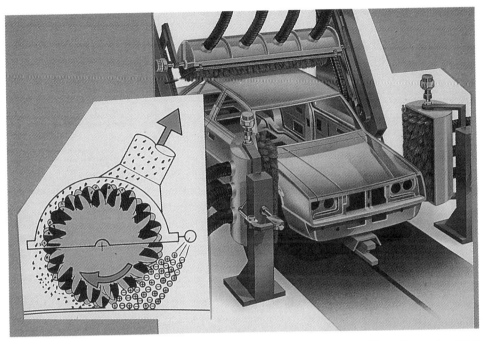

Figure 19-2. Feather Duster Using Destaticizing Air and Vacuum to Clean Body Just Prior to Painting

negatively charged airborne ions with a weak radioactive source inside a blower housing. When this air is blown across the part, the ions neutralize all static charges, and the air curtain gently blows away dust and dirt particles. Vacuum removal of dirt at this stage is strongly advised to prevent soil redeposition. The destaticizing step should immediately precede the painting operation, otherwise the parts can once again accumulate charges and become recontaminated.

Dust and lint is well controlled by confining the plastic painting operations inside a "clean room." Such a room contains carefully filtered air and is constructed of non-lint producing materials. Only authorized and properly attired personnel (lint-free attire, hairnets, shoe covers) should be allowed to enter.

## Mold Release

Mold-release agents are often used to facilitate separation of plastic products from their molds. Some mold release agents will greatly interfere with paint adhesion. Probably the most notorious mold release agents, as far as painting is concerned, are those containing silicone compounds. Silicone materials have a way of spreading throughout a plant, causing untold problems with paint adhesion to plastic products. They can be spread by a plant's air-conditioning system, causing widespread contamination of parts stacked just about anywhere in the plant. There are paintable silicone mold releases, however. Wax-type mold releases can sometimes be removed by solvent cleaning, but they are not recommended if parts are to be painted. Water-soluble mold releases that can be removed readily with aqueous detergent solutions are preferred.

Sometimes mold release agents are blended directly into a plastic formulation. These can migrate to the surface and cause paint delamination months or years after a part has been painted. For painting, internal mold releases must be avoided. Plastic raw materials are sometimes accidentally contaminated with mold release, either by the supplier or by the molder in regrind material.

Certain types of plasticizers used in plastic formulations can cause problems with paint adhesion similar to mold releases. Plasticizers are added to some molding resins to increase impact resistance. The plasticizer can slowly migrate to the surface and soften the interface between the plastic and the paint film, causing adhesion loss. Delamination of the paint film may occur weeks or months after the paint has been applied, even though initial adhesion tests are passed.

Tracking down the source of plastic surface contamination can require some disciplined detective work. Plants are not always willing to pursue the problem with that much diligence, for it can be time consuming. In such an investigation, every possible source must be examined until the contamination is identified and eliminated.

## Preparing Plastic Surfaces for Painting

Plastic surfaces are inherently low in surface profile and tend to give poor paint adhesion unless the surface is roughened by chemical or mechanical means. The most common way of overcoming surface smoothness is to microetch the surface with solvent to create a microroughness that provides anchoring sites for the paint. The etching is usually performed by the solvents present in the paint being applied. This solvent must be selected carefully because different solvents etch plastics at varying rates. Insufficient etching will not provide proper adhesion; excessive etching can damage the plastic, which exposes filler particles and, in a few cases, even creates sites for plastic raw materials to bleed into the coating. Some plastics—polycarbonate and polystyrene, for example—will crack or become totally surface-crazed from attack by selected solvents to which they are susceptible.

Sometimes during molding there are areas where plastic injection causes substantial heat from friction of the plastic flow. The part at those spots will form highly cross-linked glazed skin areas that tend to resist solvent microetching. Unless additional steps are taken, paint adhesion will be poor in these areas. Tumbling in mildly abrasive media or blasting with light grit can deglaze these areas enough to allow satisfactory paint adhesion. Treatment in hot solvents or in solvent vapors can also be effective.

When deglazing or solvent etching is not applicable, in particular with extremely nonpolar plastic surfaces, it may be necessary to use a chemical reaction to create polar oxidized groups on the surface. Two examples of plastics treated this way are polypropylene and polyethylene. These and similar low-polarity plastics may be quickly passed close to an oxidizing flame to form enough polarity in the surface to promote good paint adhesion. Putting parts through an electrical corona discharge has been used to cause a similar oxidation at the surface.

Plastics may have areas that are highly stressed from the molding process. Some solvents can form visible cracks in these areas due to stress-relief action. Many plastics can benefit from stress-relieving by heating parts to 150° F (66° C) for half an hour prior to painting. Overheating can distort thermally sensitive parts, however, and parts with high molded-in stress are likely to

# SPECIAL CONSIDERATIONS FOR PAINTING PLASTICS

warp unless held in a fixture during heating. Heat sensitivity of plastic will be discussed later in this chapter.

Low-polarity plastics are frequently treated with photosensitizers and then exposed to ultraviolet light. The UV decomposes the photosensitive compounds to form free radicals that combine with oxygen in air to generate polar groups on the plastic surface.

Cold gas-plasma technology can pretreat plastics and dramatically improve surface properties for paint adhesion. Gas-plasma has been described as a "fourth state of matter." If a gas is given enough energy, it becomes ionized (plasma). Examples include arc welding and fluorescent lighting, which involve a "glow" caused by excited ions falling to their lowest energy state. Plasma processes usually do not change surface appearance; even inert materials can be plasma-treated without discoloration. Plasma conditioning allows plastics to be painted with good adhesion; further, this excellent adhesion is achieved with the same paints used to coat metals. This is important to some manufacturers since assemblies that utilize both plastics and metals can be painted with the identical paint when the paint is suitable for both substrates.

The plasma reactor is typically a vacuum vessel with an access door for loading and unloading parts to be treated. Excitation is provided by a radio-frequency generator. When the reactor is started, a glow discharge can be observed. This process can use any gas or mixture of gases within the safety limitations of the system. Gases such as oxygen, nitrogen, air, helium, argon, and ammonia are commonly used. Gas-plasma surface treatment microetches and activates the surface. A brief treatment will make the surface polar with a high surface energy so that coatings will wet it completely.

Chemical oxidizing agents may be introduced into the paint to improve adhesion. They oxidize the surface to achieve some polarity—enough to provide good paint adhesion. Reflectance infrared spectroscopy has verified that these treatments produce oxygenated (hydroxyl, carbonyl, and carboxylic acid) groups on the plastic surface.

## Plastic Surface Peculiarities Affecting Painting

A number of plastics have inherent properties that tend to cause difficulties in painting. In addition to the solvent sensitivity already described, these include:

- Water sensitivity
- Heat sensitivity
- Sanding sensitivity
- Nonconductivity
- Color matching with metal
- Gloss variations with metal
- Wicking

### Water Sensitivity

Aqueous detergent solutions can do an excellent job of cleaning plastics for painting. Normally, water will neither corrode nor etch plastic. However, some plastic parts will absorb water, and this will lead to bubbles and blisters when the paint on the plastic is heat cured. Highly polar plastics such as nylon are especially susceptible to water absorption. Absorbed water can be removed by drying the plastic at a moderate heat for an extended period. It is better to dry the resin before molding to avoid part defects from moisture and then, if possible, to paint the parts promptly. In this way they get painted before they can absorb water and before they can become contaminated and require cleaning.

## Heat Sensitivity

Heat used for dryoff or baking may cause structural shrinkage, warping, twisting, or related distortion of plastic parts. The extent of the heat sensitivity will depend on the part configuration, the resin, and especially the molding parameters. Such distortion may sometimes be corrected by placing the part in a fixture for rigidity and stress-relieving the part at a slightly raised temperature. Figure 19-3 lists various plastics and their heat tolerance temperatures.

The thermal stability of many plastics is quite low. For example, the maximum temperature that styrene parts should endure is only about 140° F (60° C). Acrylic and ABS parts can withstand no more than about 180° F (82° C). Not all plastics will distort at these low temperatures; polycarbonates and acetals will not distort below 275° F (135° C). When heat stability problems are encountered, the applied paint can often be baked at a lower temperature for a greater length of time to achieve a full cure. However, longer bake times increase costs by requiring slow line speeds and/or extra long ovens.

### Heat Tolerance in °F

| Plastic | °F | Plastic | °F |
|---|---|---|---|
| ABS | 170 | Polyethylene | 175-250 |
| Acetal | 220 | Polypropylene | 250 |
| Acrylic | 180 | Polystyrene | 130-145 |
| Alkyd | 300 | Polysulfone | 350 |
| Cellulosic | 220 | PVC | 170-225 |
| Phenylene Oxide | 180 & Up | PVDC | 170-210 |
| Nylon | 300-425 | RIM | 250 |
| Phenolic | 225-500 | PRIM | 250 |
| Polycarbonate | 250 | TPR | 250 |
| Structural Foams | 140-250 | TPU | 250 |
| Polyester SMC | 400 & Up | TPO | 230 & Up |
| Xenoy | 210-225 | EPDM | 225 & Up |

Figure 19-3. Plastics and Heat Tolerance Temperatures

## Sanding Sensitivity

Air or other gases are deliberately molded into structural foam parts to economize on the amount of plastic that is used in the molding. The shapes of the gas pockets provide a strength equivalent to or even greater than a plastic without a cellular structure. Sanding to remove mold parting lines or other defects may open the hollow cells of structural plastic foam parts. Any sanding-caused porosity can lead to blisters and popping problems during paint baking. The sanded area may deglaze the plastic, which results in a dull paint area that may require the entire part to be given a new prime coat. Figure 19-4 lists various problems, probable causes, and remedies for finishing structural foam parts for business machines.

# SPECIAL CONSIDERATIONS FOR PAINTING PLASTICS 275

| Problem | Probable Cause | Remedy |
|---|---|---|
| Swirls | Compound or mold surface too hot | Adhere to recommended melt temperatures. Control mold temperatures carefully. Use high-volume solids primer filler and texture coating. |
| Pinholes/Craters | Solvent attack with accompanying release of gases | Use only mild solvent or waterborne primers. |
| Blisters | Absorption of solvent with subsequent release of gases | Air-dry, prebake or store foam part at room temperature for 72 hours before coating. |
| Bubbles in Substrate | Inadequate filling of mold, Inadequate foaming of part | Fill mold as fast as possible. Keep careful control of material-fill levels. |
| Poor Adhesion | Too much mold release | Use only minimal amounts of mold-release agents. |
| Wicking | Mold too hot, material too Viscous, solvent too active | Adhere strictly to recommended melt temperatures. Keep careful check of mold temperature through careful placement of gates. Use minimal amounts of mold release. |

Figure 19-4. Chart for Finishing Structural Foam

## Nonconductivity

The minimal electrical conductivity of most plastics precludes painting them by electrodeposition. Plastics are also more difficult to paint electrostatically than metals. Conductive precoats can be applied to enable plastics to be electrostatically coated, but this creates an extra step in the coating process. Such a conductive coat can be applied by various methods, including dipping and spraying. Fast-evaporating alcohol solutions of conductive organic salts can be used as a prep coat to give a plastic part conductivity for electrostatic application, but the alcohol must be counted as VOC. Aqueous solutions of calcium chloride and lithium chloride can also be used and have no VOC components. Neither of these conductive coatings interfere noticeably with paint adhesion in most instances. Conductive coatings are applied extremely thin and normally are not even visible; probably "conductive treatment" or "conductive prep" would be better terms to use.

# 276   INDUSTRIAL PAINTING

## Color Matching with Metal

Color matching between painted metal and painted plastics can sometimes be a vexing problem. Differences between metal and plastic surface textures, substrate heat capacities, and material densities can often cause an apparent color variation even when painted together as an assembly with the same paint. The different heat-up rates (heat capacities) of materials and different degrees of light reflection from their surfaces can produce apparent color mismatches that can be extremely difficult or impossible to overcome.

## Gloss Variations with Metal

Gloss variations between different plastics and between plastics and metals can occur, often for the same reasons that cause color differences. Spots that are dull are sometimes called *dive-in* or *sink*. They are the result of different plastic densities from one location to the next within a single plastic part. Density changes are caused by variations in the internal mold pressures across the part. Primer painting of such parts becomes necessary if uniform gloss across the entire surface is required.

## Wicking

Air entrapment or solvent absorption along fiberglass/plastic interfaces can occur with fiberglass-reinforced plastics. This *wicking* of the solvent deeper into the plastic results from the greater thermal expansion and contraction coefficients of plastic resins compared with the thermal coefficient of glass fibers. When the hot plastic in the mold cools, it contracts to a greater degree than the fiberglass. The plastic shrinks away from the fiberglass, leaving microvoids that can act as small capillaries. When paint is applied, the solvents in the paint can be drawn into the body of the part and away from the surface by capillary action. Solvent travels inward along the gap at the fiberglass/plastic interface. Because the solvents have been pulled deep into the part, they often do not evaporate during baking until after a skin has formed on the paint. By the time the plastic part is finally warm enough to evaporate the "wicked-in" solvent, the paint film has solidified to some extent. The solvent in the capillary is forced up under the skinned-over paint, forming a "pop" or blister.

Sometimes low–molecular-weight polymer molecules will cause an identical problem when volatilized late in the cure process after the paint has already formed a skin. Molding resins free of volatile materials should be used if the parts are to be painted.

## Paint Film Laminates for Plastics

Prepared dry sheets of both primers and topcoats can be applied to plastics in an injection or vacuum forming mold or as part of extrusion molding. Sheets of specially designed paint film laminate are inserted into the mold or applied to extrusions as they leave the die. The paint sheet becomes affixed to the plastic by heat and pressure. Two main advantages of this process are the virtually total elimination of VOC emissions from the plant at which the "painting" is done and the absence of wet overspray to dispose. A significant but less essential advantage is that any rejected parts can be recycled by regrinding since the paint is a thermoplastic resin. Appearance and durability of the paint sheet can exceed that of conventionally applied paint films. Figures

# SPECIAL CONSIDERATIONS FOR PAINTING PLASTICS 277

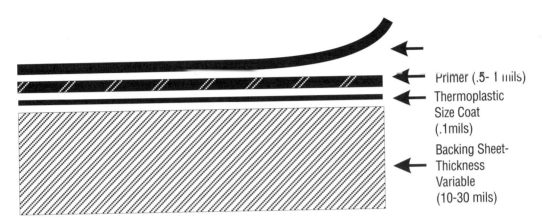

**Figure 19-5. Dry Primer Laminate for Plastics**

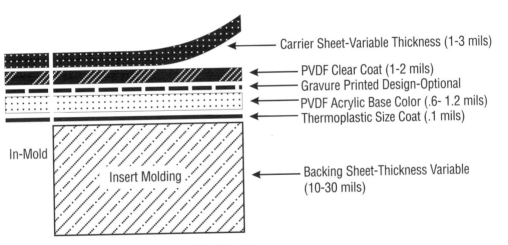

**Figure 19-6. Dry Topcoat Laminate for Plastics**

19-5 and 19-6 show the typical layered composition of laminate sheets for primer and base/clear topcoats.

The paint sheet is prepared by rollcoating a wet paint onto a flexible carrier sheet, and then curing the paint. The chemical nature of the carrier sheet must allow it to bond well to the base plastic of the part during application. The size and shape of the part has to be rather flat or at least suitable to the addition of a layer of dry paint film. Vinyl siding is a good example of a plastic part well suited for this process; one manufacturer guarantees its woodgrained vinyl siding for five decades. Some small- and medium-sized automotive parts have been painted this way, and study is being done on coating large exterior pieces with laminate sheet paint.

## Plastic Paintability Information Resources

The paintability of various plastics can be acutely different. It is unlikely that a truly helpful chart could be prepared that would give specific instructions for painting all types of plastics. It is not sufficient to specify that the plastic is a generic material such as phenolic, nylon, or acetal. Even a trade-named plastic such as Noryl (General Electric's polyphenylene oxide) has over 100 different grades available. Any chart must be too general to be of much use to painters.

In each case, the specific plastic blend must be known to predict paintability. Even then, some trial-and-error experimentation will be needed to determine which coatings are suitable for a particular plastic or a specific part. Paint companies are almost always willing to test their coatings if the plant supplies them with a sample of the plastic part(s) to be painted.

# Chapter 20

# Conveyors for Painting

## The Need for Conveyors

Industrial painting is done either by batch processing or with a conveyor. In the typical batch-processing method, relatively small quantities of products to be painted are moved manually from station to station and handled manually at each station. If the parts are small, they are picked up and repositioned by hand. Heavy items require special lifting equipment such as a hoist, crane, or forklift. A short monorail conveyor with manually moved parts is sometimes used for low volumes of large heavy items.

When large quantities of parts of any size are to be painted, the work is more likely to be moved automatically along the paint line by a power conveyor. The objective of all paint line conveyors is to move parts on a timely schedule to specific destinations. Conveyors can be categorized into the following general types:

- Flat conveyors
- Chain-on-edge conveyors
- Chain-in-floor conveyors
- Overhead chain and trolley conveyors
- Inverted conveyors

Additional classifications can also be made, based on other physical features of the conveyor. Any of these conveyors may be of various widths and lengths and have varying weight-bearing capacities, depending on the weight and overall dimensions of the product being handled. The overhead types, for example, may support the conveyor chain or cable with the trolleys riding an I-bar or being held within an enclosed track. The enclosed track in different installations may be built with the opening up, down, or to one side. (Up- and side-openings with enclosed style tracks will prevent much of the potential dirt from falling down on parts.)

## Flat Belt, Roller, or Track Conveyors

The flat conveyor is probably the earliest general category of paint line conveyor, and includes many different styles such as polyurethane belt, wire mesh, dual chain with cradles, cross slat, pin, perlon filament, and flatbed. A motor-driven, beltlike device may carry parts that are set on it along a pathway, which allows various finishing steps to be performed. Motor-driven and manual roller conveyors are other variations of this conveyor style. Powered flat conveyors are used for finishing single sides of coiled paper and metal and for painting flat stock such as doors, partitions, and wood paneling. Spray, roll, or curtain coating methods are usually used.

The parts need not be finished on the conveyor. They can be removed (by hand or with lifting equipment) from the conveyor, taken to a nearby finishing station, and then returned to the conveyor. The main advantage is simplicity. The main disadvantages are generally a low product weight limitation and the requirement to have operators remove parts for processing and then return them to the conveyor.

## Chain-on-edge Conveyors

Chain-on-edge conveyors consist of a light-duty conveyor chain equipped with vertically mounted spindles to carry parts to be painted, as shown in Figure 20-1. These conveyors are used in disc, bell, and spray-painting booths and to transport the parts through an oven. They are not likely to be used to carry parts through aqueous cleaning and conversion coating operations.

Chain-on-edge conveyors can be arranged in either a horizontally laid out configuration or in an over-and-under vertical loop arrangement. In the horizontally arranged loop, all of the spindles remain in the upright position at all times. In the over-and-under configuration, half of the time the spindles are inverted as they travel the lower section of the vertical loop. The over-and-under configuration is similar to a belt conveyor with spindles.

A chain-on-edge conveyor is not suitable for heavy objects so parts to be coated are, of necessity, rather small and lightweight. Parts must also be of rather uniform cross-section so all areas to be painted can be covered during a pass through the paint application booth. The loaded spindles passing in front of the application device usually rotate to allow paint coverage on all surfaces of the parts. Each spindle is equipped with a sprocket (pinion) that engages a rack near the spray guns, which forces the spindle to rotate. The main chain-on-edge advantages are quick load and unload, simplicity, and high efficiency. Their chief limitation is that they are useful only for the painting and baking operations; they are not very suitable for use in cleaning or pretreatment operations.

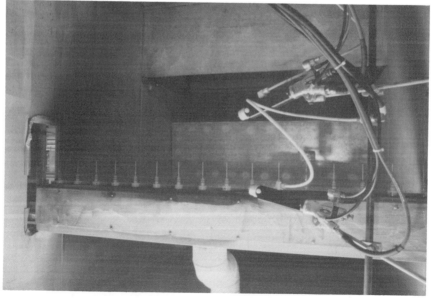

Figure 20-1. Chain-On-Edge Conveyor

## Chain-in-floor Conveyors

A chain-in-floor conveyor consists of a driven chain embedded in a channel slightly below the level of the plant floor. The conveyor's function is to pull large wheeled items or skids and trolleys bearing the weighty parts. These items are manually hooked onto and off of the conveyor chain as needed to carry items to successive finishing stages. The chain is large-gauge, designed to handle heavy loads. This type of conveyor is often used to pull heavy tracked and wheeled agricultural or earth moving machinery and large pieces of military equipment through various cleaning, pretreatment, sanding, and spray painting stations. In addition to its strength, the conveyor's main advantage is simplicity. Its basic disadvantage is its restriction to use with equipment having wheels or those items that can be placed on wheeled carriers.

## Overhead Conveyors

The overhead power conveyor consists of a motor-driven steel cable or chain to which trolley wheels and hooks are attached at regularly spaced intervals. The chain or cable is supported by the trolleys that ride on a continuous supporting track such as a "C" or "U" style channel or an I-beam. Figure 20-2A through C shows examples of the most common types. The overhead chain-and-trolley power conveyor is the workhorse of the finishing industry. It can carry large volumes of parts through all finishing steps: sanding, washing, drying, painting, and baking. It can also be manually powered when only small numbers of parts need to be moved through the manufacturing sequence.

The overhead trolley support track may be open or enclosed. The enclosed type has a box enclosure around it, except for a slot opening that runs its length to permit hanger passage. The opening may be on the bottom, top, or sides. Less debris can fall down onto painted parts with the slot located on the side of a "C" or the top of a "U" channel. Dirt-catching continuous trays (called sanitary trays) can be placed under the conveyor track to serve the same purpose. C-shaped hangers that curve around the tray are used to position the hanger under the sanitary tray.

Overhead conveyors are driven by electric motors with gear boxes and sprockets. A take-up station allows tightening the slack on the conveyor from time to time as the cable or chain stretches and the components wear. Conveyors are usually equipped with automatic oilers that periodically spray lubricant on the trolley wheel bearings.

Many conveyors have automatic brush-type cleaning equipment that runs continuously to keep the chain and trolleys clean and free of oversprayed paint. It is vitally important in electrostatic painting to have a clean I-beam surface (on which the trolley wheels ride) and clean trolley wheels to complete the electrical circuit to ground. Without continuous cleaning, the I-beam riding surface and wheels tend to collect dust and dirt, and consequently lose some of their electrical grounding ability.

Overhead conveyors should be protected in spray washers, spray booths, and ovens. A conveyor running unprotected inside a spray washer will be saturated continuously with cleaner, phosphate, and sealer rinse solutions, which will wash away lubricant and bring quick failure to wheel bearings. Chain-on-edge and inverted conveyors are not suitable for use in most pretreatment stages for this reason. An unprotected overhead conveyor inside a spray booth can collect an exorbitant amount of overspray, which fouls the trolleys and chain and eliminates a good

282 INDUSTRIAL PAINTING

Figure 20-2. Three Examples of Overhead Conveyors

electrical ground. The conveyor inside an oven may have the lubricant driven out by heat. Sometimes an overhead conveyor runs above a washer; a slot in the roof of the washer allows hangers to pass. Plastic brushes close off the gap to minimize solution and moisture escape onto the chain and trolleys, as shown in Figure 20-3.

One type of overhead conveyor protection when the chain is run inside the washer consists of a shroud device. A steady flow of air into the shroud maintains a positive air pressure that prevents moisture from reaching the conveyor. The same type of shroud can also protect a conveyor in a spray booth from overspray.

Overhead trolley conveyors can be operated at a wide speed range. The top running speed in practice is limited by the slowest operation on the paint line. For example, if a part must be in a 24-ft (7.32-m) spray booth for 3 min to allow it to be completely painted, the top conveyor speed would be 24 ft divided by 3 min, or 8 ft/min (2.4 m/min).

Overhead power conveyors can be simply power conveyors or power-and-free type conveyors. The power-and-free feature allows separate trolleys with or without parts on the hangers to be stopped at any point on the line, although the conveyor chain continues to move. Two tracks are used: a free track supporting the parts and a power track in which the chain moves continuously. Each trolley is pushed by "dogs" on the conveyor. The dogs are fixtures that can be made to extend and catch the moving chain and thus move the trolleys. The dogs can also be made to

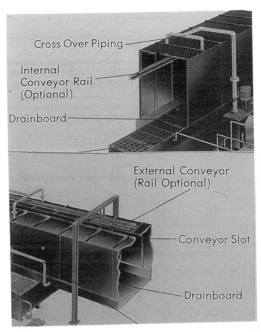

**Figure 20-3. Nylon Brush Seals in an External Overhead Conveyor Keep Cleaners and Conversion Coatings Away from Trolley Wheels and Chain Lubricants**

retract so that the trolley is not pushed along, but stays where it is at various stations for desired time intervals. Power-and-free conveyors allow individual hooks to be disengaged from the conveyor line for accumulating parts, such as in an oven for longer dwell times or for multiple routing of a variety of parts requiring non-uniform processing to and from side tracks.

The overhead conveyors can be linked with various transfer stations that allow one conveyor loop to pass parts to another conveyor loop. The conveyor can also be used with elevators to lower and raise parts for various processes such as pretreatment baths, electrocoating solutions, and batch-type ovens. Power-and-free conveyors cost about 4-5 times as much as a straight power conveyor, but they may quickly pay for themselves in handling ease and convenience.

About the only disadvantage of an overhead conveyor is that it tends to collect dust and dirt, which can fall onto parts being finished. Figure 20-4 shows car bodies being moved through a spray booth by an overhead conveyor equipped with a sanitary tray and coverings to prevent conveyor dirt from falling onto the painted bodies. Note the "C" portion of the hanger.

## Inverted Conveyors

These conveyors are "flipped-over" versions of overhead conveyors riding at either floor level or a few feet above floor level. They are equipped with regularly spaced mounting devices. They are often used to transport parts for which dirt-free finishes are critical, especially those that will be clearcoated. They are used for painting items such as car bodies or outboard motor covers to carry them through spray booths and ovens. A major advantage of this type of conveyor

284  INDUSTRIAL PAINTING

Figure 20-4. Car Bodies Moving Through Spray Booth Using Covered Overhead Conveyor

is that the conveyor mechanism—which tends to collect and distribute dirt—is below the part being finished, thus dirt cannot fall onto parts. A big disadvantage of inverted conveyors is that they restrict access inside the plant. Personnel can usually walk under an overhead conveyor line, but an inverted conveyor mounted above the floor is difficult to cross. Unless a bridge is constructed, personnel may be forced to walk around rather than over the inverted conveyor line. Inverted conveyor loops are used alone and in concert with overhead conveyor loops and power-and-free systems.

## The Importance of Good Conveyor Design

Modern finishing systems feature advanced automation, which is made possible by conveyors and computerized control. Each paint line conveyor system is custom-designed for a particular product. The conveyor system is required to move products through each finishing station. Banks of products may need to accumulate at various locations. The conveyor needs to be at the proper height above the floor for part loading and unloading.

Considerable thought and planning should be devoted to new conveyor system selection and line layout. The impact it has on manufacturing is felt every minute of operation, and changes are sometimes difficult and costly once it is in place. Poor conveyor design, inadequate conveyor cleaning, and improper lubrication can virtually guarantee a short and trouble-plagued conveyor life, plus continual serious maintenance problems. Fortunately, with the proper conveyor design, augmented by continuous brush cleaners and automatic lubrication systems, plants can expect long conveyor life with few problems and normal maintenance requirements.

# Chapter 21

# Finishing Robots

## Industrial Robot Programming

Unfortunately, for many people the term "robot" conjures up an image of a humanlike mechanical device that can walk (in zombie fashion), talk (always in a brittle monotone), and do various simple tasks (such as doing housecleaning work or bringing its owner a cold drink on a hot day). This image has been fostered by numerous puerile movies and television comedy cartoons over the years. Such a device would have little or no economic value in an industrial setting. Industrial robots have virtually no resemblance to their stilted Hollywood cousins. They don't walk or talk, and rather than do widely varied tasks, they are designed for limited numbers of similar, specific, repetitive jobs.

An industrial robot is a device that can be mounted to the floor, wall, or ceiling and is capable of moving about various axes to perform specified functions. It usually will contain an arm that mechanically resembles a human arm with a wrist, and it also performs handlike functions such as aiming or gripping. It likely will have a control center and an electronic memory so that it can be instructed to perform any selection of preprogrammed instructions.

The robot arm can be designed to have movement along (or about) three axes:

- Translation—an axis along a horizontal base
- Elevation—a vertical axis perpendicular to the base
- Reach—a horizontal axis perpendicular both to the base and to the vertical axis

The wrist part of the arm can be designed to have movement along (or about) three additional axes:

- Yaw—angular right and left
- Pitch—angular up and down
- Roll—angular around

Three additional axes of motion can be achieved by moving the entire robot as follows:

- Parallel to conveyor travel (to track a moving target)
- Perpendicular to conveyor travel (to move toward and away from the conveyor)
- Up and down

A robot with full arm, wrist, and base motions would therefore have nine axes of motion.

The electronics to control a robot's motions can be programmed point-to-point or continuous-path. *Point-to-point programming* moves a robot in a series of straight lines that connect a series of points relatively widely separated in space, which requires only a limited memory. Point-to-point programming is not used for intricate movement and extensive control. *Continuous path programming* is like point-to-point but with a major difference. In continuous path programming, the movements are between points that are very close together. Constant feedback allows the motion to follow the series of points to simulate moving along a continuous curve. Continuous path programming requires extensive memory and computing capacity. This type of programming is useful where intricate contoured paths need to be followed, such as in spray painting parts with complex shapes.

Robotic motion is usually achieved with servo controls for each axis. A *servo* is an electrical/mechanical device that provides an output motion according to a given input signal. When equipped with electrical feedback circuits, a servo's output motion can be continuously monitored and adjusted (corrected).

Robots are powered either directly through an electric motor or indirectly through a hydraulic or pneumatic motor. Characteristics of such robots include:

- Electric robots tend to be fast, highly accurate, and can carry light-to-medium payloads.
- Hydraulic robots are characterized by an ability to carry heavy-to-medium payloads.
- Pneumatic robots are often used in fast-velocity applications with light payloads.

Industrial robots are categorized according to their design function. The categories include assembly, material handling, welding, and spray painting. Assembly robots are programmed to perform one or more tasks on an assembly line. Material-handling robots are programmed to move various products from one location to another. Welding robots are programmed for use in automatic welding applications. Painting robots (see Figure 21-1) are almost always programmed to manipulate a spray gun or atomizing rotary bell through a series of motions to paint products, either while the target is stationary or while parts are being moved by a conveyor.

The four categories of robots are very similar in appearance and design. With minor modification, most robots could operate in any category. Material-handling robots vary in size according to the weight of the product being handled. A material-handling robot in a laboratory would likely be bench-mounted and be very small, perhaps with an arm about 12 in long. A material-handling robot operating in a foundry to move heavy castings would be very large and be designed with a large, sturdy arm capable of lifting the heavy parts. Gantry robots (having a bridge-like overhead frame) are capable of lifting heavy items also. Long reaches may best be handled by a robot mounted on a gantry rather than by a robot having a very lengthy arm.

About 35% of industrial robots are used for assembly, 35% for material handling, 20% for welding, and 10% for spray painting. Applying paint robotically with a rotary bell is being done on an increasing basis, but is not as common as robotic paint spraying.

## Rotary Bell and Spray Painting Robots

Robots can be used for painting operations that are uncomfortable or of marginal safety. For example, robots are being used to apply with absolute safety many two-component polyure-

# FINISHING ROBOTS 287

**Figure 21-1. Parts Rotating on a Table are Sprayed by a Stationary Robotic Arm**

thane paints that have a toxic isocyanate component. Robots are suitable to apply dirty, malodorous materials and to work in hot, humid, or other unpleasant environments. They can also perform physically demanding contortions and mechanical manipulations that would prove unendurable to a human operator if done for any length of time. Long reaches that require an extension or pole-gun for a manual sprayer can be accomplished easily with a robot.

Devices that operate in paint spray booths either must not generate sparks, or if they do, the sparks must be enclosed in a fail-safe system to prevent the possibility of the spark igniting flammable material in the booth (such as solvent). Early-model painting robots were hydraulic-powered and spark-free, averting the possibility of igniting a flammable atmosphere. However, more accurate electric-powered spray painting robots were put into service after a fail-safe system was devised to prevent the possibility of sparks, especially in the electric motors. All potential spark-generating equipment is housed in an enclosed compartment that is filled with a nonflammable gas at a positive pressure, which prevents entry of flammable solvent fumes. If the compartment should lose its positive pressure, the electric power becomes inoperative.

Electric spray painting robots became popular because of the elimination of potential hydraulic leaks and because the electric robot is a modular package. (A hydraulic robot needs a separate source of hydraulic power.) Also, a hydraulic-powered robot's ability to handle large payloads is usually not useful in painting, where a spray gun payload may weigh only several pounds and a bell less than 20–25 lbs. However, the increased accuracy capability of electric-powered robots isn't much of an advantage in most painting operations, which frequently don't require critical accuracy.

288  INDUSTRIAL PAINTING

The number of axes required in a painting robot depends on the product being painted. A simple product such as a flat object (nonconveyorized) could perhaps get by with a robot having only three axes. A complex object such as a conveyorized automobile body could require a robot with all nine axes. The robot package itself could have six axes. The axis to track the moving car body would be the seventh axis; the capability to have the entire robot move to and away from the car body would be the eighth axis. Finally, the capability for the entire robot to be raised and lowered would be the ninth axis. Figure 21-2 shows a a relatively simple multi-axis robot in use.

Robots are most often used with liquid paint spray guns or liquid paint bells, but robots have been shown to be able to successfully paint with powder guns as well. At the date of this writing, no full-scale use was being made of a robot painting with a powder bell, yet there is no reason why powder bells cannot be used this way.

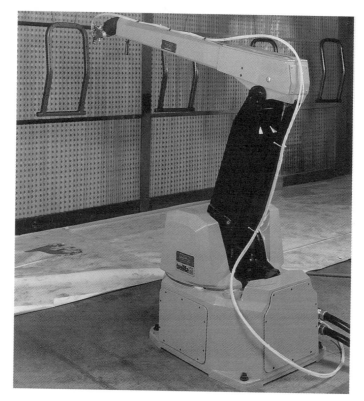

Figure 21-2. Robot Sprayer has Auxiliary Arm for Painting Interiors

FINISHING ROBOTS 289

## Programming for Painting

Both bell and spray painting robots for liquid or powder paints are usually programmed in combinations of point-to-point and continuous path, depending on the shape and size complexity of the product being painted. A growing tendency is to program the painting robots *off-line* (see Figure 21-3). In this technique, an operator in a remote location, equipped with a computer and appropriate software, can program the robot.

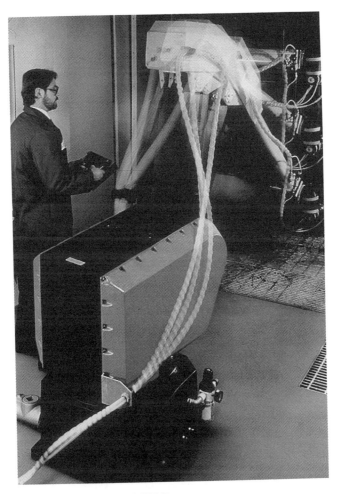

Figure 21-3. Robot Being Programmed Off-line

290  INDUSTRIAL PAINTING

Off-line programming can be used in combination with conventional teaching systems where a robot is programmed by an operator by moving a teach pendant through the desired paint applications motions, as shown in Figures 21-4 and 21-5.

All robotic painting programs need to be synchronized with conveyor loading and conveyor speeds. The programs also need to include start-stop functions and color-change operations. The part rack will activate a limit switch to signal the robot when to begin its program. This switch is located just outside the spray booth, and the appropriate delay is built into the program if the conveyor movement is continuous. Programming is simpler if the conveyor pauses (indexes) during the time the robot paints because the target is not moving. The work envelope (total reach volume) of the robot can also be smaller for an indexed conveyor.

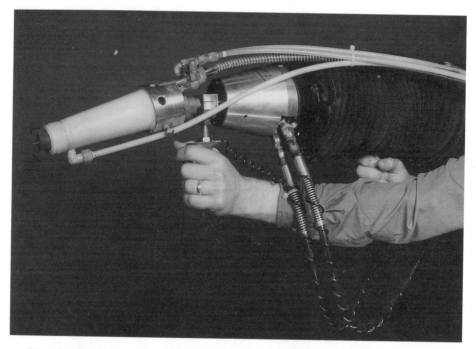

Figure 21-4. Operator Teaching Robot Desired Spray Motions

Because a robot cannot see, parts need to be hung on a conveyor the same way every time. No variation in the alignment of the hanging parts is permitted. Crooked paint hooks and hangers and parts that are swaying or swinging as they move past the robot may drift outside the programmed path of the robot, resulting in poorly painted products. Double-point hanging is often used with robotic painting to secure the parts and reduce the likelihood of sway or misalignment of parts on the conveyor line, as shown in Figure 21-6.

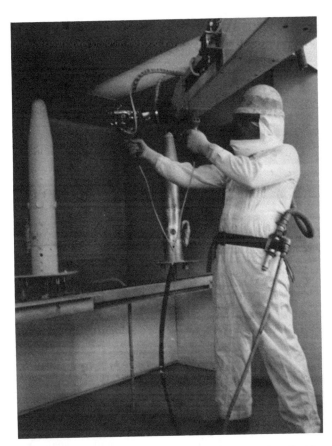

**Figure 21-5. Robot Being Programmed for Two-Part Urethane Application by Operator Wearing Special Uniform and Air-Supply Headgear**

Machine vision can be used with painting robots. In *machine vision*, video cameras record the shape of the parts on the conveyor approaching the spray booth. If a shape change is observed, the computerized vision control center can recognize the new shape and instruct the robot's computer control center to shift the painting programs accordingly. Machine vision systems tend to be expensive; the same result can usually be obtained with photocell detectors, which have been triggering spray painting systems for years. This type of system is being used in automotive sealer applications, where a bead of sealant is applied automatically to cover corrosion-prone seam-weld areas.

However, painting machine vision applications tend to be relatively simple in scope, being limited to shape recognition and location misalignment. Full-bloom vision systems that could replicate human eyesight and brain function are a long way down the road.

292  INDUSTRIAL PAINTING

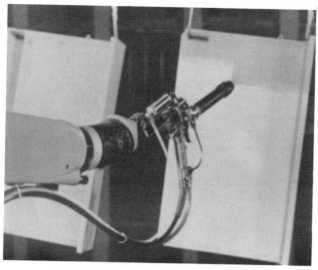

Figure 21-6. Robot Spraying Desk Parts Utilizing Double Hanging Hooks to Minimize Swaying

## Robot Operation Requirements

Before installing a painting robot, a great deal of planning must be done. The space requirements for a painting robot tend to be more demanding than for a human. Most painting robots are larger than humans and have a much bigger work envelope. The robot's physical reach must not exceed the constraints of the booth. The robot will have to be housed in a booth large enough to confine overspray and remove VOC vapors.

A spray painting robot requires considerable peripheral equipment. An off-line teaching booth in close proximity to the paint line for the preparation of new paint application programs will eliminate the need to interrupt production runs to perform robot teaching operations.

The large mass of the robot will require an overhead lifting device to relocate the robot or to move it for repairs. Easy access must be provided for servicing electrical and hydraulic components.

For ease in maintaining a painting robot, the equipment should be designed to include:

- A computer with self-diagnostics
- Interface electronics with only one printed circuit board per axis of motion
- Precalibrated board backups for each drive axis
- Easy calibration procedures

To protect people working nearby from being struck by the robot, restraining guard rails, gates, fences, or other protective devices must be installed. A number of other important safety measures are needed to protect people and equipment in the area.

In case of malfunction, runaway protection is needed so that an out-of-control robot does not cause injury. A warning bell or light can alert employees to this condition. Alarms for runaway robot conditions should be loud enough to be heard readily above plant noise. Fail-safe brakes are necessary on robot arms to prevent wild movement in the event of a hydraulic or electrical failure. Emergency shut-off switches must be located to provide high visibility and quick access.

Pressure-relief valves are needed in the hydraulic system to prevent overload of the pump and motor. A velocity "fuse" in each hydraulic supply line can prevent high flow rates. Software limits to ensure safe stoppage of the robot arm in all major axes should be backed up with electrical limit switches and mechanical shock absorbers. Robot arms can be nylon-covered to preclude any sparking in the event of a collision with a metallic object.

Perhaps most importantly, a bell or spray painting robot's cost will vary considerably, depending on the size and performance capability desired. A low-requirement unit may cost in the mid "five figures." A high-performance painting robot may cost in the low to mid "six figures." There are too many variables to give more precise dollar figures.

Added costs include robot installation and training for personnel to learn to operate and maintain the device properly. In addition, all paint shop personnel must be trained in general safety precautions associated with the robot. The average robot requires an annual maintenance expenditure of roughly $4,000, including end-of-shift cleanup and periodic preventive maintenance.

A preliminary study should be undertaken to determine the feasibility of having a robot for paint application. Payback time is the critical economic point to consider. If the payback time meets the company's financial guidelines, authority to purchase the robot can be sought. A formula to determine a robot payback can be expressed as follows:

$$Y = \frac{Rc}{Ws + Ps + Rd - (Mc + Sc)}$$

where Y = years payback, Rc = total robot purchase and installation costs, Ws = replaced worker salary for a year, Ps = production savings for a year (paint savings, energy savings, etc.), Rd = robot depreciation for a year, Mc = maintenance costs for a year, and Sc = staffing costs for robot for a year.

In a hypothetical case, suppose that:

Rc = $100,000
Ws = $30,000
Ps = $25,000
Rd = $25,000
Mc = $10,000
Sc = $20,000

Then, $$Y = \frac{\$100,000}{\$30,000 + \$25,000 + \$25,000 - (\$10,000 + \$20,000)}$$

$$Y = \frac{\$100,000}{\$50,000} = 2 \text{ years payback}$$

## Spray Painting Robot Advantages

The advantages of spray painting robots can be categorized into two areas: cost reductions and quality improvements.

The most obvious cost reduction involves labor. A typical robot installation will displace one to three painters due to its speed capability. If the robot is used on more than one shift a day, the savings are multiplied. The robot arm can simply move faster than a human arm—and without tiring.

Another cost reduction that can be substantial involves paint savings. The average replacement of a semiautomatic system by a robot will reduce paint consumption by 15%-20%. This savings accrues from repetitive and accurate painting motions, finely tuned gun and bell triggering, and multiaxis movement that can closely follow part contours. The actual material savings will depend on the volume of production, the shape of the parts, and the type of paint applied.

Another important savings is in energy. Because no human operators need be present in a robot-equipped paint booth, the amount of air exhausted from the booth can be reduced. This can result in a considerable energy savings, especially in cold climates during winter months when makeup air needs to be heated. Ventilation cost is reduced nearly 50% for robotic spraying compared with manual operations.

As for quality, robots provide absolute part-to-part uniformity of coating. This improves quality and reduces or eliminates rework and rejects (as well as possible field recalls on products due to paint problems). Some plants that switched to spray or bell painting robots have reported a 75% reduction in rejects.

A fringe benefit associated with robotic spray painting is that reduced paint use savings also translate directly into reduced VOC and HAP emissions. Paint that is not applied, naturally, will not emit VOC. Since lowered emissions are being required almost universally, any finishing system that reduces HAP and/or VOC emission totals will help a company obtain (and retain) their operating permits from jurisdictional regulatory agencies.

## Case History

The following case history describes how a company saved an appreciable amount of money by switching to spray painting robots. The plant was manually spraying plastic auto parts on two separate conveyor lines. To meet production, it had been necessary to paint reworked parts on overtime. One line was altered so that all painting was done by four robots and four stationary electrostatic guns. The second line was left unchanged and used just for spraying repair parts; it operated on straight time only. The switch to robots on one line allowed eight sprayers per shift to be reassigned. Three gun technicians were added. The net reduction of 13 positions resulted in an annual savings of over $500,000. The savings of an hour a day overtime per operator for each shift gave an additional yearly savings of over $330,000. The annual paint savings was estimated to be roughly $333,000. Although no exact dollar value can be assigned to quality and VOC/HAP reductions, both VOC and HAP emissions were reduced and paint finish quality was improved by robotic application. The total implementation cost was high—slightly over $1 million—but the annual savings was close to $1.25 million.

## Future Developments

Spray painting robotic research and development will likely focus on the following areas:

- Sensing devices
- Painting speed
- Precision of operation
- Repeatability
- Minicomputer controls
- Voice command recognition
- Artificial intelligence

Improved sensing devices could increase the efficiency of color changes, improve product identification, detect and correct misaligned parts, and identify/select parts from mixed products. Precise machine vision is available now but costs are prohibitively high. Increased painting speed could reduce robot costs and thus lower payback time. Improved precision of operation and repeatability would increase quality.

Minicomputer controls, voice command recognition, and artificial intelligence are steps toward giving the spray painting robot the capability of the human brain.

# Chapter 22

# Spray Booths for Liquid Painting

## Spray Booth Basics

Spray booths for applying powder finishes are described in Chapter 6; this chapter will deal with spray booths for liquid paint application. While similar in concept and in function as an enclosure for depositing powder coatings, spray booths for liquid painting are unique in design. Incidentally, they are all called spray booths, even though paint may be applied in them by other means, such as by rotary atomization.

A *spray booth* for liquid painting can be defined as an enclosure equipped with a means of safely capturing overspray paint, diluting and exhausting solvent vapors, and replacing the exhausted air with clean makeup air. This definition makes a spray booth sound exceedingly simple. Actually, a spray booth is quite a complex piece of equipment. In an enclosed booth, makeup air may be supplied to the inside of the booth by fans and ducts; in an open booth, plant air may be allowed to replace air exhausted from the spray booth.

A spray booth almost always has openings on both side walls to allow entry and exit of the conveyed products to be painted. A spray booth without such conveyor openings is sometimes used for low-production batch spraying. Access for product entry and exit is through the opening at the front of the spray booth. For greater cleanliness, the front opening may be fitted with doors. Replaceable air filters held in the doors will hold out some of the contaminant particles that may be present in shop air drawn into the spray booth.

The need for forced air exhaust and forced air makeup in an enclosed spray booth presents an engineering problem. The exhaust and intake air must sufficiently lower the concentrations of paint solids and solvents enough to satisfy health and insurance underwriter regulations. Air throughput requirements are even greater when humans are present in the booth. Specifically, air makeup must do two things: (1) replace the exhausted air, and (2) supply an amount of additional air to maintain a slight positive pressure. A positive booth pressure will prevent drawing plant air that might contain oil mist, dust, lint, or other particulates into the booth. The extra amount of air makeup cannot be too large, or excessive amounts of overspray might be forced out of the conveyor openings and into the plant, instead of out through the exhaust system.

Makeup air is drawn into the booth from outside the plant to avoid disturbing the rest of the plant's air balance. If the spray booth air were taken from inside the plant, a negative plant pressure would be created. The booth makeup air is filtered to introduce clean air into the booth.

## 298 INDUSTRIAL PAINTING

In northern climates, the air makeup is heated and sometimes humidified during cold weather. Figure 22-1 shows the filters and humidifiers for a spray booth makeup air housing. During summer temperature extremes, it may be air-conditioned or passed through humidifiers to cool the air, even though this also increases the relative humidity. Adjusting the temperature of booth makeup air downwards requires a great deal of energy and so is rarely done.

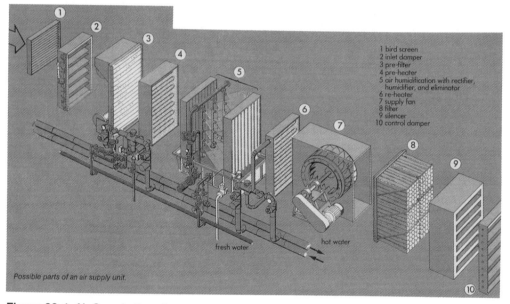

**Figure 22-1. Air Supply Housing**

Plants should also maintain buildings at a slight positive pressure with the plant air circulation system. The major advantage of keeping the building at a slightly positive inside air pressure is that outside air-borne contaminants such as dirt, dust, and pollens are not able to enter.

The amount of water vapor in the air has a significant effect on the rate at which paint solvents (and water) evaporate from wet paint films. If the air is too dry, the solvents evaporate too rapidly and the finish is less smooth and glossy as a result. Sometimes moisture is introduced into dry makeup air to prevent excessively fast evaporation of volatiles from the applied paint film. This extra moisture may be especially needed in frigid weather because the extremely cold outside air, drawn in and heated for use as makeup air in the spray booth, would otherwise contain just a tiny amount of moisture.

If the primary movement of air makeup and exhaust is from the booth ceiling to the floor, the booth is termed "downdraft." Figure 22-2 shows a downdraft booth; fresh makeup air enters through the filters at the ceiling, and overspray is exhausted through the floor grating to a

# SPRAY BOOTHS FOR LIQUID PAINTING 299

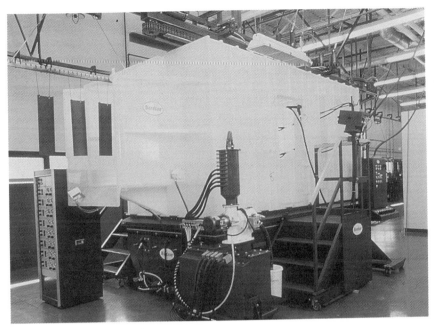

**Figure 22-2. Downdraft Spray Booth**

filtering system. (If the air movement is mainly horizontal, the booth is termed "sidedraft.") In both types, the air movement permits a human spray operator to work in a safe environment without a separate fresh-air supply for breathing. However, where toxic chemicals are sprayed, a separate fresh-air supply needs to be supplied for human breathing, usually to a hood that fits around the operator's head. OSHA specifies the maximum allowed concentrations in the breathing air for each solvent to which workers may be exposed. In automated paint booths where humans are not working, the air flow rates may be reduced.

The oversprayed paint cannot be exhausted directly from a spray booth to outside the plant because of exhaust ductwork problems and the possibility of paint drifting onto cars and buildings in the vicinity of the exhaust outlet. Unfiltered spray booth exhaust sends overspray paint into the exhaust ductwork and onto fan blades, causing a fire hazard. The overspray paint accumulation can unbalance fan blades, which causes fan vibration and wear and possible damage to the exhaust housing. Overspray paint exhausting directly outside plants and accumulating on nearby buildings and cars is not exactly a rare occurrence. A sizable number of insurance claims are filed annually in North America for the accidental painting of cars in the vicinity of a plant with poor spray booth exhaust filtration.

The method used to filter the paint overspray from the exhaust air provides another way of categorizing spray booths. If the overspray is removed by air filters, it is termed a "dry-filter booth." If the overspray is removed by a turbulent air/water mixing system, it is called a "water-wash booth" or a "wet booth."

Before we describe the dry-filter and water-wash booths, it is necessary to describe the nature of liquid paint overspray. Atomized paint exiting a spray gun consists of various sizes of paint droplets in flight toward and past the vicinity of the part to be painted. As the droplets move through the air, solvent evaporates from them continuously, leaving solvent vapors and particulate (resin, pigments, and additives) components. The evaporated solvent, both from atomized particles and from the deposited paint film, is comprised of single, extremely tiny molecules. Solvent vapors thus totally escape filtration and move directly into the exhaust air stream. The overspray paint droplets, which are relatively large, also are carried in the air exhaust stream; all but the tiniest are captured by a filtering medium before they leave the booth.

## Dry-filter Booth

In a dry-filter booth, the overspray (evaporated solvent molecules and atomized paint particulates) are directed by air circulation to a network of air filters in which particulates are trapped, but air and solvent molecules pass through into the exhaust duct system. Nearly all dry-filter booths are sidedraft type, although other variations are possible. Figure 22-3 shows a sidedraft dry-filter booth. Air enters the booth from behind the spray guns or bells and exits the booth through the filters into the exhaust plenum. Clean, filtered air can enter the booth through a ceiling plenum and descend behind the spray guns and parts to be painted, through filters on the back wall, and out the exhaust stack. It is much preferred, however, that air flow into and out of the booths be totally laminar (linear), because straight air flow results in a more efficient and clean booth operation than turbulent air flow.

**Figure 22-3. Sidedraft Dry-Filter Booth**

Dry-filter booths use easily replaceable, disposable filters that vary in size, composition, and particulate capture efficiency. The filters may be of the strainer type or the baffle type. The baffle

# SPRAY BOOTHS FOR LIQUID PAINTING

type can take heavy loading levels, but are not highly efficient at removing all of the overspray particulates. The strainer type are effective at removing most particles but tend to "face-load" or "blind" (clog) quickly. For some time, manufacturers have been making variable-density filters or dual-material strainer/baffle "sandwich" filters. A baffle-type filter is therefore often placed ahead of a strainer-type filter to obtain the advantages of both. These avoid the heavy face-loading of the strainer type but can still achieve high capture efficiencies.

When filters load up with paint to the point that air flow is significantly impaired, the used filters must be replaced and disposed. Some plants must bake their used booth filters before they are allowed to discard them in conventional ways. Baking expels residual solvents from the trapped overspray on the filters. Landfill disposal of the spent filters is usually straightforward when no toxic substances are present in the paint. The exceptions are paints that can ignite spontaneously when the used filters are stacked together. Stacking does not permit the escape of heat generated by oxygen reacting with the finely divided organic materials on the dirty filters, and spontaneous combustion can result. When this problem is encountered, the filter makers recommend the filters be plunged into a drum of water, and the filled drums then tightly sealed. This is effective in preventing combustion, but it adds considerably to overall weight and to disposal costs.

Getting rid of filters that contain toxic paint can be expensive. Depending on the applicable regulations, used filters laden with toxic paint components, such as lead or chromium compounds, may need to be disposed in facilities licensed to handle toxic wastes. Somewhat surprisingly, in at least three states, if water leach tests show that only low amounts of toxics are released, the filters do not require handling and disposal as toxic waste. Most states do not allow any materials classified as toxic to be discarded this casually.

### *Dry-filter Booth Advantages and Disadvantages*

Among the advantages of dry-filter booths is the lack of water or sludge to handle. Also, booth installation is straightforward and easy, and the initial capital investment is modest.

However, due to paint buildup on filters, booth air flow velocity and volume decreases, so filters must be replaced. The labor to replace the filters, the replacement filters themselves, and the disposal of used filters are expensive, and the replacement process may cut into production output. Storage space is needed for used and replacement filters, and the spontaneous combustion of used filters is a safety concern. The replacement process also dislodges overspray "dirt" in the booth.

## Water-wash Booths

In water-wash booths, the overspray is exhausted through one or more curtains of water or through overlapping pressurized sprays of water that are designed to entrap and remove the atomized paint overspray. Water curtains or pressurized sprays are usually located on the side of the booth behind the parts being painted. Some booths have the water-impingement system located below the floor grating. Effective water-wash booths can be of either sidedraft or downdraft design. Figure 22-4 shows a sidedraft water-wash booth, and Figure 22-5 shows a downdraft water-wash booth.

The water curtain or spray is pumped from a reservoir tank and returned to the tank. The volume of the reservoir tank depends on the size of the spray booth and may range from several

302 INDUSTRIAL PAINTING

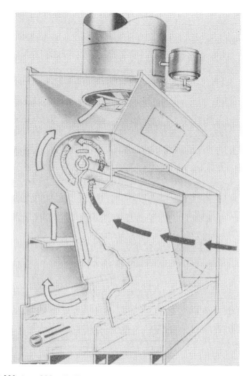

**Figure 22-4. Sidedraft Water-Wash Booth**

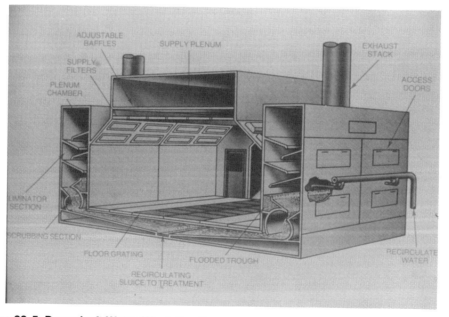

**Figure 22-5. Downdraft Water-Wash Booth**

hundred gallons for a small booth to thousands of gallons for a very large booth (see Figure 22-6). A number of booths may use a common reservoir of recirculating booth water, frequently a large "pit" tank.

If the collected overspray paint particles in the water circulation system were not chemically treated, they would stick to the sides of the reservoir and clog piping, pumps, headers, and spray nozzles. The chemicals added to prevent this are called *detackifying agents*. Figure 22-7 illustrates the detackifying process commonly referred to as "killing" the paint. Chemicals can be used either to float the detackified paint for skimming or to sink it to the bottom of the reservoir for separation from the supernatant water. Other chemical systems are used to form a "killed" material that either can be filtered out or removed by centrifuging.

Figure 22-6. Conveyorized Water-wash Booth

# 304 INDUSTRIAL PAINTING

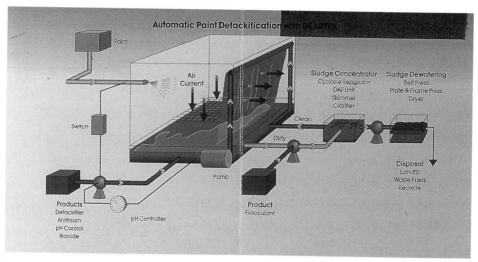

**Figure 22-7. Automatic Paint Detackification**

To kill the paint, detackification agents may combine one or more of the following methods:

- Use absorbent materials such as clay.
- Use alkaline materials that chemically degrade many paint resins. High alkalinity can "sink" the sludge; lesser amounts float the killed paint, which permits removal by skimming.
- Use polyelectrolyte materials that chemically encapsulate the paint particles, rendering them nonsticky.

Detackifier formulations are available in solid form or as concentrated liquids. Best results in separating the killed paint from the water are obtained when paint particles are agglomerated by coagulants into large particles, a result that facilitates either floating or settling.

Antifoamers or defoamers may be added to prevent foaming onto the booth floor, pump cavitation (pumping air), and impeller erosion. Foam interferes with the paint/water separation, which delays paint settling and produces excessively wet sludge that makes disposal costly because of the extra handling weight and transportation expenses. Additional environmental restrictions may also apply to sludge having a high water content.

Water that contains organic material can be potential food for fungi and bacteria unless antimicrobial agents are used. These agents prevent slime, odors, and corrosion pitting. To further minimize these problems, the water is kept mildly alkaline, i.e., at a pH of about 7.5–9.0.

Some paints are easy to kill, others are not. For example, some waterborne and a few high-solids paints tend to be rather difficult to detackify. The detackified paint sludge must be separated from the booth water and collected, which is generally done by one of four major ways. Sludge can be collected:

- By skimming if it is made to float
- By centrifuging the mixture of water and killed paint material

- By filtering
- By allowing the solids to settle and then pumping off the water; the sunken sludge in the bottom of the reservoir is next scraped up and disposed.

With all these methods, any toxic materials and solvents remaining in the booth water after sludge removal must be eliminated by water treatment before releasing it to drain. When solventborne paints are applied in a wet booth, the sludge normally contains about 1%-3% solvent.

Disposal of wet-paint sludge has become an increasingly expensive problem, especially if toxic components or excess amounts of solvents are present. In light of so many health hazards being traceable to chemicals that were improperly disposed, most areas are exceedingly reluctant to allow transport of any manufacturing wastes, much less burial or incineration, in their communities. Many regions either have or are considering laws to ban all hazardous chemical disposal except on the original plant usage site, and then only by approved and certified methods.

Some efforts have been made to recycle paint sludge, but it is usually less costly for small- to moderate-sized paint users simply to dispose of it and buy new paint. If no toxics are present, sludge can be pressed into briquettes and burned for its heat value. Pressing to remove water is necessary because wet sludges burn poorly and may require almost as much heat to evaporate the water as heat generated by combustion. At least one company makes equipment to heat and dry paint sludge and then crush it to a free-flowing powder. Where to reuse the powder is not always easy to determine.

### Water-wash Capture Booth Advantages and Disadvantages

Booth air flow velocity is constant, so no down time is required to change filters and no costs for labor and materials are incurred.

On the down side, the equipment and installation are more costly and complex. Water contamination occurs, so water treatment, "kill" chemicals, and paint sludge disposal costs are incurred. Finally, floor loading from the weight of the water is considerably higher than for a dry-filter booth.

## Dry-filter or Water-wash Booths?

Whether a plant should install a dry-filter or a water-wash booth depends largely on the amount of painting to be done. Equipment and operational costs of a dry-filter booth tend to be lower than for a water-wash booth until at least 75–100 gal (296-395 L) of paint are applied each day. If having to stop painting for 10–20 min to replace booth filters causes dirt or impedes production, or if filter changes are necessary more than 2–3 times per shift, this impediment can be avoided by switching to a wet-type spray booth. The particular paint being sprayed has an effect on the relative costs as well; how readily the paint is chemically detackified, the cost to dispose of wet sludge, and the method of paint application are important factors. The floor weight of a booth water reservoir may also be a factor to consider for a booth planned in a restricted-load portion of a building.

## High-solids Paint Overspray Recovery

Recovering and recycling high-solids paint overspray, by capturing it before it reaches the dry filters or the water-wash, is very slowly increasing in popularity. For many high-solids, the overspray remains tacky and sticky because of lower molecular weight resins and because they do not cross-link at room temperatures. It may also be helped by their relatively low amount of fast evaporating solvents. These characteristics can, in some instances, make paint recovery and recycling fairly convenient, but the viscosity of the overspray must not be too high for this process to work well. The high viscosity of the overspray of most paints renders them unsuitable candidates for recycling. Low-solids solventborne paint overspray cannot be recycled because it dries too quickly. The same type of difficulty exists with recovering waterborne coating overspray.

Most recovery and recycling methods utilize vertical baffles stacked behind parts being painted in a sidedraft booth. A large portion of the overspray is drawn toward the flat overlapping or spiral-channeled baffles. Paint collecting on them gradually flows downward into a trough. The trough slopes into a collection container, and much of the overspray paint that would have been caught up either by dry filters or by water-wash systems is recovered. Augers are also used to move the collected overspray because the highly viscous material itself flows so slowly.

The collected overspray paint needs to be filtered and adjusted to a lower viscosity by solvent addition before it can be reused. Color correction is required in all cases where several colors are jointly collected; single color paints may also need an adjustment to restore the correct color.

High-solids overspray recovery is producing meaningful cost reductions for several large manufacturers in a number of ways. Most obvious is that paint purchases are reduced, but recovery of overspray also lowers filter replacement cost in dry booths. In wet booths, both water-treatment and sludge-disposal requirements are lessened.

## VOC/HAP Reduction Methods

Attempts to devise filter or water additives that will catch or somehow destroy organic volatiles have had few commercial successes. Because solvent vapors readily pass without being captured through dry-filter and water-wash booths, VOC and HAP emission regulations for industrial painting are structured to require low-VOC coatings to be used. For example, a category of finishing (automotive, appliance, etc.) may require that coatings contain no more than 2.8 lbs of VOC/gal (0.36 kg VOC/L). As long as this requirement is met, the VOC given off during coating operations is within compliance. However, EPA regulations require that if coatings are used that exceed the VOC maximum, "add-on" VOC abatement methods must be used. These are most commonly incineration systems which oxidize solvents directly or carbon adsorption systems which capture solvents for either incineration or recovery.

### Incineration Devices

In the incineration of VOCs, the exhaust streams from paint ovens or spray booths are thermally oxidized at around 1400° F (760° C) or higher, or catalytically oxidized at about 800° F (425° C), to convert the VOCs to carbon dioxide and water vapor. Strictly speaking, both are thermal oxidizers and both need a combustion burner, but the catalytic types operate at moderately lower combustion temperatures than noncatalytic oxidizers. In noncatalytic oxidizers, the burner is always turned on during operation.

The thermal and the catalytic incineration processes are both inherently energy inefficient, unless the energy of the hot exhaust can be used elsewhere, either on the finishing line or in the plant, as shown in Figure 22-8. Often, the exhaust gases are passed through heat exchangers of varying designs. This process can be used to preheat air makeup for ovens and other burners and/or to provide heat for dry-off ovens and for washer baths. Frequently the heat exchanger is used to raise the temperature of the inlet exhaust stream which is about to undergo combustion. For economy, attempts are made to reduce the temperature of the incinerator exhaust as low as possible before it is vented from the plant.

Figure 22-8. Hot Exhaust From a Regenerative Thermal Oxidizer is Transported for Use Elsewhere in the Plant

# 308　INDUSTRIAL PAINTING

Fume incinerators are used extensively on coil coating lines, where low-solids coatings are often used and, as a result, VOC emissions from the coatings application stations and bake ovens are high. The VOC concentration in the exhaust air often is great enough to serve as fuel for a catalytic incinerator, allowing a cutback in natural gas use. Incinerator use on coil coating lines has proved to be very cost efficient. Incinerators are also common in automotive assembly plants, but these are mainly direct thermal oxidizers. These industries are major exceptions to the fact that thermal and catalytic incinerators, while not rare, are really not abundant in finishing. When they are utilized, most paint fume incinerators more often tend to be the thermal type because catalysts are expensive and are also susceptible to poisoning.

Three types of incinerators are employed for VOC abatement in painting operations: catalytic oxidizers, recuperative thermal oxidizers, and regenerative thermal oxidizers.

1. Catalytic Oxidizers. Catalytic oxidizers require a burner in order to raise the catalyst up to its operating temperature. At that point, the burner can be shut off. The catalyzed VOC oxidation reaction will become self-sustaining once that temperature is reached.

    The oxidation catalysts are normally composed either of noble metals, such as platinum and palladium, or of some more common metals like copper or manganese. Catalytic incinerators are more expensive to install than thermal incinerators due to the catalyst cost, but the lower fuel consumption will eventually offset this initial investment cost. All catalysts are subject to fouling (poisoning) by heavy metals, hydrogen halides (generated by halogenated solvent incineration), and some organic silicone compounds. The noble metals (platinum, rhodium, palladium) are more expensive than copper, manganese, and nickel, although they resist fouling more effectively. All catalytic oxidizers need a lot of care to avoid poisoning the catalyst.

2. Recuperative Thermal Oxidizers. A ceramic "shell and tube" heat exchanger (see Figure 22-9) is commonly used which will allow over 50% primary recovery of the heat generated by these VOC fume oxidizer units. While the fuel costs are high to operate recuperative oxidizers (at 1250° F–1450° F or 677° C–788° C), recovered heat can be used to run ovens or dryers or to heat water for steam or other needs. The rather elevated operating tempera-

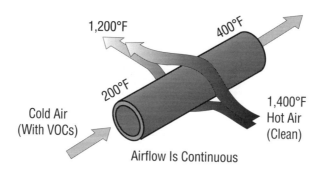

**Figure 22-9. Recuperative Thermal Oxidizer**

tures, however, do form a lot of the HAPs nitrogen oxides (NO$_x$s), far more than catalytic type oxidizers which run at about half this temperature.

3. Regenerative Thermal Oxidizers. These oxidizers employ multiple beds of closely-packed, saddle-shaped ceramic media as heat sinks and switchable ductwork which enable regenerative oxidizers to reach efficiencies above 90% in some cases. A burner ignites the solvent in the incoming air flow. The purified air is exhausted through a ceramic bed which strips the heat from the purified air. The pure air is then vented outside the plant. Once a ceramic-packed bed becomes hot, incoming solvent-laden air from booths and ovens is ducted through it to the combustion zone (see Figure 22-10). This air, preheated by passage through a hot ceramic-packed bed, ignites readily. The heat from solvent combustion is stripped out by passing the newly purified air through another bed. Continuously switching the flows optimizes the heat recovery and minimizes the fuel required for combustion of solvents. The large size and high capital and maintenance costs of regenerative oxidizers are major disadvantages, but NO$_x$ production is low.

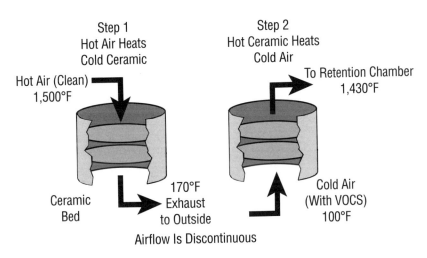

**Figure 22-10. Regenerative Thermal Oxidizer**

As a general rule, incinerator use is economically limited to destroying VOC from bake ovens where fume concentration is fairly high and where the VOC temperature is already roughly 300° F (150° C) above ambient. This obviously will require much less heat input to raise gases to the incinerating temperature. As a means of destroying VOC or HAP from a spray booth, incineration is impractical except in special instances. This is because the high air makeup and air exhaust volume requirements for spray booths produce an excessive amount of air to handle. The air is only at room temperature and it contains just a low concentration of organic vapors. Nevertheless, a number of large plants that use high volumes of paint have found they must use incineration on paint booth exhaust air for it is the least costly technique for remaining in VOC and HAP compliance. Finishers who apply only modest amounts of coatings should comply with the applicable "weight of VOC per gallon of coating" limit, and not even consider incineration. Incineration as a means to achieve EPA compliance is normally impractical for plants whose total VOC emissions are less than high (defined as more than 100 tons VOC/yr and/or more than 25 tons HAP/yr).

## Carbon-bed Solvent Adsorption Systems

Activated carbon-bed adsorption units, diagrammed in Figure 22-11, while totally effective in cleaning solvent vapors from booth exhaust air, also tend to be prohibitively expensive except for large painting operations. Huge beds of carbon are needed, plus facilities for periodically stripping solvent-saturated carbon beds with a flushing material. In most cases live steam is used for regenerating the carbon, although other fluids such as nitrogen gas can be used for this purpose. Equipment to separate both soluble and insoluble solvents from the condensed mixture of steam and organic vapors adds to the high cost of adsorption systems. These reliable and easily operated systems are very efficient (up to 99% is claimed), and they form almost no $NO_x$. But it is absolutely essential that all paint particles be removed from the exhaust flow before they reach the concentrators or the system will be fouled. Two-stage filtration is normally recommended using multiple-stage bag filters.

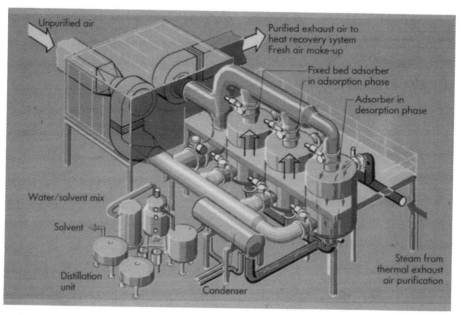

Figure 22-11. Carbon-bed Solvent Adsorption

Several newer incinerator designs use carbon adsorption to continuously strip and concentrate the solvents in a booth or oven exhaust stream prior to VOC incineration. A rotating carbon-filter wheel (see Figure 22-12) cleans the exhaust air of all solvents. Part of the hot incineration exhaust gas stream back-flushes the adsorbed solvents off the carbon wheel and concentrates them by a factor of 5- to 20-fold. The smaller exhaust volumes can be handled less expensively because smaller fans and lower diameter ductwork can be used. Additionally, the enriched

**Figure 22-12. Rotating Carbon Filter VOC Concentrating Incinerator (Interior and Exterior)**

solvent-air ratio in the exhaust mixture from the concentrator can be incinerated using noticeably less burner fuel.

Newer and smaller booth designs also attempt to reduce exhaust air volumes, but only about a maximum of 17–18% reduction has been achieved this way to date. Small booths, however, have the added benefit of requiring less treated (consecutively filtered, and possibly heated, cooled, or humidified) booth air makeup. Booths with no people in them can be operated legally at lower exhaust levels than permitted for occupied zones. This allows the use of two more air reduction tehniques: air recirculation and air cascading. A portion of the exhaust air from an unoccupied booth can be filtered free of particulate (including overspray paint), and then recirculated back through the same booth as long as the overall painting operations are not adversely affected and the fire hazards are not increased. Both bag houses and electrostatic precipitators are employed to remove particulates. If necessary this filtered air can be humidity modified and heated before reintroduction into the booth.

In air cascades, the exhaust from an occupied booth might be filtered and next passed on to a booth where paint is being applied with an automatic device such as a bell, reciprocator, or robot. Air may also be cascaded down from spray booths as makeup air for paint ovens.

Some forms of VOC are amenable to capture in aqueous alkaline scrubbers. Ultraviolet rays can then activate or even begin to decompose the VOCs in solution before they are transported to an aqueous chemical oxidation chamber. A variety of oxidants are used; ozone, chlorates, hydrogen peroxide, hypochlorites, and additional oxidizing agents have been used in this

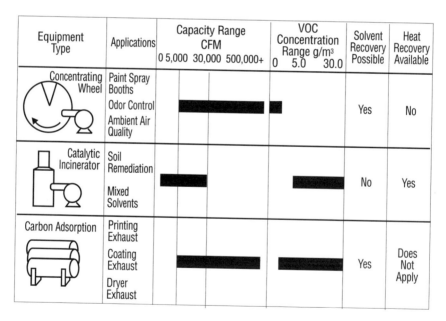

Figure 22-13. A Comparison of VOC Controls

# SPRAY BOOTHS FOR LIQUID PAINTING 313

process. The systems are called ultraviolet plus oxidation (UV/OX) emission reduction systems. VOCs that are not destroyed are adsorbed on carbon or zeolite beds and destroyed by oxidant treatment or by combustion after desorption. UV/OX is utilized most effectively for water soluble VOCs at modest concentrations, but operating costs tend to be quite high.

Some manufacturers have shown interest in biofilter systems, in which exhaust air is passed through soil materials, such as compost or peat moss. These soils contain bacteriological microorganisms that can consume solvents by converting them to water and carbon dioxide. This method forms no $NO_x$, but it cannot handle hot exhausts that would kill bacteria. Space requirements and installation costs are large, but operational expenses are small.

Figure 22-13 offers a comparison of various VOC control methods.

# Glossary

**Adhesion Loss (Adhesion Failure)**—The premature separation of a paint film from a surface. When this occurs between two coats of paint, it is termed "intercoat adhesion failure."

**Adsorption**—Some solids such as activated carbon can hold molecules of some liquids and gases on their surfaces by moderately strong secondary chemical bonds. Desorption can be accomplished by heat or displacement with other molecules. Adsorption beds can be used to purify VOC-laden air exhaust streams.

**Air Bearings**—A stream of air that supports a spinning shaft (as opposed to roller, ball, or similar mechanical bearings). Air bearings have limited load carrying capacity but require no lubricants.

**Air Dryers**—Used to remove moisture from compressed air. Dryers have three basic styles of operation: (1) deliquescent types have disposable drying agents and tend to be marginally effective for painting; (2) refrigerated dryers cool the air to condense out the water (most paint systems use this type); (3) desiccant types have a double bed dryer and are able to achieve the lowest dew point air. The beds are alternately on-stream and back-flushed to regenerate their moisture-absorbing qualities. Some plants with critical finish requirements use this style of dryer to reach dew points of -40° F (-40° C).

**Air Knife**—A slotted jet of compressed air will effectively and quickly blow superfluous water from parts after aqueous cleaning or during pretreatment before the parts enter a dryoff oven.

**Air Turbine**—(1) Electrical motor-driven fans that create volumes of relatively low-pressure atomizing air for spraying. Their output is referred to as turbine air; (2) an air-driven precision fan that is used to spin a paint atomizing disc or bell head.

**Airless Spray**—A method of paint application in which the paint's fluid pressure is used to atomize the paint into particles, and a spray gun is used to direct the particles onto the surface being painted.

**Alkali**—A substance that neutralizes acids, synonymous with caustic. Alkalis are helpful in aqueous cleaning by speeding soil removal and suspension.

**Amino Resins**—Nitrogen-containing compounds such as urea, melamine, or diethylene triamine are too rigid to be effective paint resins. The amino resins are used to cross-link polyester, epoxy, acrylic, alkyd, etc. resins to enhance their durability.

**Ampere (Amp)**—An electrodynamic unit of measure for the quantity of current in a steady electric flow.

**Anode**—The electrode at which chemical oxidation takes place. In electrodeposition (E-coating) the anode is indicated on diagrams by the positive (+) marking.

**Anodizing (of Aluminum)**—A type of conversion coating in which an aluminum oxide coating forms on the surface in a reaction forced by use of direct electrical current. The aluminum becomes an anode in the induced electrochemical cell. The process has many similarities to electroplating.

**Anolyte**—The water used to flush solubilizer molecules that form inside an electrocoating anode box. If used to flush a cathode box, it is termed catholyte.

**Aqueous Cleaning**—A system in which water plus detergent, and frequently also small amounts of acid or alkali, are used to clean surfaces prior to painting without emitting solvents.

**Arab Oil Embargo**—In 1973 a cartel of producers in Arabian countries withheld oil from the international market for a number of months in an attempt to raise prices for crude. Their efforts were devastatingly effective on all products that use petroleum, including paint products.

**ASTM**—Acronym for the American Society for Testing and Materials. The society publishes extensive standards used widely in manufacturing. Their address is 100 Barr Harbor Drive, West Conshohocken, PA 19428-2959. Phone 610-832-9585; FAX 610-832-9555.

**Atom**—Each of the 98 elements found in nature has a unique atom. Combinations of these tiny bits of matter form molecules and complex ions.

**Autodeposition (Autophoresis)**—A precipitation reaction of an organic resin occurs by the action of an acid that etches a metallic substrate. The ions of the oxidized metal codeposit with the vinyl or acrylic emulsion resin in the autodeposition coating process.

**Azeotrope**—A mixture of two miscible liquids that boil together at a specific composition at a fixed temperature lower than the boiling point of either single pure liquid.

**Barytes**—A colorless crystalline solid, a form of barium sulfate (also called "barite") used as a paint pigment.

**Bell**—A rotating head that is shaped to deliver paint forward in a circular pattern. The bell may be directed at any angle and be moved on robots or reciprocators, just as spray guns are.

**Bentonite**—A type of clay of varying composition, used as paint pigment.

**Biofilter or Soil filter**—VOC-laden air may be conducted through bacteria beds supported by soil or peat moss, etc. to biologically convert them in large measure to water and carbon dioxide.

**Bioremediation**—To destroy VOC and other wastes, they may be treated with or passed through materials containing bacteria capable of degrading the undesired wastes to harmless substances.

**Bénard Cells**—Minor circulation patterns that appear when metallic paints are applied extra wet and solvent evaporation cools the surface of a wet paint film.

**Bleeding**—A condition in which solvent migrates from the topcoat into the primer, leaching out the pigment and discoloring the topcoat.

**Blister**—A small domelike raised area in a paint film that contains, or once contained, solvent or moisture (water, water vapor, or both). Tiny blisters are called boils or pinholes.

**Blocked Isocyanates (Blocking Agent)**—Isocyanates, normally extremely reactive with water, can only be used in waterborne coatings if they can be prevented from reacting before the water is baked out of the paint film. This is done by capping or blocking the isocyanate group with a thermally decomposable chemical. In the bake oven the water evaporates, the chemical cap decomposes, and the isocyanate cross-links the paint. Blocked isocyanates are often used for E-coat curing.

**Blocking**—When freshly painted parts are stacked, they sometimes stick together if the paint is not fully cured. Sticking or blocking can be reduced by antiblock paint additives. This is also termed "printing."

**Blushing**—A defect that occurs whenever so much evaporative cooling takes place that the surface temperature of the paint film drops below the dew point, causing a whitish haze or "blush" in the paint.

**Bounce-off, Bounceback**—Paint droplets from air-atomized application have a tendency to rebound or bounce away from the surface due to the blasting effect of the air. These particles are called bounce-off or bounceback, as is the phenomenon itself.

**Brilliance**—The apparent strength of a color as perceived visually.

**Bubble**—A small domelike area in a paint film that contains, or once contained, solvent vapor. A tiny bubble is called a solvent pop.

**Bulk Coating**—The painting of large masses of small unhangable parts by a variety of possible techniques such as dip-spin, autodeposition, and hopper powder coating.

**Burn-off Ovens**—Paint stripping can be accomplished by combustion of the coating in gas-fired, burn-off ovens in which upper temperatures are controlled by injection of water spray into the oven.

**Capacitance**—The ability of ungrounded objects to retain electrical charges.

**Carcinogen, Carcenogenic**—Cancer-producing agent.

**Catalyst**—An agent that increases the rate of a chemical reaction but which undergoes no net change itself to accomplish this. Negative catalysts may be termed inhibitors.

**Catalytic Oxidizer**—A VOC abatement device in which catalysts are used to reduce fuel needed for solvent incineration.

**Cathode**—The electrode at which chemical reduction takes place. In electrodeposition (E-coating), the cathode is indicated on diagrams by the negative (-) marking.

**Caustic**—Substances that neutralize acids, synonymous with alkali. They are helpful in aqueous cleaning by speeding soil removal and suspension.

**Cellosolve**—Originally a trade name for the solvent family of mono-alkyl ethers of ethylene glycol, but the term is used generically now. A much-used solvent is butyl cellosolve, for example, which chemically is ethylene glycol monobutyl ether.

**Centrifugal Coater**—See Dip-spin Coater.

**Chemical Agent Resistant Coatings (CARC)**—Coatings used on military equipment that might become contaminated by nuclear, biological, or similar chemical substances.

**Chlorofluorocarbon (CFC)**—A general name for a multitude of halogenated hydrocarbon compounds that have either chlorine or fluorine atoms, or both, substituting for hydrogen atoms in the hydrocarbon molecule. Since many, if not all, cause ozone depletion in the earth's upper atmosphere, their manufacture and use are restricted or banned in many countries.

**Chroma**—The intensity of a hue.

**Coil Coating**—A process wherein coils of metal or polymer are unrolled and painted, then recoiled in a continuous flow.

**Cold or Dead Entry**—In E-coating, when the parts are immersed in the tank while the voltage is turned off.

**Color Mismatch**—When applying paint, a deviation in color that occurs due to misapplication, substrate variation, contamination, or batch-to-batch variation.

**Commission International de Eclairage (CIE)**—a.k.a. International Commission on Illumination; the L,a,b color scale still widely used today.

**Continuous Coater**—An enclosed automatic spray booth that recovers and reuses oversprayed paint. A continuous coater is suitable for coating large volumes of similarly sized parts.

**Conventional Air Spray**—One of the most common types of paint applications in which an air-atomizing spray gun is used to atomize the paint into tiny particles that are directed at the surface to be coated.

**Conversion Coating**—Substrates may be chemically treated prior to painting to change the chemical composition of their surface. This is most often done to enhance substrate corrosion resistance and the adhesion of applied coatings.

**Cosolvents**—Waterborne paints frequently require water-miscible organic solvents (cosolvents) in addition to water for easier manufacture and improved application properties.

**Creepback**—The spread of rust under a paint film and subsequent loss of the overlying paint. Creepback from a scribed line is often measured after accelerated corrosion testing.

**Critical Pigment Volume Concentration, CPVC**—The maximum volume of pigment that can be introduced into a paint formulation without causing a diminution of desirable film properties.

**Cross-linking**—In enamel curing, the resin molecules react to form an extensive network polymer system. The processes whereby the separate molecules unite by chemical bonds into a single macromolecule is called cross-linking.

**Cup Gun**—A spray gun used with a siphon cup or gravity-fed cup.

**Current Density**—A measure of the total electrical flow across a given surface area, frequently expressed in units of amps/square foot (A/sq ft).

**Curtain Coating**—A coating system in which paint flows from a reservoir through a slot, creating a curtain- or waterfall-like flow of paint onto the surface to be coated.

**Cyclone Separator**—A device that moves a particle-laden stream of air inside a funnel-bottomed enclosure around rapidly in a circular path. The relatively high mass of powder coating particles causes them to be thrown to the sides of the enclosure. They slide down through the funnel into a container for reuse.

**Deionized Water**—Water that has all contaminant ions removed by a double-bed ion exchanger that switches H+ (hydrogen) ions for positive impurity ions and OH- (hydroxide) ions for negative impurity ions. The hydrogen ions and hydroxide ions then combine to form HOH ($H_2O$). Deionized water is nearly equivalent in purity to distilled water but is much less costly to produce.

**Detackifying Agents**—Chemicals added to the water in a waterwash spray booth to prevent the collected overspray paint particles from sticking to the sides of the reservoir and clogging the piping, pumps, headers, and spray nozzles; also called paint kill agents.

**Diluent**—While true solvents can be added in unlimited amounts to lower paint viscosity, it may be more economical to lower viscosity with less costly diluent solvents. Diluents alone cannot dissolve resins, but when added to a prepared paint they will lower the viscosity just as effectively as a true solvent. However, if too much diluent is added, the resin will separate out of solution and the paint becomes unusable.

**Dip-spin Coater**—Bulk painting of small and unhangable parts can be accomplished by dipping a mesh basket of parts, followed by rapid rotation of the basket to remove excess paint. Parts from the dip-spin coater are dumped onto a belt for curing.

**Dirt**—In relation to paint, any and all contaminants including lint, dust, small clusters or improperly mixed pigment, tiny particles of overspray paint debris, and oil mist.

**Disc**—A relatively flat rotating head that delivers paint horizontally 360° around the rotating head. The disc is mounted horizontally, usually on a vertical reciprocator. The conveyor has an omega-shaped configuration around the disc applicator.

**Dispersion Paint**—If the paint resin is not dissolved but uniformly spread throughout the liquid formulation as a stable mixture by stirring, the coating is a dispersion paint.

**Distinction of Image (DOI)**—The measure of how well a surface acts as a reflecting mirror.

**Dive-in or Sink**—Gloss variations and dull spots in the paint film on a plastic part.

**E-Coating, Electrocoating, Electrodeposition**—In a method closely paralleling electroplating, paint is deposited using direct electrical current. The electrochemical reactions that occur cause water-soluble resins to become insolubilized onto parts that are electrodes in the E-coating paint tank. Subsequent resin curing is required.

**Eddy Current Detector**—This device, also called a magnetometer, measures paint thicknesses on metals by inducing electron flow in metals which become weaker as coating thickness increases.

# GLOSSARY

**Eductor**—Eductors (venturi nozzles) located along E-coat return headers spaced laterally at intervals across the tank bottom help to agitate the paint and prevent settling of pigments, resulting in cleaner deposited films.

**Electrons**—Sub-atomic particles of negative charge contained in all matter. Current flow in electric circuits can be thought of as a stream of electrons traveling through conducting wires.

**Electrostatic Atomization**—Electrostatic charge repulsion that pushes apart or atomizes paint into fine particles.

**ELPO**—The name is unique to General Motors where it is used to mean electrodeposition coating.

**Emulsion Paint**—A paint in which the resin is present as dispersed globules coated by an agent that prevents their agglomeration.

**Enamel**—All paints, powder or liquid, that form cross-linking chemical bonds during curing. The majority of industrial finishes fall into this category.

**Face Rust**—The appearance of rust on a continuous film of coating. The name differentiates this type of corrosion from rust at places where the paint film is broken or missing.

**Faraday Cage Effect**—Electrostatic application causes paint particles to be attracted to the nearest grounded object. The attraction force is often strong enough to pull paint particles out of their intended flight direction. Recessed areas on parts, since they require a slightly longer path for paint particles, often receive insufficient paint coverage. As a result, these Faraday Cage areas may need touch-up painting with nonelectrostatic spray.

**Fatty Edge**—An excess bead of paint that forms on the bottom edges of parts when they are in the drippage zone following dipping or flow coating.

**Fisheye**—A defect in a paint film that resembles a small depression (the eye) with a mound or dome in the center (the pupil).

**Flat Spotting or Striking In**—The appearance of gloss-deficient patches in a paint film.

**Fluidized Bed**—Finely divided powders made into a fluidlike state by passing air through a porous plate bottom of a powder hopper. This permits the fluidized bed of powder particles to be used in dip tanks and to be transported in a manner similar to liquids.

**Fluidizer**—Also known as a solvent, a fluidizer is a liquid that lowers a paint's viscosity enough to allow application.

**Flushable Electrode**—The anode in cathodic E-coating is often placed inside a semipermeable membrane enclosure so that the excess solubilizer generated at the anode can be continuously removed by water pumped into the bottom of the enclosure. Flushable electrodes in anodic E-coating can also be used (but rarely are needed) for the cathode.

**Forced Cure (Forced Bake)**—Rather than a full temperature bake needed to initiate cross linking, parts may be baked at relatively low temperatures (below 195° F or 90° C) to accelerate curing that would indeed occur, but far more slowly at lower temperatures.

**Free Radical Polymerization**—Certain organic compounds will form highly reactive electron configurations by the action of UV light (or other activation sources). These reactive species are called free radicals because, to an extent, "free" electrons are available for bonding. In free radical polymerization, the reactive electrons chemically bond to adjacent molecules and produce a cured paint film.

**Freeboard**—In a vapor degreasing tank, the freeboard is the space between the condensing coils and the top of the tank.

**Gas-liquid Chromatography**—Both qualitative and quantitative analyses of gaseous and liquid mixtures are accomplished by separating the components carried in an inert gas stream through liquid-lined chromatographic columns; sometimes simply termed gas chromatography.

**Gloss**—A measure of the capability of a surface to reflect light.

**Grain Refiners**—Finer grain zinc phosphate crystals provide superior corrosion resistance and paint adhesion. Agents used in water rinses prior to zinc phosphating or in the zinc phosphatizing bath itself to produce smaller crystals are called grain refining chemicals or grain refiners.

**Ground (Electrical Ground)**—An object so massive that it can lose or gain overwhelmingly large numbers of electrons without becoming perceptibly charged, in an electrical sense. The earth is the best electrical ground possible.

**Halogenated Hydrocarbons, CFCs, (Halogenated Solvents)**—Halogens, mainly chlorine, bromine, and fluorine, can be substituted into hydrocarbon molecules to change both the physical and chemical natures of hydrocarbon compounds. These new products are collectively called the halogenated hydrocarbons. Their nonflammability is essential for operations such as solvent degreasing. The fact that most such compounds contribute to global ozone-layer depletion curtails their use in many countries.

**Hazardous Air Pollutants (HAPs)**—VOCs which are particularly prone to cause environmental harm, and therefore, the emission of which is subject to special air quality regulations and restrictions.

**Heat of Vaporization**—The specific amount of energy required to evaporate a given quantity (one molecule) of any compound. This value is different for each substance, but is always the same for any given substance. Chemical handbooks list the values for various compounds.

**High Bake Paint**—Paints that require a cure temperature of 195° F (91° C) or greater.

**High-solids**—Solventborne coatings that are approximately 50% or greater in volume solids.

**Hot or Live Entry**—In E-coating, when the parts are immersed in the tank while the voltage is turned on.

**Hot Spray**—The process by which a paint may be heated prior to application, usually by heaters incorporated into the paint line, to reduce the paint viscosity and thus permit easier application.

**Hot Water Curing**—A particular version of autodeposition coating is cured by immersing parts in 180° F (82° C) water. Hot water curing is faster than oven curing for parts that act as a large heat sink, but the process is normally not used since it gives reduced corrosion resistance.

**Hue**—The particular shade of a color.

**Hydrophilic**—Water loving, or attracted to water.

**Hydrophobic**—Water hating, inclined to pull away from water.

**Hydroxides**—The chemical opposites of acids. Hydroxides (e.g., sodium hydroxide, potassium hydroxide) neutralize each other; also known as caustics and alkalis.

**Hygroscopic**—Describes materials that absorb water strongly. Thinly spread deposits of such materials can absorb enough water to completely dissolve.

**Ionized Air Cloud**—Around the tip of an operating electrostatic spray gun is an invisible small cloud of air molecules that have picked up excess electrons. The electrons from the power pack flow off the end of the needle electrode at the gun tip. When paint droplets pass through the ionized air cloud, they accumulate electrons that enable electrostatic attraction of the droplets to parts being coated.

**Isocyanate**—A somewhat toxic chemical compound or chemical group in a compound that is used to cross-link paint resins. It is a common component in urethane (polyurethane) coatings.

**Isolated System**—An electrostatic painting system in which the paint supply and hoses are isolated from ground in order to prevent the highly conductive paint from grounding out the paint system.

**Karl Fisher Method**—A titration procedure used to determine the water content of materials.

**Lacquer**—Any powder or liquid paint that does not cure by forming chemical cross links but cures by polar attractions among molecules of the resin.

**Latex**—An emulsion of any organic resin in water. Waterborne paints that contain emulsified resins are called latex paints.

**Low Bake Paint**—Paint that can be "forced" to cure at temperatures of 194° F (82° C) or lower.

**Low-solids**—Coatings characterized as having a low percentage of solids, both by weight and volume percent. Commonly considered as being less than 35% solids.

**Machine Vision**—An optical device used to locate, identify, and determine the orientation of parts in a continuous operational flow.

**Megaohm**—An electrical resistance unit equal to one million ohms. (An ohm is the resistance of a conductor in which 1 ampere current flows with 1 volt potential.)

**Metamerism**—The appearance of a color shift in a paint film caused by a change in the nature of the incident light. For example, an object can appear in daylight to have a color different from the color seen in fluorescent light.

**Mho**—A standard unit of electrical conductivity reciprocal to an ohm.

**Micromho**—One millionth of a mho, a unit for measuring electrical conductivity.

**Mold Release**—Materials used to facilitate the removal of molded objects, usually plastic, from the mold in which they are formed. Internal mold release is formulated into the plastic resin; external mold release is applied to the mold cavity surfaces.

**Molecules**—Compounds exist in identical, submicroscopic groups of chemically bonded elements. Each individual group is called a molecule of that compound. Water, for example, consists of molecules having 2 hydrogen atoms and 1 oxygen atom. The chemical formula, $H_2O$, indicates this molecular composition. Few molecules are that small. Polymers often have many thousands of atoms per molecule.

**Molten Salt Bath**—Mixtures of inorganic salts will melt at 650° F–900° F (343° C–482° C). Painted items and paint racks immersed in these mixtures are rapidly stripped by combustion of the paint in the molten salt bath.

**Mottle**—A defect that occurs when metallic paint is applied excessively wet, and the color pigments separate from the metallic flakes.

**Multicoat**—Two or more coatings applied to a surface, each of which has a unique function to perform.

**Nonrinsed or React-in-place Conversion Coating**—Several pretreatments for aluminum can be used when corrosion resistance requirements are not extreme that react fully enough to avoid the need for being flushed off the surface with water.

**Nubbing**—Localized over-etching of zinc alloys.

**Off-line Programming**—The programming of a painting robot from a location other than the conveyor line. The completed robot program is then used to operate the on-line painting robot.

**Ohm**—A standard unit of resistance to electrical flow equal to a conductor carrying one ampere of current from a 1-volt potential. A megaohm is equal to $10^6$ ohms.

**Ohmmeter**—A device used to measure electrical resistance in a circuit. The electrical ground of parts to be electrostatically painted can be checked with an ohmmeter.

**Omega Loop**—The conveyor for rotating disc paint applicators shaped to produce a circular path around the vertically reciprocating disc that delivers paint from all 360° of its circumference. The "omega" term is used because the shape of the conveyor loop resembles the capitalized form of the Greek letter omega.

**One K (1K) Paint**—A single component version of a paint such as a polyurethane, which can also be formulated in a two-component version.

**Orange Peel**—The surface in a paint film caused by poor paint flowout may feel smooth, but have a visibly undulating pattern.

**Paint**—All organic resin-containing materials used as decorative, protective, or functional coatings on any kind of surface. Four materials typically found in powder and liquid paints are resins, pigments, fluidizers, and additives.

**Permeate**—The resin- and pigment-free output of electrocoating bath ultrafiltration used to rinse freshly coated parts; also called flux or ultrafiltrate.

**pH**—The relative acidity or basicity of aqueous liquids is indicated by the pH scale. Numbers below the neutral pH or 7 are increasingly acidic, and above pH 7 are increasingly basic (anti-acid).

**Picture Framing**—A paint defect in which extra paint accumulates at the edges of a panel, darkening the edges.

**Pigment**—These tiny solid particles are used to enhance the appearance, color, or functional properties of a paint film.

**Plastic**—1) A broad category of materials produced from petroleum-based and agricultural plant-derived organic polymeric resins which can be molded into many shapes. 2) Adjective that denotes a material will deform, melt, and flow when heated.

**Polarity**—Every compound is made up of molecules, and all the molecules of each compound are identical and distinct in the arrangement, number, and types of atoms that comprise them. As a result, molecules have a characteristic magnet-like property called polarity, which can vary from no polarity to extremely high polarity. Compounds with uniform electron distributions are described as having low polarity; those with less uniformity have slight, moderate, or high polarity, depending on the degree of electron nonuniformity.

**Polymer**—In organic chemistry, a chain or network linkage of many repeating individual chemical structures (monomers or oligomers).

**Powder Coating**—This type of coating is comprised of resins, often with pigments, and additives, but does not contain solvents. Powder coats are thus applied dry with virtually zero VOC emissions.

**Power-and-Free Conveyor**—A power conveyor can use a separate pusher chain unattached to paint hooks riding freely on a separate support beam (as distinguished from a continuous power conveyor). This conveyor allows for variable parts spacing and for parts to be held stationary even when the pusher chain is moving.

**Power Conveyor (Continuous)**—An electrically driven cable or chain power conveyor with mechanically attached hooks, onto which parts to be painted are hung. The conveyor is used to carry parts through the painting process operations. When the line is operating, all individual hooks on the line will continue to move and maintain their spacing.

**Pressure Feed**—Paint delivered to an application device by air pressure on a paint reservoir (pot) connected to the application equipment, as opposed to gravity flow or positive displacement pumping of paint.

**Pressure Pot**—Various-sized paint tanks with delivery tubes extending to the bottom inside the tank are pressurized with compressed air to force paint to the application device. The tanks, known as pressure pots, usually have bolt-on covers.

**Primer**—A paint formulated to be directly applied to a substrate, before any other paint is applied. Primers promote better adhesion and provide protection to the substrate, among other benefits.

**Printing**—When parts with soft paint films are stacked together, or tightly packed, a permanent contact pattern is left in the paint film. This is also termed "blocking."

**Programming (Robots)**—Creation of electronically recorded instructions that tell the robot what to do mechanically; can use point-to-point or continuous path instructions.

**Radiation-cure Coatings**—Coatings that cure when exposed to ultraviolet (UV) or electron beam (EB) radiation. Infrared radiation is a heat cure method, sometimes considered to be a radiation-cure method, but paints cured by infrared are definitely not radiation-cure coatings.

**React-in-place or Nonrinsed Conversion Coating**—Several pretreatments for aluminum can be used when corrosion resistance requirements are not extreme. They react fully enough to avoid the need for being flushed off the surface with water.

**Reactive Diluents**—Substances that can be added to a polymer to reduce viscosity, that will cross-link with the base polymer during curing without producing VOCs.

**Reciprocator**—An automatic device to move a paint-applying tool in alternating directions along a straight or slightly curved horizontal or vertical path.

**Recuperative Thermal Oxidizers**—In these VOC abatement devices, a heat exchanger, normally a stainless shell-and-tube type unit, transfers heat from hot exhaust gases exiting a combustion chamber to incoming solvent-laden air as it approaches the combustion chamber. Heat recovery is about 50%. Catalysts can be used to reduce fuel costs for incineration.

**Regenerative Thermal Oxidizers**—These VOC abatement units have two or more ceramic media beds. Solvent-laden air passes through one bed to be heated near its combustion point just before it enters a gas-fired combustion chamber. The hot combustion gases, now free of up to 95% of their solvent, pass through a second bed which extracts the heat before the air is vented to the atmosphere.

**Resin**—The polymer (plastic) component that cures to form a paint film; also known by many names, such as binder or vehicle. The term binder is used because resins must bond with the substrate.

**Resin Kickout**—A condition in which resin separates from a paint formulation due to various causes.

**Resistive System**—A protective device in an electrostatic paint system designed to reduce voltage as current rises.

**Reverse Osmosis**—By the natural process of osmosis, solvent is driven through a semi-permeable membrane separating solutions of different concentrations. This is how water reaches the top leaves of even the tallest trees, for example. Water always travels from the less concentrated to the more concentrated solution. In reverse osmosis, high pressures are applied to force water out of the concentrated solution, often to obtain pure (or purer) water. It has been used to produce drinking water from the ocean and to reduce contaminants in processed water.

**Runs, Sags, or Curtains**—Scalloped or uneven paint film on vertical surfaces caused by wet paint running down the surface before curing.

**Rupture Voltage**—In E-coating, the point at which the voltage will cause excess current to flow between electrodes and form gases under the paint film, which causes the paint film to lift off the substrate.

**Sand-scratch Swelling**—A defect that occurs when the solvent from a topcoat "swells" and magnifies the sanding scratches left in an undercoat.

**Sealer**—1) A paint film used to bridge large differences in the polarity between primer and topcoat. 2) A coating used to lock in wood stains. 3) A coating used to prevent sand-scratch swelling. 4) A coating used to prevent pigment bleeding.

**Servo**—An electrical/mechanical device that provides an output motion according to a given input signal.

**Sheen**—The specular gloss of a paint film at very small angles of light incidence and reflection.

**Singlecoat**—A single paint film that can perform all the required functions alone.

**Siphon Cup (Suction Cup)**—When a special air spray tip is employed, a partial vacuum is created by the atomizing air just outside the fluid orifice. As a result, atmospheric pressure on the paint in a container connected to the fluid line (such as a siphon cup) will force paint up out of the container into the fluid line leading to the gun tip. The term siphon is actually a misnomer; suction is a somewhat more accurate description of the action.

**Slitting**—The process of cutting wide coils of roll coated materials into narrower widths.

**Soft Paint Film**—Deposit coatings cured to a hardness below the required specifications.

**Soil filter or Biofilter**—VOC-laden air may be conducted through bacteria beds supported by soil or peat moss, etc. to biologically convert them in large measure to water and carbon dioxide.

**Solids**—That portion of paint made up of resins, pigments, and additives.

**Solubilizer**—Since water is a polar solvent and resins are usually nonpolar, the resins must be treated to increase their polarity if they are to be used in waterborne paints. Water-insoluble resins used for paints will form polar (and hence water-soluble) polymer ions when mixed (reacted) with compounds known as solubilizers.

**Solution Paint**—If the resin molecules are fully dissolved by solvents, the paint is a solution paint.

**Solvent Wash**—Paint voids or areas with thin paint caused by solvent condensation. This is a common defect in dipping and flowcoating.

**Solventborne**—A paint that uses traditional, organic solvents to carry or disperse the resins, pigments, and additives.

**Specular Gloss**—The shininess or brilliance on highlighted areas of a painted part.

**Spray Booth**—An enclosure in which liquid or powder coatings are deposited on workpieces.

**Static Electricity (Electrostatic)**—Electrons temporarily removed from various items can cause static charges. Whatever has excess electrons has a negative charge; the object from which electrons have been taken will be positively charged. Electrons will tend to jump from one object to another if at all possible in order to neutralize all charges. This behavior differs from an electrical current, or electrodynamics; instead it is electrostatics and charges that are termed static electricity.

**Stripping**—The removal of unwanted cured paint film from a surface.

**Styrenated Alkyd**—Styrene is allowed to react with alkyd resin to improve the water resistance, alkali resistance, and drying speed.

**Surface Tension**—Liquids tend to reduce their surface area due to unequal intermolecular attractive forces in this region. A low degree of surface tension is preferred for liquid coatings to maximize adhesion and minimize edge-pull and fisheye effects.

**Surfactant**—Organic surface active agents used in organic acid cleaners to remove stains, soils, streaks, and related blemishes.

**Thermally Inverse Solubility**—Nearly all solids follow the pattern of becoming more water-soluble as temperature increases. A relatively few exceptions decrease in solubility with a rise in temperature.

**Thermoplastic Resin**—Resins that soften and melt with heat, and can be completely dissolved with the appropriate solvents.

**Thermosetting Resin**—Resins that will not melt or soften to any appreciable extent when heated; they will soften in some solvents but not dissolve in any of them.

**Thixotrope**—By forming loosely held three-dimensional particle networks within paint fluids, thixotropes cause temporarily high paint viscosities. Agitation of the paint by stirring, pumping, spraying, etc. quickly destroys the networks and viscosity drops sharply. When agitation is halted, the networks reform rapidly and paint viscosity again rises. Thixotropic is an adjective describing such materials.

**Toll Coater**—A trade term for custom or contract coaters who coil coat for other companies.

**Toners**—Unpigmented paints may need to be transparent but shaded to a specific color. Soluble colorants, i.e. toners, produce this effect and give a tinted clearcoat.

**Transfer Efficiency**—The percentage of applied paint used that is deposited on the parts. The balance of the paint goes onto the booth surfaces, hooks, filters, floors, etc.

**Tribo Charging or Friction Charging**—A method of charging some types of powder coating without using a charging electrode.

**Triglycidyl isocyanurate (TGIC)**—A complex chemical used to cross-link paint, especially polyester powders, to increase exterior durability.

**Two K (2K) Paint**—A dual component version of polyurethane paint. Because they react when mixed, the two parts are mixed shortly prior to application.

**Ultrafiltrate**—The output of ultrafiltration; also called permeate and flux.

**Ultrafiltration**—A process using low-pressure membrane filtration to separate small molecules from large molecules and fine particulates. For example, E-coat rinse water is extracted from the paint bath by ultrafiltration.

**Ultrasonic Cleaning**—A process in which vibrational frequencies slightly higher than those audible are used to agitate immersion cleaning tanks. Microbubble formation in the liquid accelerates dislodgement of soils.

**Unsaturated Bonds**—These carbon-to-carbon double bonds react chemically with atmospheric oxygen to form a cross-linked paint film.

**Value**—A grayness scale that varies gradually from pure white to pure black.

**Vapor Curing**—A process using a room temperature catalyst vapor to speed the curing of a two-component coating.

**Vapor Degreasing**—A method of using freshly distilled solvent vapors to clean metal parts prior to coating or painting. It requires halogenated (nonflammable) solvents that are now tightly restricted by many countries, but are still used in third-world countries.

**Viscosity**—The ratio of the shearing stress to the rate of shear of a Newtonian liquid.

**Volatile Organic Compounds (VOC)**—Most, but not all, VOC emissions are subject to governmental air pollution regulations.

**Volatiles**—That portion of paint that evaporates during curing, principally the fluidizers or solvents, but including low molecular-weight resin components as well.

**Voltage**—A measure only of the potential difference (force or pressure) in electrical systems; it does not indicate amounts of current flow.

**Waterborne**—A paint that uses mostly water to carry or disperse the resins, pigments, and additives.

**Wavelength**—When referring to radiant energy such as UV, visible light, and infrared, the tiny distance between corresponding consecutive points in the electromagnetic radiation.

**Weir**—The (often adjustable) barrier that controls the paint depth in an E-coat tank and over which the paint flows to the circulation pump.

**Wicking**—Air entrapment or solvent absorption along fiberglass/plastic interfaces in fiberglass-reinforced plastics.

**Wraparound or Wrap**—In electrostatic painting, the force that attracts paint droplets around to the back side of a part being painted.

Accelerated weathering tests, 250–252
Accelerators, 209
Acid cleaning of metal surfaces, 91–92
Acid etching cleaners, 93, 95
Acrylic resins, 28–29, 32
Additives
    categories, 13–15
    defined, 13
    introduced, 13
Adhesion/flexibility tests, 246
Adhesion loss, 232–233
Adhesion (tape) tests, 246–247
Air-assisted spray guns, 176–177
Air-atomizing spray guns
    air spray characteristics, 167
    components, 155–157
    compressed air supply, 157–158
    gun operation, 161–166
    high-efficiency low-pressure (HELP), 166
    high-volume low-pressure (HVLP), 162–165
    low-volume high-pressure (conventional air spray, LVHP), 161–162, 163, 164, 165
    low-volume low-pressure (LVLP), 166
    paint supply, 158–161
    spraying techniques, 166–167
Air bearings, 201–202
Air-drying paints, 3–4
Airless spray guns
    advantages, 174
    air-assisted, 176–177
    components, 169–171
    disadvantages, 174–176
    operation, 171
    skin injection danger, 175–176
    spraying techniques, 172–173
Alkaline detergent cleaning of metal surfaces, 92–93
Alkaline etching cleaners, 95
Alkyd resins, 29, 32
Aluminum
    anodizing, 109
    chromates for, 106–107
    cleaners, 93–95
    conversion coatings, 106–109
    phosphates for, 108–109
American Society for Testing and Materials (ASTM) guidelines, 237, 256–257

Ampere (unit), 182
*Annual Book of ASTM Standards,* 256
Anodic E-coat systems, 137
Anodizing aluminum, 109
Anolyte solutions, 140
Antiblock agents, 14
Antifoamers (defoamers), 14
Antifreeze, 14
Antimar agents, 15
Antimicrobial agents, 14
Appearance pigments, 6–7
Application methods
    autodeposition, 152–154
    coating by immersion, 133
    coil coating, 127–131
    continuous coaters, 125
    curtain coating, 126–127
    dip-spin coaters, 125–125
    dipping, 122
    electrocoating advantages, 150
        bath chemical reaction, 138–142
        bath parameters, 146–147
        curing cycle, 149–150
        disadvantages, 150–152
        electrical considerations, 135, 137
        introduced, 133–135, 136
        paint constituents, 137–138
        pigments, 146
        primers, 20
        rinsing, 142–146
        tank details, 148–149
    electrostatic painting
        advantages, 186–187
        disadvantages, 187–191
        effects of humidity, 186
        electrostatics and, 179–180
        grounding and safety precautions, 182–185
        introduced, 180–182
    flow coating, 123–125
    historical, 121
    roll coating, 127–131
    rolling, 121–122
    rotary atomizers
        advantages, 204–205
        configuration of the disc system, 201
        disadvantage, 205
        disc and bell rotation, 201–202
        hand-held bells, 203–204
        introduced, 193–196, 197, 198, 199

        paint application, 200
        rotary system operation, 202–203
        rotational speed and degree of atomization, 196, 199–200
    safety procedures and cleanliness, 131–132
        fire safety, 131
        safe spraying techniques, 132
        waste disposal, 132
    spraying. *See* Spray guns
Aqueous cleaning of metal surfaces, 89–91, 92
Aqueous solvents, 11
ASTM (American Society for Testing and Materials) guidelines, 237, 256–257
Atomization of waterborne coatings, 57
Atomizers. *See* Rotary atomizers
Atoms, 1
Autodeposition, 152–154
Azeotropic mixtures, 52

BACT (Best Available Control Technology), 52, 60
Bake primers, 19–20
Baking enamels, 4
Basecoats, 21–23
Bells. *See* Rotary atomizers
Benard cells, 228
Bend tests, 246
Best Available Control Technology (BACT), 52, 60
Beta-ray backscatter devices, 245
Binders. *See* Resins
Blasting
    by abrasive grit, 260–261
    by water, 261
    media, 260–261
Bleeding, 20
Blending aids, 14
Blisters, 219–221
Blushing, 227
Bonds, unsaturated, 4
Bounceback, 173
Brilliance, 252
Bubbles, 221–222
Burn-off ovens, 265

CAA (Clean Air Act), 54–55
    Amendments of 1990, 12
Capacitance, 56, 188

Carbon-bed solvent adsorption
   systems for VOC/HAP
   reduction, 310–313
CARC (chemical agent resistant
   coatings), 25
Catalyst vapor coatings, 210–211
Catalysts, 209
Catalytic cure enamels, 4
Catalytic oxidizers (incinerators), 308
Cathodic (cationic) E-coat
   systems, 137
Cellulosic resins, 29
CFC (chlorofluorocarbon)
   solvents, 10
CFCs (chlorofluorocarbons), 87–89
Chain-in-floor conveyors, 281
Chain-on-edge conveyors, 280
Chemical agent resistant
   coatings (CARC), 25
Chemical resistance tests, 247
Chemical stripping, 261–262, 264
Chlorinated hydrocarbon
   solvents, 10
Chlorofluorocarbon (CFC)
   solvents, 10
Chlorofluorocarbons (CFCs), 87–89
Chroma, 255
Chromates for aluminum, 106–107
CIE scale, 255–256
Clean Air Act (CAA), 54–55
   Amendments of 1990, 12
Cleaning surfaces.
   *See* Surface cleaning
Clearcoats, 23
Coating solvents and the
   environment, 11–13
Coatings. *See also* Electrocoating
   autodeposition, 152–154
   catalyst vapor, 210–211
   coating by immersion, 133
   coil coating, 127–131
   continuous coaters, 125
   conversion
      aluminum, 106–109
      dip processes, 110–113
      metal, 100–110
      other, 119
      phosphate
         troubleshooting, 113–118
      qualifications, 99–100
      sealing phosphate
         coatings, 118–119
      specialty and mixed
         metal phosphates, 110
      spray processes, 110–113

steel, 101–105
waste treatment, 120
zinc, 105–106
curing. *See* Curing methods
curtain coating, 126–127
defects. *See* Film defects
dip-spin coaters, 125–126
flow coating, 123–125
heat cross-linking, 208–209
high-solids
   advantages, 43–44
   disadvantages, 44–46
   introduced, 40–41
   resins, 41–42
   surface tension and, 42
low-solids, 39–40
moisture-cure, 208
oxidizing, 208
paint versus, 2
powder
   advantages, 74–78
   defined, 61
   disadvantages, 78–83
   electrostatic spray
      application, 64–71
   fluidized bed application,
      72–73
   introduced, 61–63
   isocyanate cross-linked
      polyesters, 31
   properties, 63
   specialty application
      methods, 73–74
   transfer efficiency, 81, 83
   triglycidyl isocyanurate
      (TGIC) cross-linked
      polyesters, 31
radiation cure, 211–213
reactive catalytic, 209–210
roll coating, 127–131
ultraviolet-cure (UV-cure),
   211–212
waterborne
   advantages, 59–60
   agitation and, 58
   application problems, 57–58
   atomization of, 57
   cleanup, 58
   cost, 55
   defined, 47–48
   dip tanks and, 59
   disadvantages, 59–60
   dispersion, 51
   electrostatic application,
      56–57

emulsion, 49–51
fire hazard of, 59
foam and, 59
introduced, 47–48
modifications, 55
odor, 58
operation permits, 54–55
pretreatment, 55
process commonality, 55
resin availability, 55
solution, 48
solvents in, 51–53
storage of, 59
wood finishes, 53–54
Coil coating, 127–131
Cold (dead) entry, 149
Cold solvent strippers, 262–265
Color match tests, 253–256
Color matching plastics
   with metal, 276
Color measurement scales, 255–256
Color mismatch, 222–223
Composition of paint, 2–3.
   *See also* Additives; Pigments;
   Resins; Solvents (fluidizers)
Concrete paints, 24
Conductivity, 240, 270, 275
Consumer paints, 17
Continuous coaters, 125
Continuous path programming, 286
Convection ovens, 213–215
Conventional air spray.
   *See* Air-atomizing spray guns
Conversion coatings
   aluminum, 106–109
   dip processes, 110–113
   metal, 100–110
   other, 119
   phosphate troubleshooting,
      113–118
   qualifications, 99–100
   sealing phosphate coatings,
      118–119
   specialty and mixed metal
      phosphates, 110
   spray processes, 110–113
   steel, 101–105
   waste treatment, 120
   zinc, 105–106
Conveyors
   chain-in-floor, 281
   chain-on-edge, 280
   flat belt, 279–280
   importance of good design, 284
   introduced, 279

# INDEX 335

inverted, 283–284
overhead, 281–283, 284
roller, 279–280
track, 279–280
Corrosion of paint, 18
CPVC (critical pigment volume concentration), 8–9
Craters, 221–222
Creepback, 250
Critical pigment volume concentration (CPVC), 8–9
Cross-linkers, 209
Cryogenic stripping, 260–261
Cups. See Rotary atomizers
Cure mechanism categories, 37–38
Curing agents, 14
Curing methods
  coatings that cure by cross linking
    catalyst vapor coatings, 210–211
    catalytic coatings, 209–210
    heat cross-linking coatings, 208–209
    introduced, 208
    moisture-cure coatings, 208
    oxidizing coatings, 208
    radiation-cure coatings, 211–213
  coatings that cure by solvent evaporation only, 207–208
  heat-of-condensation curing, 218
  induction heating, 217
  introduced, 207
  microwave curing, 217
  ovens
    convection, 213–215
    infrared (IR), 215–216, 217
    introduced, 213
    radio frequency (RF) curing, 217
Curtain coating, 126–127
Curtains, 230–232

Dead (cold) entry, 149
Defects. See Film defects
Defoamers (antifoamers), 14
Deionized water (DI) rinse, 96–97
Detackifying agents, 303, 304
Detackifying process, 303, 304
DFT (dry-film thickness), 241, 242–245
DI (deionized water) rinse, 96–97
Dip processes for conversion coatings, 110–113

Dip-spin coaters, 125–126
Dipping, 122
Dirt in paint, 223–225
Discs. See Rotary atomizers
Dispersion of pigments, 8–9
Dispersion paints, 34, 36
Dispersion waterborne coatings, 51
Distinction of image (DOI), 252–253, 254
Dive-in, 276
DOI (distinction of image), 252–253, 254
Downdraft booths, 298–299
Downdraft water-wash booths, 301, 302
Dry-film thickness (DFT), 241, 242–245
Dry-filter booths
  advantages and disadvantages, 301
  defined, 299
  introduced, 300–301
  water-wash booths versus, 305
Dryoff ovens, 119

E-coat. See Electrocoating
EB-cure (electron-beam-cure) coatings, 212–213
Eddy current detectors, 244–245
Eductors, 145
Electrical conductivity, 240, 270, 275
Electrical ground, 182–185, 241
Electrocoating advantages, 150
  bath chemical reaction, 138–142
  bath parameters, 146–147
  curing cycle, 149–150
  disadvantages, 150–152
  electrical considerations, 135, 137
  introduced, 133–135, 136
  paint constituents, 137–138
  pigments, 146
  primers, 20
  rinsing, 142–146
  tank details, 148–149
ELectrodeposition of POlymers (ELPO). See Electrocoating
Electrodeposition primers, 20
Electromagnetic radiation, 253–255
Electron-beam-cure (EB-cure) coatings, 212–213
Electrostatic application
  of powder coatings, 64–71
  of waterborne coatings, 56–57
Electrostatic atomization, 196, 198
Electrostatic painting

advantages, 186–187
disadvantages, 187–191
effects of humidity, 186
electrostatics and, 179–180
grounding and safety precautions, 182–185
introduced, 180–182
Electrostatic voltage, 241
Electrostatics, 179–180
ELPO (ELectrodeposition of POlymers). See Electrocoating
Emulsifiers, 36, 49
Emulsion cleaners, 93–95
Emulsion paints, 36
Emulsion waterborne coatings, 49–51
Enamels, 3–4, 37–38, 208
Environment and coating solvents, 11–13
Environmental Protection Act, 12–13
Environmental Protection Agency (EPA), 10, 11–13
Epoxy resins, 29–30, 32
Etch (wash) primers, 19
Evaporation of solvents, 13
Extent of cure tests, 248

Face rust, 250
Faraday cage effect, 189, 190
Film defects
  adhesion loss, 232–233
  blisters, 219–221
  bubbles, 221–222
  color mismatch, 222–223
  craters, 221–222
  curtains, 230–232
  dirt, 223–225
  fisheyes, 225–226
  gloss variations, 226–228
  mottle, 228
  orange peel, 229–230
  root cause of, 219
  runs, 230–232
  sags, 230–232
  soft paint films, 233–234
  solvent pops, boils, and pinholes, 234–235
  solvent wash, 236
Finishing robots
  advantages, 294
  case history, 294
  future developments, 295
  introduced, 285–287
  operation requirements, 292–293
  payback time for, 293
  programming, 286, 289–292

Finishing robots *(cont.)*
  rotary bell, 287–288
  spray-painting, 287–288
Fisheyes, 225–226
Flash primers, 19
Flash rust, 116–118
Flat belt conveyors, 279–280
Flat spotting, 226
Flow coating, 123–125
Flow control agents, 14
Fluidized bed application of
  powder coatings, 72–73
Fluidized sand stripping, 266
Fluidizers. *See* Solvents (fluidizers)
Flux, 143
Forced-curing (forced-dry)
  primers, 19–20
Freeboard length, 87, 88
Friction charging, 71
Functional pigments, 7–8

Gloss modifiers, 14
Gloss tests, 252
Gloss variations, 226–228, 276
Grayness (value), 255
Ground, electrical, 182–185, 241

Halogenated resins, 30
HAPs. *See* Hazardous Air
  Pollutants (HAPs)
Hardness tests, 247
Hazardous Air Pollutants
  (HAPs). *See also* Volatile
    Organic Compounds (VOCs)
  carbon-bed solvent
    adsorption systems for
    VOC/HAP reduction,
    310–313
  CFC solvents, 10
  chlorinated hydrocarbon
    solvents, 10
  coating solvents, 11–13
  high-solids coatings, 46
  incineration devices for VOC/
    HAP reduction, 306–310
  National Emission Standards
    for HAPs (NESHAP), 88
  operation permits, 54–55
  solvents, 88–89
  spray-painting robots and, 294
  waterborne coatings, 60
  wood finishes, 54
Heat cross-linking coatings, 208–209
Heat-of-condensation curing, 218

Heat of vaporization, 58
Heat sensitivity of plastics, 274
High-efficiency low-pressure
  (HELP) air spray guns, 166
High-solids coatings
  advantages, 43–44
  disadvantages, 44–46
  introduced, 40–41
  resins, 41–42
  surface tension and, 42
High-solids overspray recovery, 306
High-volume low-pressure
  (HVLP) air spray guns,
  162–165
Hot (live) entry, 149
Hot alkaline strippers, 261–262, 264
Hot fluidized sand stripping, 266
Hue, 255
Humidity resistance tests, 248–249
HVLP (high-volume low-pressure)
  air spray guns, 162–165
Hydophilic binders, 47
Hydrogen embrittlement, 92
Hydrophobic binders, 47

Impact tests, 246
Incineration devices for VOC/
  HAP reduction, 306–310
Induction heating, 217
Industrial paint types
  basecoats, 21–23
  chemical agent resistant
    coatings (CARC), 25
  clearcoats, 23
  concrete paints, 24
  cure mechanism categories, 37–38
  marine finishes, 25
  multicoats, 26
  other end-use
    classifications, 25–26
  peel coats, 25
  physical makeup categories
    introduced, 34
    liquid fluidization
      methods, 34–36
  primers, 18–20
  resin categories
    acrylics, 28–29, 32
    alkyds, 29, 32
    cellulosics, 29
    epoxies, 29–30, 32
    halogenated, 30
    introduced, 27–28
    oleoresinous (oil-based)
      coatings, 30

phenolics, 30
polyesters, 31
resin mixtures, 34
silicones, 31
urethanes, 31–33
vinyls, 33
sealers, 20–21
singlecoats, 25
surfacers, 21
topcoats, 23–24
wood finishes, 24
Industrial robots. *See* Finishing
  robots
Infrared (IR) ovens, 215–216, 217
Inverted conveyors, 283–284
Ionized air clouds, 181
IR (infrared) ovens, 215–216, 217
Iron phosphating, 102–104
Isocyanate cross-linked
  polyesters, 31
Isolated systems, 184

Karl Fisher method, 240
Kickout, 48, 49

Lacquers, 3, 37, 207
LAER (Lowest Available Emission
  Requirements), 52, 60
Latex, 36, 49
LEL (lower explosive limit), 214
Light spectrum, 253–255
Liquid fluidization methods, 34–36
Live (hot) entry, 149
Low-solids coatings, 39–40
Low-volume high-pressure
  (conventional air spray,
  LVHP) guns, 161–162,
  163, 164, 165
Low-volume low-pressure
  (LVLP) air spray guns, 166
Lower explosive limit (LEL), 214
Lowest Available Emission
  Requirements (LAER), 52, 60
LVHP (conventional air spray,
  low-volume high-pressure)
  guns, 161–162, 163, 164, 165
LVLP (low-volume low-
  pressure) air spray guns, 166

Machine vision, 290–291
MACT (Maximum Available
  Control Technology), 60
Magnetic banana gauges, 243–244
Magnetic pencil gauges, 243
Magnetometers, 244–245

# INDEX  337

Maintenance paints, 17–18
Marine finishes, 25
Maximum Available Control
    Technology (MACT), 60
Mechanical cleaning
    of metal surfaces, 86
    solvent cleaning, 87–89
Mechanical stripping, 259–260
Media blasting, 260–261
Metal
    blisters in paint on, 220
    cleaning surfaces
        acid cleaning, 91–92
        alkaline detergent
            cleaning, 92–93
        aqueous cleaning, 89–91, 92
        emulsion cleaning, 93–95
        introduced, 85–86
        mechanical cleaning, 86
        solvent cleaning, 87–89
    conversion coatings, 100–110
Metallic paint, 190–191
Metamerism, 222
Microwave curing, 217
Milling pigments, 8–9
Moisture-cure coatings, 208
Mold-release agents, 271–272
Molecules, 1
Molten salt stripping, 265
Monomers, 3
Mottle, 228
Multicoats, 26
Munsell scale, 255

National Emission Standards
    for HAPs (NESHAP), 88
Nonconductivity of plastics, 270, 275
Nonetching (silicated alkaline)
    cleaners, 93
Nonsilicated alkaline cleaners, 93

ODSs (Ozone Depleting
    Substances), 87
Off-line programming, 289–290, 291
Oil-based (oleoresinous) coatings, 30
Oleoresins, 30
Omega loops, 201
Operation permits, 54–55
Orange peel, 229–230
Organic acid cleaners, 93
Overhead conveyors, 281–283, 284
Oxidizers (incinerators), 308–309
Oxidizing coatings, 208
Oxygenated solvents, 10

Ozone Depleting Substances
    (ODSs), 87

Paint
    air-drying, 3–4
    applying. *See* Application
        methods
    cleaning surfaces for.
        *See* Surface cleaning
    coatings versus, 2
    composition of, 2–3. *See*
        *also* Additives; Pigments;
        Resins; Solvents (fluidizers)
    consumer, 17
    corrosion of, 18
    defects. *See* Film defects
    defined, 2
    dirt in, 223–225
    dispersion, 34, 36
    emulsion, 36
    industrial types. *See*
        Industrial paint types
    introduced, 1–2
    maintenance, 17–18
    metallic, 190–191
    preparation of, 2–3
    removing. *See* Stripping
    shear/viscosity relationship, 15
    solids portion of, 2
    solution, 34
    solventborne, 34
    testing. *See* Paint tests
    trade sale, 17
    volatiles in, 2
    waterborne, 34
Paint tests
    American Society for Testing
        and Materials (ASTM)
        guidelines, 237, 256–257
    *Annual Book of ASTM*
        *Standards* for, 256
    applied paint film
        accelerated weathering,
            250–252
        adhesion (tape), 246–247
        adhesion/flexibility, 246
        chemical resistance, 247
        color match, 253–256
        distinction of image
            (DOI), 252–253, 254
        dry-film thickness
            (DFT), 241, 242–245
        extent of cure, 248
        gloss, 252
        hardness, 247

        humidity resistance, 248–249
        introduced, 242
        stain resistance, 248
        water immersion, 248
        wet-film thickness, 242
    choosing, 256–257
    introduced, 237
    paint "in the bulk," 237–240
    parts and paint application
        equipment, 241
    selecting, 256–257
Payback time, 293
Peel coats, 25
Permeates, 142–144
Petroleum hydrocarbon solvents, 10
pH scale, 94
Phenolic resins, 30
Phosphates
    for aluminum, 108–109
    flash rust and, 116–118
    loose, powdery, and
        nonadherent coatings, 115
    mixed metal, 110
    salt deposits, acid spots, and
        water spots from, 116
    sealing coatings, 118–119
    specialty, 110
    streaky, blotchy, thin, or
        discontinuous coatings,
        113–114
Phosphating
    iron, 102–104
    troubleshooting, 113–118
    zinc, 104–105
Picture framing, 228
Pigment volume concentration
    (PVC), 8–9
Pigments
    appearance, 6–7
    defined, 6
    dispersion of, 8–9
    functional, 7–8
    introduced, 6
    milling, 8–9
Plastic resins, 267–268
Plastics
    chemistry of, 267–268
    cleaning before painting, 270–272
    color matching with metal, 276
    common, 267–268
    conductivity, 270, 275
    defined, 267
    gloss variations with metal, 276
    heat sensitivity, 274

Plastics (cont.)
    introduced, 267–268
    nonconductivity, 270, 275
    paint film laminates for, 276–277
    paintability information
        resources, 278
    preparing surfaces for
        painting, 272–273
    reasons for painting, 268–270
    sanding sensitivity, 274–275
    surface peculiarities affecting
        painting, 273–276
    water sensitivity, 273
    wicking into, 276
Point-to-point programming, 286
Polarity of materials, 1–2
Polyester resins, 31
Polymers, 3
Pot life, 209
Powder coatings
    advantages, 74–78
    defined, 61
    disadvantages, 78–83
    electrostatic spray
        application, 64–71
    fluidized bed application, 72–73
    introduced, 61–63
    isocyanate cross-linked
        polyesters, 31
    properties, 63
    specialty application
        methods, 73–74
    transfer efficiency, 81, 83
    triglycidyl isocyanurate
        (TGIC) cross-linked
        polyesters, 31
Pressure-feed, 159–160
Pressure pots, 159
Primers, 18–20
Printing, 233
Programming robots, 286, 289–292
Pull-off gauges, 242–244
PVC (pigment volume
    concentration), 8–9

Radiation cure (radcure)
    coatings, 211–213
Radiation cure (radcure) enamels, 4
Radio frequency (RF) curing, 217
Reactive catalytic coatings, 209–210
Reactive diluents, 4, 212
Reciprocators, 195–196
Recuperative thermal oxidizers
    (incinerators), 308–309

Regenerative thermal oxidizers
    (incinerators), 309
Removing paint. See Stripping
Resin kickout, 48, 49
Resins
    acrylic, 28–29, 32
    alkyd, 29, 32
    availability
        for powder coatings, 78
        for waterborne coatings, 55
    cellulosic, 29
    defined, 3
    enamels, 3–4, 37–38, 208
    epoxy, 29–30, 32
    halogenated, 30
    high-solids, 41–42
    introduced, 3, 27–28
    lacquers, 3, 37, 207
    mixtures, 34
    oleoresins, 30
    phenolic, 30
    plastic, 267–268
    polyester, 31
    properties, 28, 32
    qualities, 5–6
    silicone, 31
    soft-to-hard ranges, 28
    thermoplastic, 62, 208, 267
    thermosetting, 62, 267
    urethane, 31–33
    vinyl, 33
Resistive systems, 185
Reverse osmosis, 142
RF (radio frequency) curing, 217
Rinsing after cleaning, 95–98
Robots. See Finishing robots
Roll coating, 127–131
Roll-off gauges, 242–244
Roller conveyors, 279–280
Rolling, 121–122
Rotary atomizers
    advantages, 204–205
    configuration of the disc
        system, 201
    disadvantage, 205
    disc and bell rotation, 201–202
    hand-held bells, 203–204
    introduced, 193–196, 197,
        198, 199
    paint application, 200
    rotary system operation, 202–203
    rotational speed and degree of
        atomization, 196, 199–200
Rotary bell robots, 287–288
Runs, 230–232

Rupture voltage, 139–140

Safety procedures
    cleanliness, 131–132
    fire safety precautions, 131
    safe spraying techniques, 132
    waste disposal, 132
Sags, 230–232
Salt bath stripping, 265
Sand stripping, 266
Sanding sensitivity of plastics,
    274–275
Sandscratch swelling, 20
Sealers, 20–21
Servos, 286
Shear/viscosity relationship of
    paint, 15
Sheen, 252
Shopcoat primers, 19
Shopplate primers, 19
Sidedraft dry-filter booths, 300
Sidedraft water-wash booths,
    301, 302
Silicated alkaline (nonetching)
    cleaners, 93
Silicone resins, 31
Singlecoat, 25
Sink, 276
Siphon cups, 158–160
Soft paint films, 233–234
Softening agents, 14
Solids portion of paint, 2
Solution paints, 34
Solution waterborne coatings, 48
Solvent cleaning of metal
    surfaces, 87–89
Solvent pops, boils, and
    pinholes, 234–235
Solvent stripping, 262–265
Solvent wash, 236
Solventborne paints, 34
Solvents (fluidizers)
    aqueous, 11
    carbon-bed solvent adsorption
        systems for VOC/HAP
        reduction, 310–313
    chlorinated hydrocarbon, 10
    chlorofluorocarbon (CFC), 10
    coating solvents and the
        environment, 11–13
    defined, 10
    evaporation of, 13
    introduced, 10
    oxygenated, 10
    petroleum hydrocarbon, 10

INDEX 339

properties, 12
selecting, 11, 12
terpene, 11
in waterborne coatings, 51 53
Spectrum of light, 253–255
Specular gloss, 252
Spray booths
  defined, 297
  downdraft, 298–299
  dry-filter
    advantages and
      disadvantages, 301
    defined, 299
    introduced, 300–301
    water-wash booths versus, 305
  high-solids overspray
    recovery, 306
  introduced, 297–300
  VOC/HAP reduction methods
    carbon-bed solvent
      adsorption systems,
      310–313
    comparison of methods, 312
    incineration devices, 306–310
    introduced, 306
  water-wash
    advantages and
      disadvantages, 305
    defined, 299
    dry-filter booths versus, 305
    introduced, 301–305
Spray guns
  air-atomizing
    air spray characteristics, 167
    components, 155–157
    compressed air supply,
      157–158
    gun operation, 161–166
    high-efficiency low-
      pressure (HELP), 166
    high-volume low-pressure
      (HVLP), 162–165
    low-volume high-pressure
      (conventional air spray,
      LVHP), 161–162,
      163, 164, 165
    low-volume low-pressure
      (LVLP), 166
    paint supply, 158–161
    spraying techniques, 166–167
  airless
    advantages, 174
    air-assisted, 176–177
    components, 169–171

disadvantages, 174–176
operation, 171
skin injection danger,
  175–176
spraying techniques,
  172–173
Spray-painting robots, 287–288
Spray primers, 20
Spray processes for conversion
  coatings, 110–113
Stain resistance tests, 248
Staining, 20
Static electricity, 180
Steel conversion coatings, 101–105
Storage stabilizers, 14
Striking in, 226
Stripping
  blasting
    by abrasive grit, 260–261
    by water, 261
    media, 260–261
  burn-off ovens, 265
  chemical, 261–262, 264
  hot fluidized sand, 266
  introduced, 259, 260
  mechanical, 259–260
  molten salt, 265
  selecting a stripping process, 266
  solvent, 262–265
Surface cleaning
  introduced, 85
  metal surfaces
    acid cleaning, 91–92
    alkaline detergent
      cleaning, 92–93
    aqueous cleaning, 89–91, 92
    emulsion cleaning, 93–95
    introduced, 85–86
    mechanical cleaning, 86
    solvent cleaning, 87–89
    other considerations, 95
    rinsing after cleaning, 95–98
Surface tension, 42
Surfacers, 21
Surfactants, 93

Tape tests, 246–247
TDS (total dissolved solvents), 96
TE (transfer efficiency), 81, 83, 241
Terpene solvents, 11
Testing paint. *See* Paint tests
TGIC (triglycidyl isocyanurate)
  cross-linked polyesters, 31
Thermal oxidizers (incinerators),
  308–309

Thermoplastic resins, 62, 208, 267
Thermosetting resins, 62, 267
Thixotropes, 14, 15
Topcoats, 23–24
Total dissolved solvents (TDS), 96
Track conveyors, 279–280
Trade sale paints, 17
Transfer efficiency (TE), 81, 83, 241
Tribo charging, 71
Triglycidyl isocyanurate (TGIC)
  cross-linked polyesters, 31
Two-part reactive enamels, 4

Ultrafiltrates, 143
Ultrafiltration, 142–144
Ultraviolet (UV) stabilizers, 14–15
Ultraviolet-cure (UV-cure)
  coatings, 211–212
Ultraviolet plus oxidation
  (UV/OX) emission reduction
  systems, 312–313
Unsaturated bonds, 4
Urethane resins, 31–33
UV (ultraviolet) stabilizers, 14–15
UV-cure (ultraviolet-cure)
  coatings, 211–212
UV/OX (ultraviolet plus
  oxidation) emission reduction
  systems, 312–313

Value (grayness), 255
Vapor curing, 210–211
Vapor degreasing, 87–88
Vapor injection curing (VIC),
  210–211
Vinyl resins, 33
Viscosity, 15, 237–238, 239
Volatile Organic Compounds
  (VOCs). *See also* Hazardous
  Air Pollutants (HAPs)
  basecoats, 22–23
  carbon-bed solvent adsorption
    systems for VOC/HAP
    reduction, 310–313
  coating solvents, 11–13
  high-solids coatings, 41, 43, 46
  incineration devices for VOC/
    HAP reduction, 306–310
  lacquers, 207
  operation permits, 54–55
  powder coatings, 75
  radiation cure (radcure)
    enamels, 4
  spray-painting robots and, 294
  waterborne coatings, 52–53, 59

340  INDUSTRIAL PAINTING

VOCs *(cont.)*
  weight percent solids/weight percent VOC, 238, 240
  wood finishes, 54
Volatiles in paint, 2
Voltage
  defined, 182
  electrostatic, 241
  rupture, 139–140
Volume percent solids, 240

Wash (etch) primers, 19
Waste disposal considerations for solvent strippers, 263–265
Waste treatment for conversion coatings, 120
Wastewater treatment for cleaning and rinse solutions, 97–98
Water immersion tests, 248
Water sensitivity of plastics, 273
Water-wash booths
  advantages and disadvantages, 305
  defined, 299
  dry-filter booths versus, 305
  introduced, 301–305
Waterborne coatings (WBCs)
  advantages, 59–60
  agitation and, 58
  application problems, 57–58
  atomization of, 57
  cleanup, 58
  cost, 55
  defined, 47–48
  dip tanks and, 59
  disadvantages, 59–60
  dispersion, 51
  electrostatic application, 56–57
  emulsion, 49–51
  fire hazard of, 59
  foam and, 59
  introduced, 47–48
  modifications, 55
  odor, 58
  operation permits, 54–55
  pretreatment, 55
  process commonality, 55
  resin availability, 55
  solution, 48
  solvents in, 51–53
  storage of, 59
  wood finishes, 53–54
Waterborne paints, 34

Wavelength of light, 253–255
WBCs. *See* Waterborne coatings (WBCs)
Weathering tests, 250–252
Weight per gallon, 237
Weight percent solids/weight percent VOC, 238, 240
Weirs, 147
Wet booths. *See* Water-wash booths
Wet-film thickness, 242
Wicking into plastics, 276
Wood, blisters in paint on, 221
Wood finishes, 24, 53–54
Wraparound (wrap), 187

Zinc
  conversion coatings, 105–106
  phosphating, 104–105